The Handbook of Environmental Chemistry

Volume 126

Founding Editor: Otto Hutzinger

Series Editors: Damià Barceló • Andrey G. Kostianoy

In over four decades, *The Handbook of Environmental Chemistry* has established itself as the premier reference source, providing sound and solid knowledge about environmental topics from a chemical perspective. Written by leading experts with practical experience in the field, the series continues to be essential reading for environmental scientists as well as for environmental managers and decision-makers in industry, government, agencies and public-interest groups.

Two distinguished Series Editors, internationally renowned volume editors as well as a prestigious Editorial Board safeguard publication of volumes according to high scientific standards.

Presenting a wide spectrum of viewpoints and approaches in topical volumes, the scope of the series covers topics such as

- local and global changes of natural environment and climate
- anthropogenic impact on the environment
- water, air and soil pollution
- remediation and waste characterization
- environmental contaminants
- biogeochemistry and geoecology
- chemical reactions and processes
- chemical and biological transformations as well as physical transport of chemicals in the environment
- environmental modeling

A particular focus of the series lies on methodological advances in environmental analytical chemistry.

The Handbook of Environmental Chemistry is available both in print and online via https://link.springer.com/bookseries/698. Articles are published online as soon as they have been reviewed and approved for publication.

Meeting the needs of the scientific community, publication of volumes in subseries has been discontinued to achieve a broader scope for the series as a whole.

Groundwater Quality and Geochemistry in Arid and Semi-Arid Regions

Volume Editors: Shakir Ali · Abdelazim Negm

With contributions by

I. Abd-Elaty · S. J. Adams · M. Ahmed · H. Allali · S. Ali ·
A. M. Armanuos · M. Bahir · P. Bhattacharya · S. Chander ·
S. Difi · O. el Mountassir · Y. Elmeddahi · M. El-Rawy ·
M. K. Elshaarawy · M. G. Eltarabily · H. Fathi ·
K. S. Gemail · A. Gupta · F. Hallouz · M. M. Khalil ·
M. M. Ladyka · M. Meddi · M. B. Okyere · R. Ragab ·
V. M. Starodubtsev · E. D. Sunkari · M. S. Taha · S. Yadav ·
M. S. Zango · B. A. Zeidan

Springer

Editors
Shakir Ali
Department of Geology
University of Delhi
Delhi, India

Abdelazim Negm
Department of Water and Water Structures Engineering
Faculty of Engineering, Zagazig University
Zagazig, Egypt

ISSN 1867-979X ISSN 1616-864X (electronic)
The Handbook of Environmental Chemistry
ISBN 978-3-031-53776-9 ISBN 978-3-031-53777-6 (eBook)
https://doi.org/10.1007/978-3-031-53777-6

This Springer imprint is published by the registered company Springer Nature Switzerland AG
The registered company address is: Gewerbestrasse 11, 6330 Cham, Switzerland

Prof. Dr. Teresa Rocha-Santos
University of Aveiro, Aveiro, Portugal
Prof. Dr. Yolanda Picó
Desertification Research Centre - CIDE, Moncada, Spain

Preface

The use of groundwater can be traced back to a millennium ago. However, the research on groundwater quality has seen an exponential increase in the last few decades. This is due to increased groundwater pollutants through various geogenic or anthropogenic means. In arid and semi-arid regions, groundwater is the most important/only resource fulfilling the demands of millions of populations. Therefore, an in-depth understanding of groundwater resources is required through science-based studies. Thus, a book covering various facets of groundwater contamination, particularly in arid and semi-arid areas, is imperative to understand and combat the pollutants.

This book is the outcome of hard work from both the editors and selected worldwide authors who contributed their high-quality chapters on groundwater quality in arid and semi-arid regions. Both the editors worked closely with the authors, who implemented several revisions in response to the comments from editors, reviewers, and the Springer team to ensure high-quality contributions and to ensure that the readers would find this book handy. All authors have extensively presented their contribution concerning groundwater quality, highlighted various challenges in the arid and semi-arid regions and concluded with concrete conclusions and a set of recommendations.

This book contains 12 chapters covering various domains of groundwater quality. Broadly, this book is divided into two themes: the first is about the assessment of groundwater quality and is written in nine chapters, while the second is about major global contaminants in groundwater and includes three chapters.

The first chapter in this book is entitled "*Unveiling the Hidden Depths: A Review for Understanding and Managing Groundwater Contamination in Arid Regions*". The authors conducted a comprehensive review of groundwater contaminants and explored various protection techniques for safeguard of groundwater quality in arid regions. The second chapter entitled "*Risk Assessment of Potential Groundwater Contamination by the Agricultural Drainage Water in the Central Valley Watershed, California, USA*". The authors used various statistical methods to analyse nitrate contamination in the groundwater of the Central Valley of California in the

USA and recommended various measures to minimize the impacts of the agricultural contaminants. The next chapter titled "*Impact of Bugun Reservoir on Groundwater and Soil: A Case Study from South Kazakhstan*" presents the results of investigating the effect of reservoir on groundwater quality and soils in the vicinity of reservoir in Kazakhstan. The authors highlighted various environmental consequences of the reservoir.

The authors of the chapter titled "*Impact of Climate Changes on Seawater Intrusion in the Nile Delta Aquifer (Egypt)*" applied and documented various numerical models and investigated how climatic change vis-à-vis. over-exploitation along coastal area influence seawater intrusion in the Nile Delta Aquifer in Egypt.

Furthermore, the chapter titled "*Groundwater Quality Prediction in Upper and Middle Cheliff Plain, Algeria Using Artificial Intelligence*" applied machine learning to improve the prediction of water quality in Cheliff plain in Algeria, while the authors of the chapter titled "*Evolution of Groundwater in the Cheliff and Mitidja Aquifers (North Algeria) in Qualitative and Quantitative Terms*" investigated two potential plains of Algeria i.e., Cheliff and Mitidja and presented status of both of these aquifers concerning their deterioration both quantitatively and qualitatively. Consequently, the authors proposed numerous recommendations to combat the quality and quantity of the aquifers. The chapter titled "*Groundwater Pollution Sources and Its Quality in Kingdom of Saudi Arabia- State-of-the-Art*" presents the state-of-the-art review on groundwater quality issue in various aquifers throughout Saudi Arabia. Additionally, the chapter entitled "*Isotopic and Chemical Composition of Egypt's Groundwater Resources*" presents the compiled and published historical water isotopes and chemical (salinity) compositions for seven important aquifer systems in Egypt. The compositions were used to understand country-wide groundwater resources and investigated the recharge and contamination sources. The authors further provided sustainable management scenarios for the seven potential Egyptian aquifers. The last chapter in the first theme is titled "*Understanding Seawater Intrusion by Hydrochemical Parameters and Stable Water Isotopes along the Coastal Alluvial Aquifers of the Essaouira Basin, Morocco*". The authors used hydrochemical parameters along with stable water isotopes to investigate salinization processes along coastal aquifers of Morocco. The authors successfully identified the aquifers that are mostly exposed to sea-water intrusion.

On the other hand, the first chapter in the second theme is titled "*Geochemical Controls on Fluoride Enrichment in Groundwater of a Geologically Heterogeneous Part of Ghana: Implications for Human Health Risk Assessment*". The authors investigated hydro-geochemistry of groundwater in the northern region of Ghana and evaluated the human health risk assessment due to consumption of fluoride contaminated groundwater. On the other hand, the chapter titled "*Uncovering Fluoride Contamination in Groundwater of Arid and Semi-Arid Regions: Stigma to Solutions*" documented and investigated the fluoride contamination in arid and semi-arid regions. The author highlighted various challenges and suggested potential solutions in providing sustainable fluoride-free water in the regions. The last chapter is entitled "*Nitrate Contamination in Groundwater of Arid and Semi-Arid*

Regions: Ecotoxicological Impacts and Management Strategies". The authors presented the review results of nitrate contamination in the groundwater of arid and semi-arid regions. The chapter also focuses on health and environmental impacts due to the presence of nitrate contamination in groundwater. The authors also compiled and highlighted various management strategies and feasible options for safe water supply.

The efforts done to finalize this book indicated that more books on groundwater quality and management in arid and semi-arid regions are needed to benefit the understanding of the scientific communities concerning various dimensions, challenges, and opportunities.

Last but not least, the editors thank all the authors for their invaluable, high-quality contributions. Thanks are extended to Springer's team, editors, and the entire editorial board of the HEC series, for their kind pieces of advice and critical and constructive comments that helped in producing this high-quality book.

Delhi, India
Zagazig, Egypt
November, 2023

Shakir Ali
Abdelazim Negm

Contents

Part I Assessment of Groundwater Quality

Unveiling the Hidden Depths: A Review for Understanding and Managing Groundwater Contamination in Arid Regions 3
Khaled S. Gemail and Ismail Abd-Elaty

Risk Assessment of Potential Groundwater Contamination by Agricultural Drainage Water in the Central Valley Watershed, California, USA 37
Mohamed Galal Eltarabily and Mohamed Kamel Elshaarawy

Impact of Bugun' Reservoir on Groundwater and Soil: A Case Study from South Kazakhstan 77
Vladimir M. Starodubtsev and Maryna M. Ladyka

Impact of Climate Changes on Seawater Intrusion in the Nile Delta Aquifer (Egypt) 97
Asaad M. Armanuos, Mohamed Samir Taha, and Bakenaz A. Zeidan

Groundwater Quality Prediction in Upper and Middle Cheliff Plain, Algeria Using Artificial Intelligence 165
Yamina Elmeddahi, Salah Difi, Hemza Allali, and Ragab Ragab

Evolution of Groundwater in the Cheliff and Mitidja Aquifers (North Algeria) in Qualitative and Quantitative Terms 185
Faiza Hallouz and Mohamed Meddi

Groundwater Pollution Sources and Its Quality in the Kingdom of Saudi Arabia: State of the Art 215
Mustafa El-Rawy and Heba Fathi

Isotopic and Chemical Composition of Egypt's Groundwater Resources 237
Mohamed Ahmed and Mahmoud M. Khalil

Understanding Seawater Intrusion by Hydrochemical Parameters and Stable Water Isotopes Along the Coastal Alluvial Aquifers of the Essaouira Basin, Morocco 267
Mohammed Bahir, Otman el Mountassir, and Shakir Ali

Part II Major Global Contaminants in Groundwater

Geochemical Controls on Fluoride Enrichment in Groundwater of a Geologically Heterogeneous Part of Ghana: Implications for Human Health Risk Assessment 297
Emmanuel Daanoba Sunkari, Moses Boakye Okyere, Salaam Jansbaka Adams, Musah Saeed Zango, Prosun Bhattacharya, and Shakir Ali

Uncovering Fluoride Contamination in Groundwater of Arid and Semi-Arid Regions: Stigma to Solutions 327
Shakir Ali

Nitrate Contamination in Groundwater of Arid and Semi-Arid Regions, Ecotoxicological Impacts, and Management Strategies 339
Subhash Chander, Sangita Yadav, and Asha Gupta

Part I
Assessment of Groundwater Quality

Unveiling the Hidden Depths: A Review for Understanding and Managing Groundwater Contamination in Arid Regions

Khaled S. Gemail and Ismail Abd-Elaty

Contents

1 Introduction 4
2 Aquifers Contamination 7
3 Assessment of Aquifers Contamination Methods 9
3.1 Geophysical Investigations 9
3.2 Hydro-Geochemical Investigations 11
3.3 Experimental Studies 14
3.4 Mathematical Studies 15
4 Aquifers Protection Techniques 16
4.1 Operation Management 17
4.2 Hydraulic Barrier Management 17
4.3 Material Management 21
4.4 Physical Barriers Management 23
5 Conclusion 27
6 Recommendations 27
References 28

Abstract Groundwater, as a vital freshwater source in arid regions, faces significant risks of contamination from various natural and anthropogenic sources such as agricultural and industrial activities, sewage ponds, open drains, and saltwater intrusion. This chapter provides a comprehensive review of the contamination

K. S. Gemail (✉)
Environmental Geophysics Lab (ZEGL), Geology Department, Faculty of Science, Zagazig University, Zagazig, Egypt
e-mail: khaledgemail@zu.edu.eg

I. Abd-Elaty
Water and Water Structures Engineering Department, Faculty of Engineering, Zagazig University, Zagazig, Egypt

Shakir Ali and Abdelazim Negm (eds.), *Groundwater Quality and Geochemistry in Arid and Semi-Arid Regions*, Hdb Env Chem (2024) 126: 3–36, DOI 10.1007/698_2023_1049,

Published online: 15 December 2023

sources affecting groundwater aquifers in arid regions and explores diverse protection techniques employed to safeguard groundwater quality. The review encompasses an assessment of different investigative methods used to understand aquifer contamination, including geophysical, hydro-geochemical, experimental, and mathematical approaches. Furthermore, the study highlights proven methods for effectively protecting groundwater in arid regions, such as aquifer operation, hydraulic barriers, material management, and physical barrier management. By comprehending the sources and protection techniques associated with groundwater contamination, we can implement effective measures to prevent contamination and ensure the availability of clean groundwater for future generations. Thus, the future application of these protection methods holds immense promise in ensuring the availability of pristine freshwater and safeguarding human health, particularly in regions facing high water stress such as Egypt, and the adverse effects of climate change, notably arid and semi-arid regions. By adopting and integrating these approaches, water resources management can significantly address the challenges posed by groundwater pollution and secure sustainable water supplies. This chapter serves as a valuable resource for researchers and practitioners involved in the management and conservation of groundwater resources.

Keywords Aquifer operation, Arid regions, Experimental, Geophysical, Groundwater contamination, Groundwater management, Hydraulic barriers, Hydro-geochemical, Material management, Mathematical, Physical barriers

1 Introduction

Freshwater scarcity in arid and semi-arid regions, particularly in relation to groundwater, is a significant environmental and socio-economic challenge in many countries around the world. Arid regions are characterized by very low rainfall and high evaporation rates due to the limited surface water resources. As a result, groundwater becomes a critical water source for domestic, agricultural, and industrial purposes in these regions [1, 2]. In addition, the mounting water demand resulting from rapid population growth and evolving living standards has engendered a multitude of environmental, economic, and social problems. These predicaments have underscored the significance of effective water resource management, rendering it an intricate and imperative undertaking, particularly in arid and semi-arid regions (Fig. 1a). Accordingly, numerous regions across the globe are grappling with the pressing issue of water scarcity as seen in Fig. 1b. In these complex environments, the groundwater is a vital source of drinking water, providing approximately 31% of the world's drinking water supply. As a result, water supply engineers and managers are increasingly focused on the exploitation of fresh groundwater resources [1, 5]. Figure 2 illustrates a comprehensive global map of groundwater resources. The depicted map utilizes a colour scheme to convey valuable information; the predominant blue hues indicate major groundwater basins, categorized based on their recharge rates. The green areas correspond to hydrogeological environments

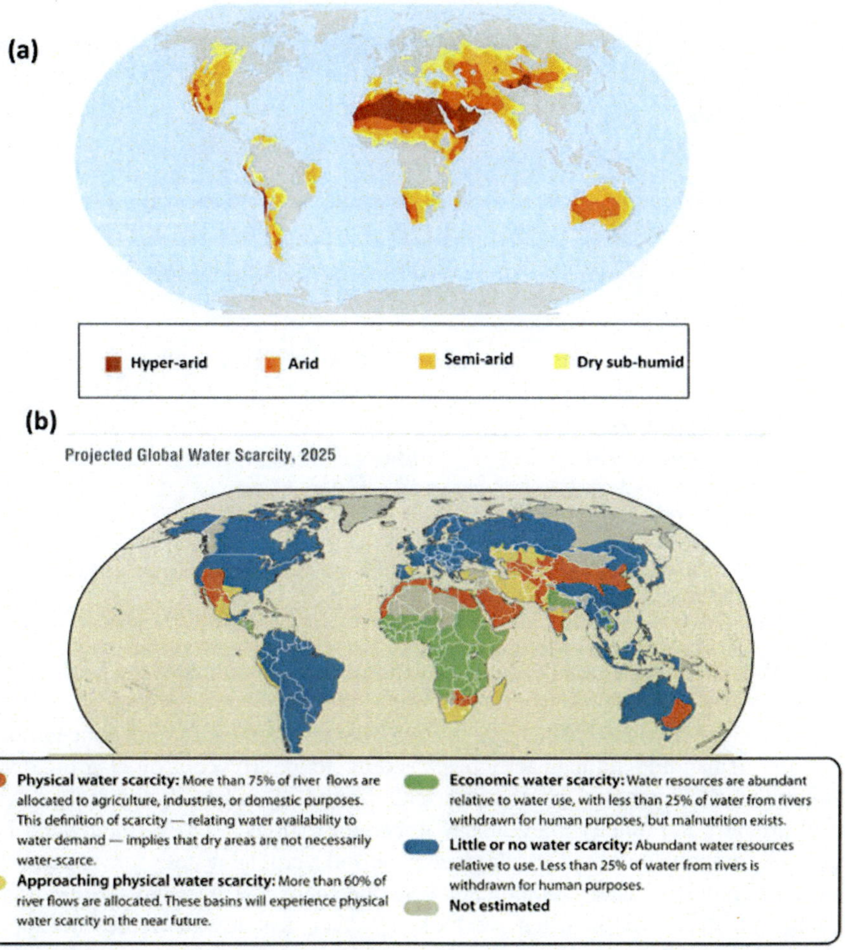

Fig. 1 World maps show the distribution of (**a**) aridity index and arid regions [3], and (**b**) projected global water scarcity [4]

encompassing aquifers situated within or proximate to non-aquifer strata. Finally, the brown shades indicate local and shallow aquifers characterized by the presence of dense rock formations in close proximity to or at the surface.

The over-extraction of groundwater, resulting from increased water demand in arid regions and a lack of sustainable management practices, poses one of the most significant challenges, leading to the depletion of groundwater resources. Over-pumping can result in declining water tables, reduced well yields, and even the drying up of wells, causing severe water scarcity. In coastal arid regions like the Nile Delta, over-pumping groundwater can cause saltwater intrusion. Excessive groundwater extraction from coastal aquifers creates a pressure gradient that draws in saline

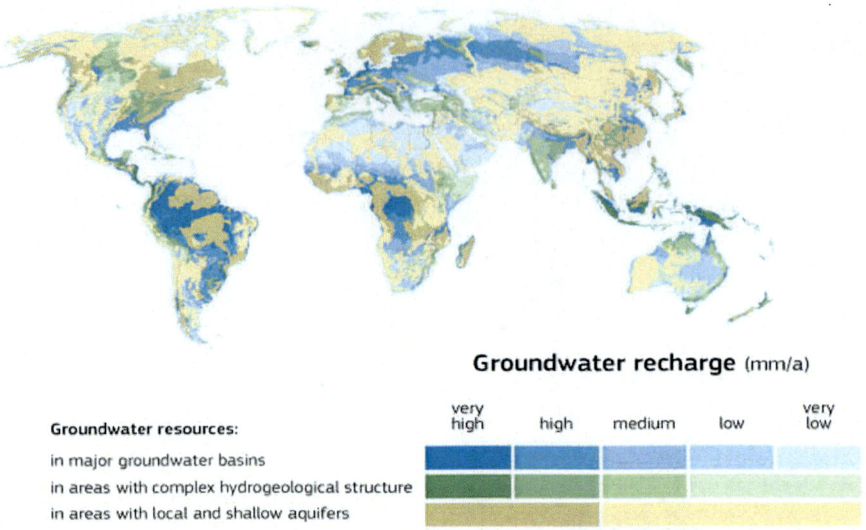

Fig. 2 World map of groundwater distribution displaying groundwater aquifers and the recharge rates with different hydrogeological conditions [6]

water from the adjacent sea or ocean. This intrusion contaminates the fresh groundwater, making it unsuitable for use without costly desalination processes.

In the Mediterranean basin, there is strong development with high concentrations of water-demanding human activities, leading to poorly controlled groundwater withdrawal and increased groundwater salinization. This has significantly deteriorated groundwater quality, particularly in certain regions [7, 8]. Groundwater is an essential component of the hydrological cycle and a crucial water source for meeting domestic, industrial and commercial water demands [9]. Despite its importance, groundwater resources are often overlooked. In fact, groundwater represents approximately 22% of all freshwater on Earth, with polar ice accounting for nearly 77% and freshwater in rivers and lakes representing just 0.30% [10]. Therefore, it is critical that we recognize the importance of groundwater as a vital resource and take measures to protect and manage it effectively. This will require a combination of effective policies, management strategies, and technological solutions to ensure this valuable resource's sustainable use and protection for future generations. According to the World Health Organization (WHO), more than 50% of the developing countries' population is suffering from health problems due to a lack of clean drinking water [11]. In arid regions, the sustainable management of natural water resources and food security is crucial to mitigate the adverse effects of changing climate conditions on coastal communities. Increasing inundation and groundwater contamination will likely adversely affect agricultural production [12].

Globally, groundwater resources provide about 50% of drinking water and 40% of industrial needs [13]. Groundwater pollution is usually traced back and damaged by four sources: environmental, domestic, industrial, and agricultural [14]. Water resources in hyper-arid and arid regions are limited compared with humid and wet regions [15]. Effective management strategies are required to preserve fresh groundwater resources [16]. Several methods can be employed to understand the pollutants' transportation mechanism and recharge processes in the vadose zone. For instance, the high-resolution geophysics, hydro-geochemical, and analytical and numerical models techniques can be used as a monitoring tool for detecting the flow of pollutants and migration patterns in the subsurface from disposal systems (e.g. [17–22]). In addition, the chemical characteristics, multivariate statistics, and spatial analysis can be applied to study the origin of TDS and contaminants in the groundwater ([23–30]). These integrated approaches have been applied in different environments to obtain direct quantitative information about the source and level of pollution.

The objective of the current chapter is to provide a comprehensive overview of groundwater as a vital source of drinking water, and the challenges it faces in arid regions due to population growth, climate change, and pollution from various sources. This chapter highlights the importance of understanding groundwater pollutant's sources and transportation mechanisms and recharge processes. It also discusses the various techniques and strategies that can be employed to protect and manage groundwater resources effectively, including high-resolution geophysics, hydro-geochemical, analytical, and numerical models, as well as chemical characteristics, multivariate statistics, and spatial analysis. By understanding these challenges facing groundwater sustainability and the techniques and strategies that can be employed to manage them effectively, we can take steps to protect this vital resource and ensure its availability for years to come.

2 Aquifers Contamination

Groundwater quality is an important issue in developing and managing water resources in many parts of the world. Groundwater pollution is usually traced back to seven sources as seen in Fig. 3 and could be summarized after Bear [10], and Zaporozec and Miller [31] as follows:

Natural and other sources: including lakes and spills, natural leaching, water from fault zones, acid rains, and volcanic origin [32]. Saltwater intrusion (SWI) is due to the interface between the freshwater and saline water in the coastal regions. This occurs due to the high density of saline water heads compared with the freshwater. The water resources salinity is a worldwide problem, particularly in coastal aquifers [33]. SWI into freshwater coastal aquifers is due to the over-pumping of freshwater and sea level rise, it causes profound changes in the biogeochemistry of groundwater and the dynamic systems changes called subterranean estuaries [34]. Storm water can be one of the main sources of pollutants (heavy metals, hydrocarbons, and other organic compounds) produced by cities. As a result, conventional urban drainage systems have given rise to numerous technical and environmental challenges, with a notable concern being the contamination of surface receptor media [35].

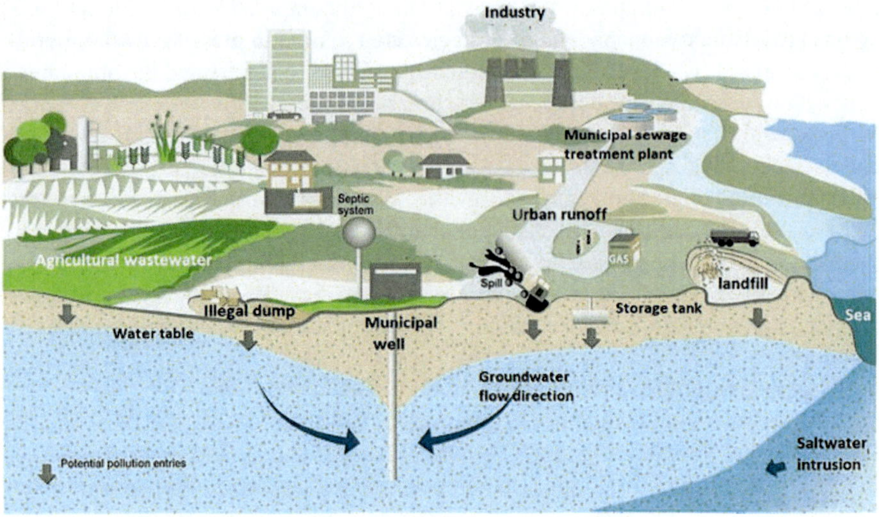

Fig. 3 Potential sources of groundwater contamination [31]

(i) Environmental factors that influence the flow of groundwater. For example, the characteristics of the surrounding environment play a significant role in groundwater contamination. This can occur when wastewater passes through fractured carbonate or permeable cap zones, introducing pollutants into the groundwater. Additionally, seawater intrusion and the infiltration of brackish water from adjacent aquifers can further contribute to groundwater pollution [36–39].

(ii) Human sources of water pollution can arise from various factors. These include the accidental breaking of sewer lines, which can release untreated sewage into the environment. Additionally, percolation from septic tanks and cesspools is commonly used in areas without centralized sewage systems. Rainwater infiltration through sanitary landfills is another potential domestic source of water pollution, as heavy rainfall can cause leachate-containing harmful substances to seep into nearby water sources [12, 17, 40].

(iii) Industrial pollution originates primarily from the disposal of sewage generated by industrial activities, encompassing the presence of heavy metals, non-degradable compounds, radioactive materials, and industrial landfills [12, 41, 42].

(iv) Agricultural pollution arises from various sources, including the dissolution and transportation of fertilizers, salts, herbicides, and pesticides by irrigation water and rainwater. Additionally, the return flow from irrigation systems can transport solutes, contributing to groundwater contamination. Animal waste, dryland farming practices, vegetation evapotranspiration, and feedlots contribute to agricultural pollution. The literature supports these findings, as highlighted by Bouderbala and Gharbi [43], Dugga et al. [44], and Abd-Elaty et al. [12].

(v) Oil and mining fields serve as significant sources of water pollution, primarily attributed to activities such as petroleum production and development, the presence of abandoned oil wells and test wells, buried pipelines and storage tanks, the improper disposal of oil field brines, as well as mining activities encompassing mine drainage and mine tailings [45, 46].

(vi) Hydraulic activities contribute to water pollution through various mechanisms, including the disposal, drainage, and abandonment of wells, excessive pumping of water resources, inadequate well construction practices, river infiltration, and the encroachment of seawater into freshwater aquifers [27, 32, 47, 48].

3 Assessment of Aquifers Contamination Methods

3.1 Geophysical Investigations

Geophysical methods play a crucial role in the identification and mapping of areas affected by soil and groundwater contamination [49, 50]. These methods enable the differentiation of lithological units and their respective water contents, thereby aiding in the assessment of subsurface conditions. Geophysical techniques are essential tools for measuring the spatial distribution of physical properties within the Earth's subsurface, including the electrical conductivity of geological materials. Moreover, geophysics is highly valuable for the detection, delineation, and monitoring of ground pollution resulting from spills, leaks, or unauthorized discharges, which can significantly alter the conductivity/resistivity of the affected area [49, 51, 52].

For instance, Gemail et al. [36] utilized the DC resistivity method to evaluate the vulnerability of an aquifer to industrial wastewater in the central Nile delta, as illustrated in Fig. 4. The study focused on mapping sand bodies within the protective clay cap, forming a vital component in assessing the aquifer vulnerability index [41]. This approach involved interpreting layer resistivity and thickness values above the aquifer zone to derive the integrated electrical conductivity (IEC). The IEC serves as an indicator of groundwater vulnerability, with the impermeable upper zone playing a pivotal role in controlling potential contamination.

Apart from the DC resistivity technique, other geophysical methods can also be employed for hydrogeological investigations. Surface geophysical methods, including induced polarization (IP), electromagnetic (EM), and ground penetrating radar (GPR), offer valuable insights into the subsurface characteristics. These methods aid in delineating aquifer boundaries, assessing water quality, and determining flow direction [10]. Furthermore, borehole logging tools such as Electric logs, Gamma-ray, and SP logs, provide additional information for understanding aquifer properties and dynamics at various depths. These logging tools contribute to the characterization of aquifer boundaries, evaluation of water quality, and assessment of flow patterns [10]. By employing a combination of these geophysical methods, hydrogeologists can enhance their understanding of aquifer vulnerability and gain

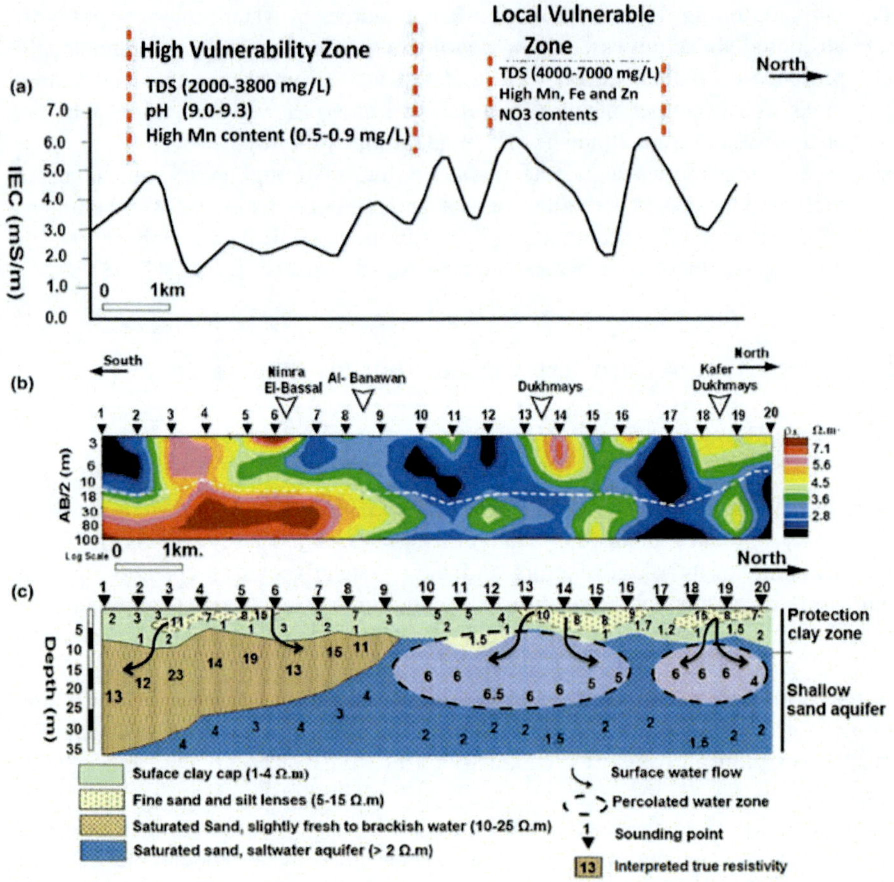

Fig. 4 Application of DC resistivity for mapping the heterogeneities in the clay cap in the enteral Nile Delta: (**a**) the estimated vulnerability index from DC resistivity soundings (IEC), (**b**) the apparent resistivity distribution along a profile closed to an open industrial wastewater drain, and (**c**) the infiltration of wastewater in the upper zone of the aquifer layer [36]

insights into the subsurface hydrogeological conditions, enabling informed decision-making for sustainable groundwater management.

The integration between multidisciplinary techniques to investigate hazardous waste and groundwater pollution sites is now at a rapid increase and cost-effective means of preliminary evaluation. The information obtained from a geophysical investigation can be used to determine the subsurface site conditions. Such conditions include; hydro-stratigraphic framework, depth to bedrock, the extent of concentrated groundwater contaminant plumes, the location of voids, and faults or fractures. In recent years, the growing demand for groundwater pollution investigations has coincided with notable advancements in geophysical data resolution, acquisition, and interpretation. It is worth noting that this process is continuously evolving, and as improvements are made in instrumentation and interpretation algorithms, the outlines of geophysical techniques and procedures are subject to revision [53]. In a study conducted by Benabdelouahab et al. [54] in northern Morocco, geo-electrical investigations were employed to characterize the aquifer and assess the geo-environmental conditions. The findings of the study revealed that the coastal aquifer under investigation consists of extensive deposits of coarse sand, gravel, and pebbles, overlaying substrata with varying hydrogeological and electrical properties.

It is evident that the integration of geophysical techniques has proven to be instrumental in enhancing our understanding of aquifer systems and their geo-environmental characteristics. As research in this field continues to evolve, advancements in instrumentation and interpretation algorithms further contribute to the refinement and optimization of these techniques, enabling more accurate assessments of groundwater pollution and aquifer characterization.

The use of multidisciplinary geophysical, hydro-geochemical, microbiological, and statistical approaches can be considered a good practice for revealing the migration of surface pollutants within the vadose zone and their flow paths in urban regions with local scale and high-resolution imaging. In this context, Abu Salem et al. [17] have used this integrative approach to validate the electrical resistivity tomography (2D-ERT) for mapping the permeable zones in the hazardous groundwater pollution sites in the Nile Delta where the wastewater from open sewage drain preferentially flows towards the shallow aquifers as illustrated in Fig. 5.

3.2 *Hydro-Geochemical Investigations*

The distinction of different salinization mechanisms is vital to the evolution of water origin, pathways, rates, and future salinization. Several geochemical criteria could be used to identify the origin of salinity in coastal aquifers [10]. Table 1 presents groundwater quality standards based on the guidelines set by the WHO and Egyptian drinking water standards.

The weighted arithmetic index method described in a research publication [56] was used for the evaluation of the water quality index (WQI). The quality rating was evaluated using:

$$q_n = \frac{100[V_n - V_{10}]}{[S_n - V_{10}]} \tag{1}$$

Water quality index (WQI) is calculated by taking aggregates of the products of determined parameter qualities and the unit weights divided by aggregates of the unit weight using:

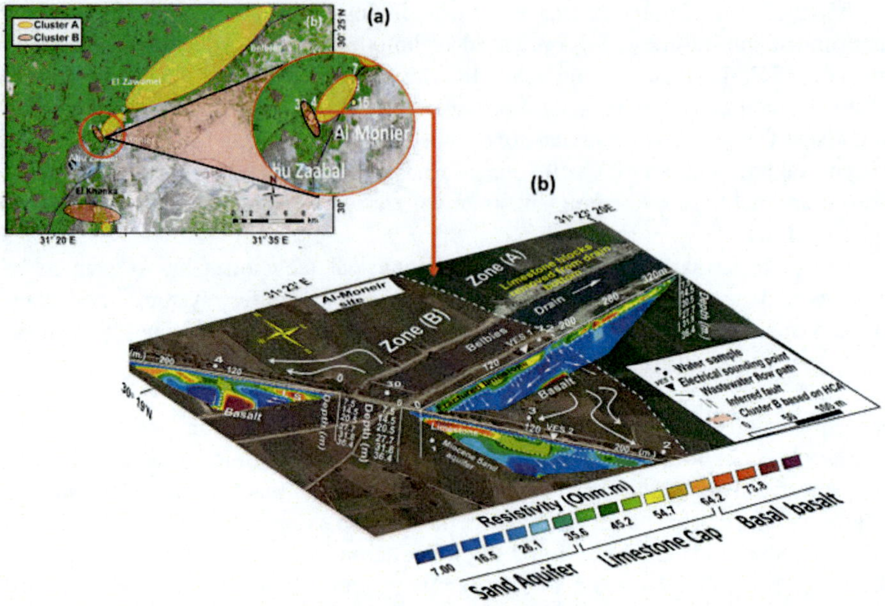

Fig. 5 Application of electrical resistivity tomography for mapping the flow paths of wastewater in the southeastern part of the Nile Delta aquifer: (**a**) the cluster map hierarchical cluster analysis (HCA) shows the rapid changes in the groundwater quality based on the statistical analysis of the total coliform, fecal coliform, and nitrate concentrations, and (**b**) the inverted ERT sections and the suggested wastewater flow paths projected onto the Landsat image of the boundaries cluster A (moderate vulnerability) and cluster b (high vulnerability) [17]

$$\mathrm{WQI} = \frac{\sum q_n\, W_n}{\sum W_n} \tag{2}$$

where q_n: is the quality rating for the water quality parameter, V_n is the estimated value of the water quality parameters of collected samples, S_n is the standard permissible value of the water quality parameter, V_{10} is the ideal value of the water quality parameter impure water, and W_n is the unit weight (1/parameter value). Table 2 lists the classification of water according to the water quality index (WQI).

In a recent study, Ismail et al. [58] employed a comprehensive approach to investigate groundwater quality in the west Assiut and El-Minia districts of Egypt. Their investigation encompassed geo-electrical, hydro-geochemical, and stable isotope (Oxygen-18 and Deuterium) analyses. The results revealed that the collected groundwater samples were suitable for irrigation purposes. Furthermore, the oxygen and hydrogen isotope values indicated that the Eocene aquifer received recharge from both surface water and the Nubian aquifer.

Similarly, Fallatah and Khattab [59] conducted an assessment of groundwater quality and its suitability for irrigation and human consumption in Saudi Arabia.

Table 1 Physical and chemical parameters [55]

		Egyptian standard		WHO [11] Standard	
Item	Unit	Drinking	Irrigation	Drinking	Irrigation
Temperature	c°	Max 5°C	Max 3°C	Max 5°C	Max 3°C
PH	–	6.5–8.5	6.5–8.5	6.5–8.5	6.5–8.5
EC	mS/cm	0.4	0.4	0.4	0.4
Turbidity	NTU	–	–	–	–
TDS	mg/L	500	1,000	300	1,500
TSS	mg/L	–	–	10	50
COD	mg/L	10	50	10	50
BOD	mg/L	6	30	6	30
CO_3	mg/L	1.25	2.5	–	–
HCO_3	mg/L	50	400	–	–
TA	mg/L		–	–	–
Ca	mg/L	100	200	100	200
Mg	mg/L	30	50	30	50
Na	mg/L	20	200	20	175
K	mg/L	10	12	10	12
Cl	mg/L	25	250	25	200
SO_4	mg/L	200	–	–	–
NO_3	mg/L	2	45	25	50
NO_2	mg/L	0.02	0.02	–	0.1
PO_4	mg/L	0.4	5	0.4	5
S	mg/L	10	10	–	0.01
Cr	mg/L	0.05	0.05	–	0.05
Cu	mg/L	0.01	1	0.1	3
Fe	mg/L	0.5	3	0.05	0.2
Mn	mg/L	0.2	2	0.02	0.05
Ni	mg/L	0.02	0.1	–	0.05
Pb	mg/L	0.01	0.1	–	0.05
Zn	mg/L	0.01	5	0.1	5
Total coliform	CFU/100 ml		N/D		
Fecal coliform	CFU/100 ml		N/D		

Table 2 Water quality index and water quality status [57]

Water quality index	Water quality status
0–25	Excellent water quality
26–50	Good water quality
51–75	Poor water quality
76–100	Very poor water quality
>100	Unsuitable for drinking

Their findings demonstrated that, following appropriate treatment processes, approximately 92.5% of the collected groundwater well samples were deemed suitable for drinking and irrigation purposes.

In another study, Giao et al. [60] evaluated the quality of groundwater for drinking purposes in the Mekong Delta, Vietnam. The researchers utilized the irrigation water quality index (IWQI) to assess the groundwater samples. The results indicated that 53.1% of the samples were classified as excellent, 25% as good, 9.4% as poor, 4.7% as very poor, and 7.8% as undrinkable. These variations in quality were primarily attributed to coliform contamination. These studies highlight the significance of assessing groundwater quality using multidisciplinary approaches and provide valuable insights into the suitability of groundwater for different purposes in various regions.

3.3 Experimental Studies

A number of experimental studies investigate groundwater contamination. Physical models (e.g. a sand tank) replicate physical processes, often on a smaller scale than encountered in the field [61]. A sandbox model consists of a tank filled with an unconsolidated porous medium through which water is induced to flow. A major drawback of sandbox models is the problem of scaling down a field situation to the dimensions of the laboratory model. It is not a practical tool to simulate the real situation under different boundary conditions and heterogeneous media [62]. Remya et al. [9] presented experiments to investigate the fate and transport of two significant anions through the soil to explore their potential as groundwater contaminants using by use of an adsorption test and an adsorption-diffusion column test. The results indicated that chloride ions and sulphate ions can serve as reliable chemical indicators of groundwater contamination from various sources. The results from the batch adsorption test indicated a maximum chloride adsorption capacity of 3.7 and 1.16 mg/g in different characteristics of soil-1 (high clay) and soil-2 (high sand) respectively while the corresponding values for sulphate were 24.09 and 13.83 mg/g in the two soils.

Sharma and Jyoti [63] applied an experimental study of groundwater quality improvement by recharging the rainwater. The results showed an improvement of groundwater quality using this method to the developed aquifer strata in a laboratory with the controlled laboratory setup. Ohta et al. [64] applied an experimental and numerical study of the groundwater quality in the Hakkouda Tunnel in the Sankaku-dake Mountains, Aomori prefecture, northern Japan. The test results demonstrate that the chemical composition of the leaching water from the altered rock sample closely resembles that of the groundwater present within the same rock mass. Abdoulhalik et al. [65] investigated how layered heterogeneity affects transient saltwater up coning in a laboratory-scale coastal aquifer using an experimental study. The results highlighted the increased vulnerability of such layered aquifer systems to saltwater up coning compared to homogeneous systems. Zhu et al. [66]

applied sand tank experiments for studying the intermittent abstraction on SWI, the study showed that this method enhances the mixing zone width relative to constant pumping. Also the constant pumping causes a larger seawater in the aquifer than using the intermittent abstraction.

3.4 Mathematical Studies

Analytical and numerical models have recently been developed to investigate groundwater contamination in polluted aquifers. The modelling process is translating the physical properties into mathematical terms and the groundwater flow model is used to calculate head and flow characteristics, but the transport model is used to calculate concentration [67]. Groundwater models were used and continued in use to answer specific questions or to achieve a specific objective. Modelling objectives and methods vary depending on the nature of the question being asked and the site's or system's characteristics. The necessary level of detail or accuracy of results can vary, depending on the objective. Within the natural resource industry, groundwater models are used for many purposes. In terms of environmental effects, models are used for environmental assessments or other permitting requirements. Typical uses of groundwater models include the following [68]:

- Conceptualize and quantify current conditions (synthesize existing information)
- Understand system dynamics to identify and quantify controlling and significant processes (e.g. surface water – groundwater interactions, recharge areas, seepage rates, transport dynamics, etc.)
- Predict a future change or impact in response to a planned or potential stress, such as water table drawdown related to a planned extraction well or construction of a given mine plan component (e.g. inflow to open pits, seepage from tailings facilities, etc.)
- Evaluate the sensitivity of the system to model uncertainty and/or magnitude of stresses
- Identify capture zones or source protection areas (for groundwater resource projects)
- Assess mitigation options (e.g. seepage interception, pump and treat, etc.) to guide future data collection
- Improve the design of monitoring networks (e.g. determine aquifer units and/or specific areas requiring additional monitoring)
- Act as a management tool (e.g. assess different proposed management scenarios in managing a multiple-use aquifer)
- Evaluate engineering designs (e.g. phreatic surface in a tailings dam, mine dewatering systems, detailed mitigation designs)

In their study, Ger et al. [69] developed a program in the C++ programming language specifically designed to address groundwater contamination problems. This program offers a range of capabilities, including the visualization of groundwater head, contamination distributions, and velocity fields within the discretized aquifer domain. Additionally, the program provides users with multiple output file options to save and analyse the obtained solutions. Further details about the program can be found in their publication available at (https://open.metu.edu.tr/handle/11511/53670). Abd-Elaty et al. [70, 71] developed a numerical study to investigate the effect of climate changes on groundwater quality and quantity. In a simulation conducted by Chen et al. [72] on groundwater contamination in Tainan City, Taiwan, the study findings revealed a significant 72% increase in the total mass of pollutants within the aquifer over a span of 10 years. Li and Yin [73] developed an analytical solution for the convection-diffusion equation considered by the two-dimensional and inverse Fourier transform. The study provided a quick and intuitive decision-making basis for water resources protection, especially in dealing with water pollution emergencies. Abd-Elaty et al. [74] conducted a comprehensive investigation on soil and groundwater contamination resulting from fertilizer usage in arid and semi-arid regions with intensive pumping. The study revealed that excessive pumping activities led to a decline in groundwater levels, thereby reducing the transfer of fertilizers to the groundwater and increasing soil contamination. Abd-Elaty et al. [75] investigated and evaluated groundwater salinity related to groundwater abstraction and desalination brine deep injection on a coastal aquifer. Bahrami and Zarei [76] conducted an assessment and modelling of groundwater quality in a fissured aquifer in Iran, focusing on its suitability for various purposes such as drinking, irrigation, and industrial use. The study employed water quality indices and leveraged the GIS technique to provide comprehensive insights into the groundwater quality dynamics in the region. The results showed that reduction of water quality from the dry season to the wet season shows the destructive effect of anthropogenic activities such as agriculture on groundwater quality. In a recent study, Khiavi et al. [77] applied machine learning algorithms (MLAs) to assess groundwater quality (GWQ) across the Central Plateau region of Iran. Their analysis of water quality parameters revealed that approximately 42.71% of the area had poor GWQ. Additionally, they found that 18.93% and 38.36% of the region fell within the moderate and high GWQ classifications, respectively. Eid et al. [78] applied support vector machine regression (SVMR), geographic information systems (GIS), and the irrigation water quality index (IWQI) to evaluate the groundwater quality Souf Valley, Algeria; the study showed that the combination of IWQIs, SVMR, and GIS was effective and an applicable technique for interpreting and forecasting the irrigation water quality used in both arid and semi-arid regions.

4 Aquifers Protection Techniques

The growing environmental awareness and the recognition of the need for protection have led to a heightened focus among researchers on studying solute transport and its relationship to groundwater contamination [79]. Abd-Elaty and Straface [80] presented a numerical study to manage groundwater salinity in coastal aquifers.

4.1 Operation Management

Groundwater aquifers are recharged from rainfall and seepage from canals and drains. The discharge occurs by irrigation and drinking water abstraction. Over-pumping is considered the main cause of groundwater salinity [10]. Ezzeldin et al. [81] employed an artificial jellyfish search algorithm to develop an optimal management model for the coastal El-Arish Rafah aquifer in Egypt. Their model developed under constraints to control saltwater intrusion into pumping wells successfully simulated the management of the unconfined coastal aquifer system. Through the application of this computational approach, the study demonstrated the potential of the artificial jellyfish search algorithm for determining groundwater abstraction strategies that could sustainably balance freshwater demands while minimizing intrusion risks over both the short- and long-term. With further refinement and validation using local aquifer data, models of this nature may offer national water regulators and local authorities an effective tool to guide evidence-based groundwater governance into the future.

Vahdat-Aboueshagh et al. [82] applied machine learning for multi-objective optimization of aquifer storage and recovery (ASR) operations under uncertainty via surrogates. The results showed that optimal ASR operations are highly influenced by hydraulic conductivity and longitudinal dispersivity. The high values of hydraulic conductivity lead to a higher number of active stress periods during storage and recovery phases, which requires a large volume of extraction to recover the dispersed injectate while the higher ratios of longitudinal dispersivity to hydraulic conductivity adversely impact the injectate recovery efficiency. Zhu et al. [66] developed a new technique for aquifer operation using intermittent pumping on SWI; the study showed that intermittent pumping increases the freshwater storage to be extracted before well salinization. This method could be good for non-pumping which allows the aquifer to recover.

4.2 Hydraulic Barrier Management

To safeguard the quality of freshwater within the aquifer, it is imperative to effectively manage the irrigation and wastewater drainage systems. Additionally, the modification of boundary conditions pertaining to head and concentrations in these networks can be accomplished through three specific measures, as depicted in Fig. 6 and outlined by Abd-Elaty et al. [71].

4.2.1 Reuse of Agricultural Drainage Water

The agricultural wastewater is initially utilized separately from domestic-industrial wastewater. Subsequently, it undergoes a blending process with fresh water,

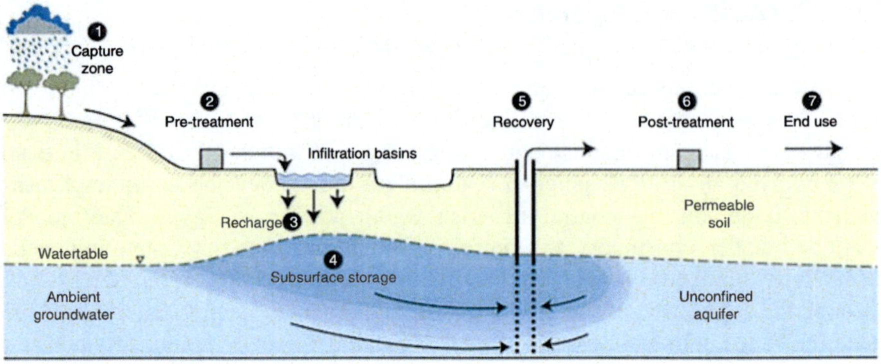

Fig. 6 Managed aquifer recharge cycle [83, 84]

considering the required salinity level for water supply systems and irrigation canals. This integration helps in maintaining the water level within irrigation networks, ensuring adequate support for irrigation purposes. Abdelhalim et al. [85] conducted a study on this subject, providing valuable insights into this practice. The study carried out a numerical modelling for groundwater management applied to the Samalut area in the Minia Governorate, Egypt where the River Nile acts as a drain in the area, while El-Ibrahimiya Canal and Bahr Yusef act as a source of aquifer recharge. The proposed scenarios showed that surface water is important in recharging the aquifer during increasing groundwater extraction. The findings of the study indicate a notable decrease in aquifer storage change, transitioning from +48,125 m^3 day^{-1} in the current state (2013) to +27,134 m^3 day^{-1} and −869 m^3 day^{-1} when groundwater extraction is increased by 25% and 50% respectively. In a separate investigation by Ashu and Lee [86], focused on the reuse of agricultural drainage water within a mixed land-use watershed situated in the Osan River watershed of central Korea, it was observed that the quantities of reused water accounted for 77.2% and 49.8% of the available ADW during the summer and winter seasons respectively. These figures corresponded to 49.1% and 54.5% of the seasonal canal delivery in the study area.

4.2.2 Constructed Wetlands (CWs)

Constructed wetlands (CWs) have emerged as effective methods for mitigating pollution in waterways and water bodies. These engineered systems mimic the natural processes that occur in wetlands, utilizing a combination of vegetation, soil, and microorganisms to treat and improve polluted water quality. CWs are designed to receive and treat various sources of pollution, including urban runoff, industrial effluents, and agricultural runoff. As polluted water flows through the wetland system, it undergoes a series of physical, chemical, and biological processes that remove or transform contaminants [87].

4.2.3 Natural and Artificial Recharge

Natural and artificial recharge methods play a crucial role in replenishing aquifers and mitigating the pollution load in waterways. One effective approach involves utilizing the hydraulic method, which entails mixing domestic-industrial wastewater with submerged porous media that are lightweight and capable of overlapping. This process facilitates water turnover and enhances the dissolved oxygen content, thereby aiding in the reduction of pollutants. Recognizing the need for alternative water resources, managed aquifer recharge (MAR) utilizing treated wastewater (TWW) has emerged as a viable solution [88]. MAR constitutes an essential component in addressing water scarcity, safeguarding water security, combating declining water quality, stabilizing falling water tables, and protecting groundwater-dependent ecosystems. It is considered a cost-effective, environmentally friendly, resilient, and socially acceptable solution. However, the widespread implementation of MAR has been hindered by factors such as limited awareness, inadequate understanding of aquifers, incomplete risk perception, and insufficient policies for integrated water management, including the integration of MAR with demand management strategies [89]. Recharging aquifers can be achieved through various sources, including surface water from rivers, storm water, and treated wastewater [90]. The following technologies can be employed for MAR:

(i) *Natural recharge* by infiltration ponds or tanks into the aquifer's unsaturated zone. The water structures with Ponds infiltration dams, reservoirs, storm water management or runoff, and individual wastewater treatment units [90, 91].
(ii) *Artificial recharge* can be accomplished through injection wells, employing methods such as aquifer storage (AR), aquifer storage and recovery (ASR), and aquifer storage transfer and recovery (ASTR). This approach proves particularly effective for recharging confined or semi-confined aquifers. Another technique, known as induced recharge or riverbank filtration, involves the installation of pumping wells near the shoreline of rivers...

The injection of treated wastewater (TWW) into the aquifer serves the purpose of creating a hydraulic barrier to prevent seawater intrusion while simultaneously enhancing water quality in the vicinity through the implementation of soil aquifer treatment (SAT) techniques [88]. Chocat et al. [35] demonstrated that the management method of utilizing managed aquifer recharge (MAR) is widely adopted by large cities, including notable examples such as Berlin, Paris suburbs, Lyon, Dunkirk, and Geneva. This technique effectively manages storm water by capturing surface runoff and directing it into infiltration basins for recharge into the aquifer. Iván et al. [92] used managed aquifer recharge (MAR) by proposing the recharge of island aquifer by deep wells with regenerated water in Gran Canaria (Spain). The study revealed that the implementation of MAR holds the potential to complete the water cycle in the Canary Islands. By utilizing reclaimed water as an alternative resource, the dependency on the aquifer can be alleviated, thereby reducing the strain on its reserves. Furthermore, MAR gains significant momentum in the face of a changing island climate. Page et al. [93] showed that the suitable location for the MAR scheme is typically in areas of low topographic relief; and the organizational capability, institutional arrangements, and supportive policies to operate the scheme sustainably and economically. Furthermore, several factors contribute to the feasibility of MAR implementation. These factors include the ongoing demand for water in high-value agriculture, the availability of water for recharge, and the presence of a suitable aquifer capable of storing and recovering water. In their research, Perzan et al. [94] examined the controls on flood-managed aquifer recharge within a heterogeneous vadose zone, focusing on the infiltration rate at the surface. Their findings suggest that the method can lead to an increase in saturated-zone storage beneath a specific site, as well as an enhancement in flux recharge. Moreover, it can result in an augmented lateral discharge away from the site, occurring beneath the water table.

4.2.4 Positive and Negative Hydraulic Barriers

Pool and Ramírez [95] studied the dynamics of negative hydraulic barriers to prevent seawater intrusion, the study showed three cases for the pumping rate, the first is a high pumping rate in which drawdowns cause saltwater to flow along the aquifer bottom around the seawater well, contaminating the freshwater well. The second is low pumping rate at the seaward well leads to insufficient desalinization at the freshwater well. The third is the critical pumping rate at the seawater well is defined as that which produces optimal desalinization at the freshwater well. In their study, Ozaki et al. [96] highlighted the widespread utilization of a negative hydraulic barrier approach, which involves the implementation of barrier wells that extract

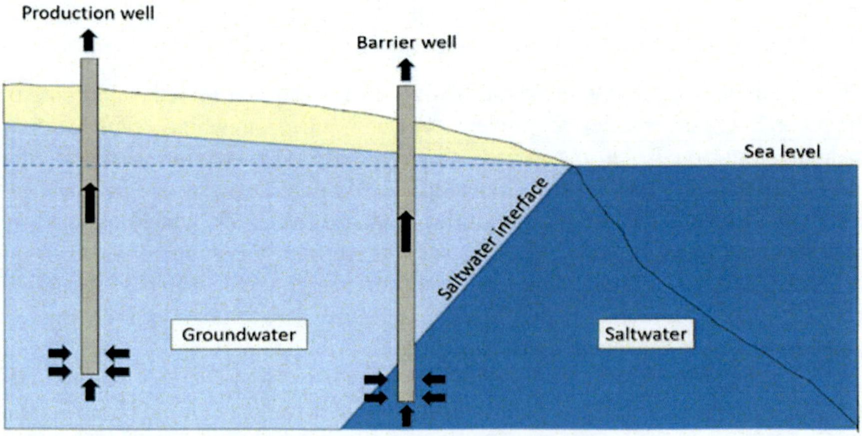

Fig. 7 Conceptual diagram of saltwater pumping by a barrier well [96]

saline or brackish water (Fig. 7). By maintaining pumping rates below the critical ratio, the study demonstrated that the critical pumping ratio between the barrier and the production well can effectively control saltwater intrusion. Saad et al. [97] developed optimal management of mixed hydraulic barriers in coastal aquifers using multi-objective Bayesian optimization; the results showed that the injection barriers have higher control over the remediation system while the abstraction barriers are useful as an alternative source of water. In addition, the average amount of water required for abstraction ranged from 1.5% to 25% of that for injection.

4.3 *Material Management*

The material management (MM) method has been devised to assess the influence of lining materials on groundwater and surface water contamination. This method involves modifying the conductance of polluted drains by implementing linings composed of low-permeability materials. The selection of appropriate lining materials is crucial for effectively safeguarding the aquifer against contamination. The following materials have been considered for lining purposes:

4.3.1 Geo-synthetic Materials

Geo-synthetic materials (i.e. geo-textiles, geo-nets, geo-membranes, geo-synthetic bentonite, and others) are crucial in protecting groundwater from pollution in various applications. These materials provide impermeable barriers, filtration, and separation solutions, preventing the migration of contaminants and preserving groundwater quality. Geo-synthetic bentonite and geo-membranes are commonly employed in landfill liners and caps. These materials act as impermeable barriers, preventing the leachate from surface landfills and dump sites. Geo-synthetic bentonite is composed of bentonite clay and geotextiles, and provides excellent hydraulic containment properties, while geo-membranes provide a continuous impermeable layer [98, 99]. In addition, geo-textiles and geo-membranes are employed as barriers to isolate contaminated soil or waste materials from the surrounding environment, preventing the spread of pollutants to shallow groundwater aquifers. It's important to note that the selection and design of geo-synthetics for groundwater protection depend on site-specific conditions, regulatory requirements, and the nature of the pollutants involved.

Geo-membrane can vary depending on various factors such as the specific formulation of the material, its thickness, and the testing conditions. For example, the permeability coefficients for high-density polyethylene ranged from 10^{-11} to 10^{-17} cm/s. Moreover, it boasts stable chemical properties, corrosion resistance, and a well-established track record in project implementation [100, 101]. Lambert and Touze-Foltz [102] estimated the geo-membranes' permeability is less than 10^{-6} m day^{-1}. In contrast, a typical nonwoven needle-punched geotextile has a hydraulic conductivity below 4×10^{-8} m s^{-1} under normal conditions, although it can reach approximately 1×10^{-7} m s^{-1} or higher [103, 104]. In a study conducted by Elkamhawy et al. [105], the effectiveness of different liner materials was evaluated. The findings indicated that the concrete liner exhibited the highest efficiency, followed by the geo-membrane liner, and then the bentonite liner, with efficiencies of approximately 99%, 96%, and 54%, respectively.

4.3.2 Natural Materials

To protect the shallow groundwater aquifers, natural materials (compacted clay soils, natural fibres, vegetation) are sometimes used in combination with geo-synthetic materials to enhance the performance of landfill and dumpsite liner systems [106]. For example, natural clay soils, such as bentonite or kaolin, are often used as a component in composite liners or as geo-synthetic clay liners (Fig. 8). It consists of a layer of bentonite clay sandwiched between geotextiles. The swelling properties of bentonite clay provide an additional barrier to the migration of liquids and contaminants, enhancing the overall performance of the liner system [107, 108]. Compacted clay is used as a natural soil with low permeability to manage groundwater contamination where the compacted clay liner permeability is 1×10^{-7} cm s^{-1} at a thickness of 100 cm. In particular, Abd-Elaty et al. [87] conducted research demonstrating the considerable effectiveness of clay covers in confined aquifers for mitigating groundwater contamination from polluted waterways. Their findings indicate that such protective measures are notably more successful in confined aquifers compared to unconfined ones.

4.3.3 Mixed Materials

Mixed materials can play a crucial role in landfill liner systems by combining the advantages of different components to enhance overall performance. Mixed materials can help manage hydraulic conditions within the landfill liner system. The liner system can control the flow of water or leachate by combining layers with varying

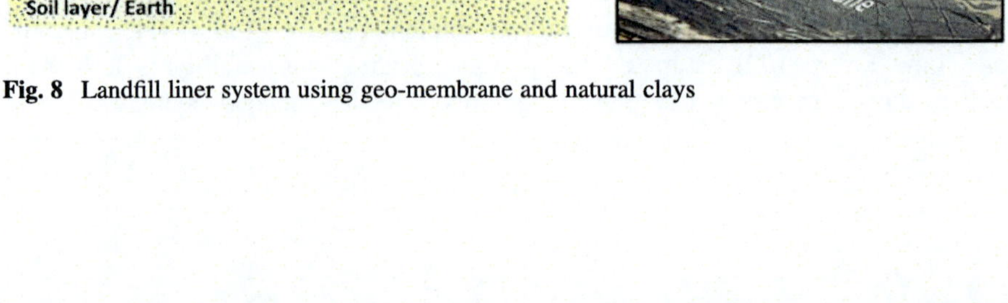

Fig. 8 Landfill liner system using geo-membrane and natural clays

permeability, such as geo-textiles, geo-composites, or geo-nets. This promotes proper drainage and prevents the build-up of hydraulic pressure that could lead to liner failure [109]. It's important to note that the selection and combination of mixed materials in landfill liner systems should be based on site-specific conditions, regulatory requirements, and engineering considerations. Professional geotechnical and environmental engineers assess these factors to design and implement liner systems that effectively contain and manage waste while minimizing environmental impacts [110].

Here are some examples of using mixing materials in landfill and dumpsite liner systems. The mixing cement and bentonite has a remarkable strength and exhibits low compressibility, rendering it particularly suitable for deployment on steep slopes characterized by unstable soil conditions. Its hydraulic conductivity measures approximately 10^{-6} cm s^{-1} [111]. Nevertheless, comprehensive research and practical insights gained across the United Kingdom and Europe have revealed the potential to attain a hydraulic conductivity of less than 9^{-10} m s^{-1} by introducing ground granulated blast furnace slag as a supplement to the cement-bentonite mixture, as demonstrated by the work of Jefferis [112]. The hydraulic conductivity of concrete and mortar materials were found; 5.67×10^{-15} cm s^{-1} for concrete and 5.87×10^{-16} cm s^{-1} for mortar [113].

4.4 *Physical Barriers Management*

The physical barrier walls are intricately designed structures, strategically positioned at different depths and intervals, customized to suit the specific characteristics of the aquifer and the contaminated watercourse. Utilizing a combination of chemical or cement grout and sheet piling, these cut-offs wall materials play a vital role in restraining the migration of polluted groundwater. These barrier walls effectively impede the flow of contaminated groundwater by employing materials such as chemical or cement grout and sheet piling. The physical barrier walls are meticulously designed, taking into account the geological properties of the aquifer (Fig. 9). For instance, in sandy aquifers prone to rapid water flow, barrier walls are strategically placed at varying depths to ensure maximum efficiency in blocking the movement of contaminants [87, 115]. The physical barrier walls include: (i) cutoff wall, (ii) subsurface dam, and (iii) earth fill.

Vertical barriers with low permeability are extensively utilized in environmental control systems to effectively manage the migration of contaminants from polluted sites. In the United States, soil-bentonite slurry wall technology is commonly preferred, whereas in the United Kingdom, cement-bentonite slurry wall technology is favoured [116]. Slurry walls are widely recognized as one of the most commonly employed subsurface barrier technologies, serving as a fundamental benchmark in barrier construction [111]. These walls have been extensively utilized as passive vertical barriers to regulate the horizontal movement of groundwater and contaminants, effectively curtailing the migration of pollutants through the subsurface

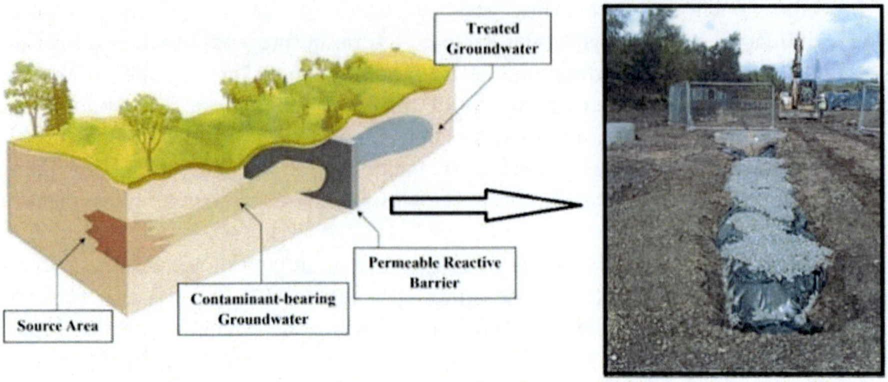

Fig. 9 The concept of permeable reactive barrier, PRB [114]

[106]. They are composed of vertical trenches filled with permeable materials (Fig. 9) exhibiting a hydraulic conductivity of 1×10^{-6} cm s^{-1} or lower [117]. Luca et al. [118] installed a vertical slurry wall around an Italian quarry lake: complications arising and simulation of the effects on groundwater flow, the study showed that the construction of the slurry wall would not cause any change in the piezometric head in the area where there are municipal wells; hence, it will not have any negative effect on the functionality of the municipal wells. *Plastic concrete barriers* are a modified type of slurry wall with permeability ranging from 10^{-6} to 10^{-8} cm s^{-1} [117]. *Sheet pile wall* is a conventional type of subsurface barrier adapted to geotechnical construction and groundwater contamination [119].

The subsurface dam, composed of an impermeable wall, is strategically constructed at the bottom of the aquifer, allowing for the discharge of fresh upstream groundwater through an opening in the upper aquifer. Conversely, the cut-off wall, depicted in Fig. 10, consists of an impermeable barrier constructed in the upper aquifer, with an opening at the bottom of the aquifer [120].

McMahon [121] evaluated the flow through sheet pile walls for vertical barriers using two studies, the first is bentonite slurry and an organic polymer was used to seal the interlocks between the panels of waterloo barrier sheet piles. With the bentonite slurry as an interlock sealant, a bulk hydraulic conductivity of 6×10^{-11} m s^{-1}, whereas the bulk hydraulic conductivity for the Waterloo sheet pile wall using the organic polymer sealant is less than 1×10^{-11} m s^{-1}. The second is a sheet pile barrier using Arbed sheet piles (European sheet piles with a different interlock system than sheet piles manufactured in the United States) that was driven through a silty sand deposit into an underlying clay deposit and a field pumping test was performed using a pumped well and four observation wells. In this case, no interlock sealant was used and the measured bulk hydraulic conductivity of the wall system was 7×10^{-9} m s^{-1}. Luyun et al. [122] showed that the subsurface dams with smaller heights could achieve faster removal of inland residual saltwater as well as more reduction of the expected increase of the saltwater wedge height along the coastline boundary than higher dams. Elkhawas [110] carried out a numerical study

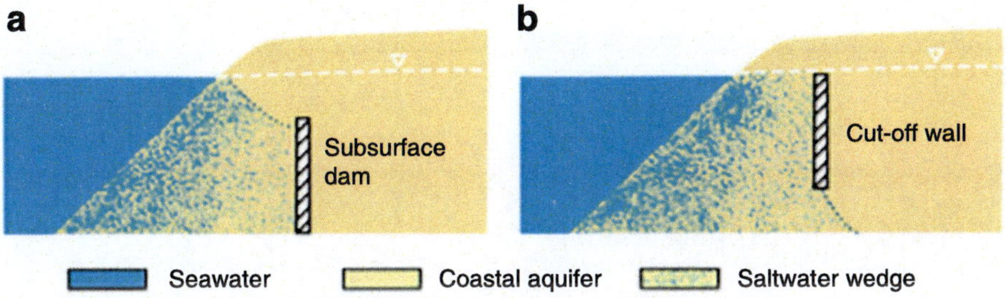

Fig. 10 (**a**) Schematics of a subsurface dam and (**b**) a cut-off wall [120]

using a 2-D finite element model, Z-Soil finite elements program, and experimental study to develop a mix, as a construction material, for a cut-off wall containing sand, silt, bentonite, and cement by-pass dust. The result of the experimental study showed that the minimum permeability was achieved for a mixture of 10% CBPD, 15% bentonite, 15% silt, and 85% sand by weight of the native soil which was represented in the laboratory program by the weight of sand and silt. The permeability value obtained using this mix was in the order of 10^{-6} cm s^{-1}. Chang et al. [123] developed the concept of minimum effective height for the subsurface dams to mitigate aquifer salinity and the influence of subsurface dams on the freshwater discharge. Abdoulhalik et al. [124] proposed the mixed physical barrier method which combines an impermeable cut-off wall with a semi-permeable subsurface dam on the seaward side. This method caused a visible saltwater lifting process whereby freshwater flowing below the wall opening with increased velocity transported a dispersive flux of salt above the subsurface dam and discharged it towards the outlet. Abd-Elaty et al. [87] found that installing cut-off walls along drainage sides led to a reduction in the spread of contamination within shallow aquifers. Abd-Elaty et al. [87] demonstrated that the cut-off walls can effectively help manage contaminant plumes in shallow underground water systems. However, the study also showed such barriers have little impact on contaminant mobility. While, Qian et al. [125] used a composite cut-off wall consisting of HDPE geomembrane combined with soil-bentonite backfill that can be effectively applied to site remediation projects. The study showed that this wall is considered to be currently the safest and most effective vertical barrier technique to block horizontal pollution migration as presented in Fig. 11.

Gao et al. [126] applied an experimental study for mixed physical barriers on residual saltwater removal and groundwater discharge in coastal aquifers. The efficiency of MPB in removing residual saltwater could be 40–100% and 0–56% higher than that of traditional subsurface dams and cut-off walls, respectively. Zheng et al. [127] studied the dynamic desalination of SWI after the construction of cut-off walls in a coastal unconfined aquifer, the results showed that the desalination performance degrades dramatically when the hydraulic conductivity of the wall is greater than a certain threshold (8×10^{-7} m/s in the simulated cases). Chang et al. [123] studied the SWI repulsion driven by groundwater level differences around

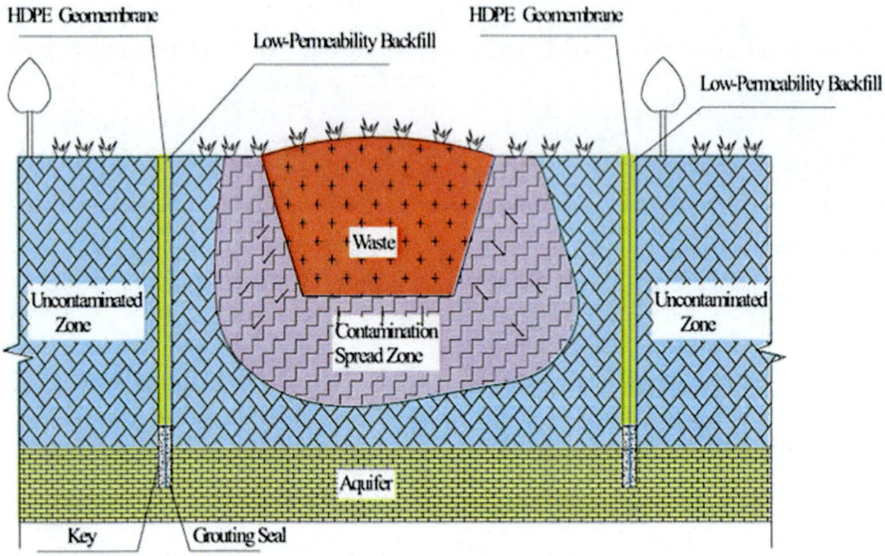

Fig. 11 Schematic diagram of the function of geomembrane composite cut-off wall [125]

cut-off walls on seawater intrusion in unconfined aquifers. The results showed that any way to increase the inland hydraulic head upstream of the wall would improve the efficiency of cut-off walls such as the freshwater recharge, the air injection, and the subsurface dam.

Oude Essink [128] showed that the landfill provides the land area required to meet growing urbanization and population increase and create a foreland that may develop a new zone for a freshwater body. Hu and Jiao [129] concluded that groundwater discharge and the head were increased after land reclamation and had a long-lasting impact on the aquifer system. Abd-Elaty et al. [12] studied cost-effective management measures for coastal aquifers affected by saltwater intrusion and climate change. The study showed that the coastal land reclamation is useful for the integrated management of coastal zones by delaying aquifer salinity, protecting fresh groundwater bodies, increasing agricultural lands, supporting surface water supplies by harvesting rainfall and flash flooding, and desalinating saline water using wave energy. Abd-Elaty et al. [130] showed that a comprehensive cost-benefit analysis is required in order to further account for the feasibility and the economic costs related to the construction of physical subsurface barriers for groundwater protection in coastal aquifers.

5 Conclusion

To face the water scarcity challenge in arid regions, effective water management policies are essential for sustainable development. These strategies may include implementing water conservation technologies, improving the traditional irrigation system, promoting the use of drought-resistant crops, and adopting efficient water demands in urban and industries. Sustainable groundwater management, such as monitoring and regulating groundwater extraction, implementing recharge projects, and promoting water recycling, can help ensure long-term water availability. In some cases, arid regions explore alternative water sources to supplement groundwater. This may include seawater desalination, wastewater treatment and recycling, rainwater harvesting, and importing water from other regions. While, these options can provide additional water supplies, they often come with high costs and energy requirements.

In arid regions, groundwater is considered an important water source and involves a variety of inputs such as precipitation, river leakage, drainage canals, and excess irrigation. However, groundwater contamination is considered a significant threat, leading to the degradation of freshwater storage due to agricultural, industrial, sewage, open drains, and saltwater intrusion.

6 Recommendations

Through the following recommendations groundwater resources management can achieve remarkable progress in effectively facing the challenges of groundwater contamination and ensuring the long-term availability of sustainable water supplies.

1. Geophysical techniques, such as electrical resistivity, seismic surveys, and ground-penetrating radar, provide valuable insights into the subsurface characteristics of a site. These methods help identify geological formations, lithology, structural features, and the presence of aquifers or pollution flow paths. A thorough understanding of site characterization is crucial for effective groundwater conservation and management.
2. Integration of hydrogeological and geophysical surveys can be employed for long-term monitoring of groundwater systems. Regular monitoring allows for tracking changes in groundwater levels, and water quality, identifying trends, and evaluating the effectiveness of conservation measures. Geophysical techniques, such as time-lapse resistivity surveys, geophysical well logging, and the GRACE satellite dataset provide valuable data for assessing aquifer recharge rates, mapping variations in groundwater storage, and detecting potential anomalies or subsidence.
3. It is highly recommended to prioritize the installation of wastewater drains, landfills, disposal dumpsites, and wastewater ponds in low permeability and thick layers, such as clays or consolidated bedrock, as a key strategy to minimize

contamination. Extensive research and practical experience have consistently demonstrated that this approach yields superior results compared to installing drains in high permeability or fractured layers, such as sandy-rich soils, where the presence of a protective clay cap is absent. By focusing on low permeability layers, the risk of contaminants infiltrating is significantly reduced, ensuring effective groundwater conservation and safeguarding the quality of this vital resource.

Acknowledgements This work was conducted as part of a comprehensive literature review for the research project titled "Roadmap for sustainable groundwater resources in the Nile Delta using integrated approaches under the impact of climate change and wastewater pollution." The project was funded by the Science, Technology, & Innovation Funding Authority (STDF), Egypt, under the grant number 46209, Applied Science 1st Call. The authors would like to express their sincere gratitude to STDF for their generous support in making this study possible. The authors thank the Department of Water and Water Structures Engineering, Faculty of Engineering, Zagazig University, Zagazig 44519, Egypt, for constant support during the study.

References

1. Ibrahim A, Gemail KS, Bedair S et al (2023) An integrated approach to unravel the structural controls on groundwater potentialities in hyper-arid regions using satellite and land-based geophysics: a case study in Southwestern Desert of Egypt. Surv Geophys 44:783–819. https://doi.org/10.1007/s10712-022-09755-8
2. Gemail K, Ibrahim A, Bedair S (2021) Geophysical groundwater exploration in arid regions using integrated land-based magnetic and DC resistivity measurements: a case study at Gilf Kebir area, South Western Desert, Egypt. In: Negm A, Elkhouly A (eds) Groundwater in Egypt's deserts. Springer, Cham. https://doi.org/10.1007/978-3-030-77622-0_5
3. UNEP World Conservation Monitoring Centre (UNEP-WCMC) (2008) Annual report 2007. Available. online: https://www.unep-wcmc.org/system/dataset_file_fields/files/000/000/214/original/Annual_Report_2007.pdf?1400677478. Accessed latest on 21st Nov 2020
4. IWMI International Water Management Institute (2012) Projected global water Scracity 2025. Poster [1210x904]. Aviable at Data & Tools:: IWMI (cgiar.org)
5. Jasechko S, Perrone D (2017) Hydraulic fracturing near domestic groundwater wells. PNAS 114:13138–13143. https://doi.org/10.1073/pnas.1701682114
6. WHYMAP (2008) Groundwater resources of the world: transboundary aquifer systems. Poster for special edition for 4th world water forum, Mexico City, March 2006. Available www.whymap.org
7. Abd-Elaty I, Saleh OK, Ghanayem HM, Grischek T, Zelenakova M (2021) Assessment of hydrological, geohydraulic and operational conditions at a riverbank filtration site at Embaba, Cairo using flow and transport modeling. J Hydrol Regional Stud 37. https://doi.org/10.1016/j.ejrh.2021.100900
8. Giovanna DF, Mauro G, Sergio N, Stefano M, Laura C, Chiara V (2014) Numerical modeling of groundwater flow in the coastal aquifer system of Taranto (Southern Italy). Geophys Res Abstr 16. EGU20 14-393-1
9. Remya N, Roshni T, Yadav RR, Shukla N (2019) Experimental investigation of groundwater contamination by surface sources: determination of adsorption capacity, diffusion, and sorption potential of selected anions in different soils. Environ Qual Manag 2019(29):139–148. https://doi.org/10.1002/tqem.21658

10. Bear J, Cheng AH, Sorek S, Quazar D, Herrera I (1999) Seawater intrusion in coastal aquifers, concepts, methods and practices. Kluwer Academic Publisher, Dordrecht. ISBN 0-7923-5573-3
11. WHO (2011) Guidelines or drinking-water quality.4th edn. World Health Organization, Geneva
12. Abd-Elaty I, Kushwaha NL, Grismer ME, Elbeltagi A, Kuriqi A (2022) Cost-effective management measures for coastal aquifers affected by saltwater intrusion and climate change. Sci Total Environ 836:155656. https://doi.org/10.1016/j.scitotenv.2022.155656
13. Udmale P, Ichikawa Y, Manandhar S et al (2014) Farmers′ perception of drought impacts, local adaptation and administrative mitigation measures in Maharashtra State, India. Int J Disaster Risk Reduct 10:250–269. Return to ref 2014 in article
14. Qian H, Chen J, Howard KWF (2020) Assessing groundwater pollution and potential remediation processes in a multi-layer aquifer system. Environ Pollut 263:114669. https://doi.org/10.1016/j.envpol.2020.114669
15. Abd-Elaty I, Straface S, Kuriqi A (2021) Sustainable saltwater intrusion management in coastal aquifers under climatic changes for humid and hyper-arid regions. Ecol Eng 171. https://doi.org/10.1016/j.ecoleng.2021.106382
16. Abd-Elaty I, Saleh OK, Ghanayem HM, Zeleňáková M, Kuriqi A (2022) Numerical assessment of riverbank filtration using gravel back filter to improve water quality in arid regions. Front Earth Sci 10:1006930. https://doi.org/10.3389/feart.2022.1006930
17. Abu Salem H, Gemail KS, Nosair AM (2021) A multidisciplinary approach for delineating wastewater flow paths in shallow groundwater aquifers: a case study in the southeastern part of the Nile Delta, Egypt. J Contam Hydrol 236:103701. https://doi.org/10.1016/J.JCONHYD.2020.103701
18. Commer M, Pride S, Donald R, Vasco W, Finsterle S, Kowalsky MB (2020) Imaging of a fluid injection process using geophysical data – a didactic example. Geophysics 85(2) 1MA-Z8. https://doi.org/10.1190/geo2018-0787.1
19. Gemail KH, Ghomimi A, Shebl S, Soliman SH, El-Sayed K, Saleh A (2023) Delineating groundwater flow-paths in fractured aquifers under hazardous environment using conceptual and geophysical modeling with a case study. https://doi.org/10.21203/rs.3.rs-2367575/v1 (preprint)
20. Rani P, Di Maio R, Piegari E, Soupios P, Milano L (2016) Time evolution of self-potential anomaly sources due to organic contaminant transport by different data inversion approaches. Adv Water Resour
21. Simyrdanis K, Papadopoulos N, Soupios P, Kirkou S, Tsourlos P (2018) Characterization and monitoring of subsurface contamination from olive oil mills' waste waters using electrical resistivity tomography. Sci Total Environ:637–638., 991–1003. ces, 28, 467–477. https://doi.org/10.1016/j.scitotenv.2018.04.348
22. Soldi M, Jougnot D, Guarracino L (2019) An analytical effective excess charge density model to predict the streaming potential generated by unsaturated flow. Geophys J Int 216(1): 380–394. https://doi.org/10.1093/gji/ggy391
23. Kolsi SH, Bouri S, Hachicha W et al (2013) Implementation and evaluation of multivariate analysis for groundwater hydrochemistry assessment in arid environments: a case study of Hajeb Elyoun–Jelma, Central Tunisia. Environ Earth Sci 70:2215–2224. https://doi.org/10.1007/s12665-013-2377-0
24. Matiatos I (2016) Nitrate source identification in groundwater of multiple land use areas by combining isotopes and multivariate statistical analysis: a case study of Asopos Basin (Central Greece). Sci Total Environ 541:802–814. https://doi.org/10.1016/j.scitotenv.2015.09.134
25. Mouser PJ, Rizzo DM, Röling WF, Van Breukelen BM (2005) A multivariate statistical approach to spatial representation of groundwater contamination using hydrochemistry and microbial community profiles. Environ Sci Technol 39(19):7551–7559. https://doi.org/10.1021/es0502627

26. Yang YH, Zhou F, Guo HC, Sheng H, Liu H, Dao X, He CJ (2010) Analysis of spatial and temporal water pollution patterns in Lake Dianchi using multivariate statistical methods. Environ Monit Assess 170(1–4):407–416. https://doi.org/10.1007/s10661-009-1242-9
27. Abu Salem HS, Gemail KS, Junakova N, Ibrahim A, Nosair AM (2022) An integrated approach for deciphering hydrogeochemical processes during seawater intrusion in coastal aquifers. Water 14(7):1165. https://doi.org/10.3390/w14071165
28. Arumugam T, Kinattinkara S, Kannithottathil S, Velusamy S, Krishna M, Shanmugamoorthy M et al (2023) Comparative assessment of groundwater quality indices of Kannur District, Kerala, India using multivariate statistical approaches and GIS. Environ Monit Assess 195(1): 29. https://doi.org/10.1007/s10661-022-10538-2
29. Bux RK, Haider SI, Mallah A, Solangi AR, Moradi O, Karimi-Maleh H (2022) Spatial analysis and human health risk assessment of elements in ground water of district Hyderabad, Pakistan using ArcGIS and multivariate statistical analysis. Environ Res 210:112915. https://doi.org/10.1016/j.envres.2022.112915
30. Jia H, Howard K, Qian H (2020) Use of multiple isotopic and chemical tracers to identify sources of nitrate in shallow Groundwaters along the northern slope of the Qinling Mountains, China. Appl Geochem 113:104512
31. Zaporozec A, Miller JC (2000) Groundwater pollution. UNESCO-PHI, París, pp 1–24
32. El Tahlawi MR, Farrag AA, Ahmed SS, Groundwater of Egypt (2008) An environmental overview. Environ Geol 55:639–652. https://doi.org/10.1007/s00254-007-1014-1
33. Bahir M, Salah O, Paula MC (2018) Geochemical and isotopic approach to decrypt the groundwater salinization origin of coastal aquifers from semi-arid areas (Essaouira basin, Western Morocco). Environ Earth Sci 77(13):485
34. Moore WS, Joye SB (2021) Saltwater intrusion and submarine groundwater discharge: acceleration of biogeochemical reactions in changing coastal aquifers. Front Earth Sci 9: 600710. https://doi.org/10.3389/feart.2021.600710
35. Chocat B, Bertrand-Krajewski J-L, Barraud S (2007) Chapitre: Les eaux pluviales urbaines et les rejets urbains de temps de pluie. Les techniques de l'Ingénieur. Edition T.I. Doc. W6 800 Ŕ 8–2007, p 17
36. Gemail KH, El-Alfy M, Ghoneim MF, El-Shishtawy AM, Abd El-Bary M (2017) Comparison of DRASTIC and DC resistivity modeling for assessing aquifer vulnerability in the Central Nile Delta, Egypt. Environ Earth Sci 76(9). https://doi.org/10.1007/s12665-017-6688-4
37. Kurwadkar S, Kanel SR, Nakarmi A (2020) Groundwater pollution: occurrence, detection, and remediation of organic and inorganic Ollutants. Water Environ Res 92:1659–1668. https://doi.org/10.1002/wer.1415
38. Mensah DO, Appiah-Adjei EK, Asante D (2023) Groundwater pollution vulnerability assessment in the Assin municipalities of Ghana using GIS-based DRASTIC and SINTACS methods. Model Earth Syst Environ 9:2955. https://doi.org/10.1007/s40808-022-01680-4
39. Xu H, Xing Y, Daqing W, Yihua H, Zijian C, Yue S, Pu Z, Liang S (2023) Multivariate and spatio-temporal groundwater pollution risk assessment: a new long time serial groundwater environmental impact assessment system. Environ Pollut 317:120621. https://doi.org/10.1016/j.envpol.2022.120621
40. Iqbal J, Chunli S, Rashid A, Yang N, Baloch MYJ, Talpur SA, Ullah Z, Rahman G, Rahman NU, Earjh, and Meer Muhammad Sajjad. (2021) Hydrogeochemical assessment of groundwater and suitability analysis for domestic and agricultural utility in southern Punjab, Pakistan. Water 13(24):3589. https://doi.org/10.3390/w13243589
41. Gemail KH, El-Shishtawy AM, El-Alfy M, Ghoneim MF, Abd El-Bary MH (2011) Assessment of the aquifer vulnerability using Geoelectrical and geochemistry measurements. Study case of El Gharbyia Main Drain, Egypt. J Appl Geophys 75:140–150
42. Gemail KS, Attwa M, Eleraki M, Zamzam S (2017) Imaging of wastewater percolation in heterogeneous soil using electrical resistivity tomography (ERT): a case study at east of tenth of Ramadan City, Egypt. Environ Earth Sci 76:666. https://doi.org/10.1007/s12665-017-7013-y

43. Bouderbala A, Gharbi BY (2017) Hydrogeochemical characterization and groundwater quality assessment in the intensive agricultural zone of the Upper Cheliff plain, Algeria. Environ Earth Sci 76:1–17. https://doi.org/10.1007/s12665-017-7067-x
44. Dugga P, Pervez S, Tripathi M, Siddiqui MN (2020) Spatiotemporal variability and source apportionment of the ionic components of groundwater of a mineral-rich tribal belt in Bastar, India. Groundw Sustain Dev 10:100356. https://doi.org/10.1016/j.gsd.2020.100356
45. Burston MW, Nazari MM, Bishop PK, Lerner DN (1993) Pollution of groundwater in the Coventry region (UK) by chlorinated hydrocarbon solvents. J Hydrol 149(1–4):137–161. https://doi.org/10.1016/0022-1694(93)90104-H
46. Wang XW, Zhong NN, Hu DM, Liu ZZ, Zhang ZH (2009) Polycyclic aromatic hydrocarbon (PAHs) pollutants in groundwater from coal gangue stack area: characteristics and origin. Water Sci Technol 59(5):1043–1051. https://doi.org/10.2166/wst.2009.050
47. Gemail K, Samir A, Oelsner C, Mousa SE, Ibrahim S (2004) Study of saltwater intrusion using 1D, 2D and 3D resistivity surveys in the coastal depressions at the eastern part of Matruh Area, Egypt. Near Surf Geophys 2:103–109. https://doi.org/10.3997/1873-0604.2004007
48. Youssef YM, Gemail KS, Sugita M et al (2021) Natural and anthropogenic coastal environmental hazards: an integrated remote sensing, GIS, and geophysical-based approach. Surv Geophys 42:1109–1141. https://doi.org/10.1007/s10712-021-09660-6
49. Gemail KMS (2012) Monitoring of wastewater percolation in unsaturated sandy soil using geoelectrical measurements at Gabal el Asfar farm, Northeast Cairo, Egypt. Environ Earth Sci 66:749–761. https://doi.org/10.1007/s12665-011-1283-6
50. Nwankwo C, Emujakporue G (2012) Geophysical method of investigating groundwater and sub-soil contamination a case study. Am J Environ Eng 2(3):49–53. Available at SSRN: https://ssrn.com/abstract=2720748
51. Gaber A, Gemail KS, Kamel A, Atia HM, Ibrahim A (2021) Integration of 2D/3D ground penetrating radar and electrical resistivity tomography surveys as enhanced imaging of archaeological ruins: a case study in San El-Hager (Tanis) site, northeastern Nile Delta, Egypt. Archaeol Prospect 28(2):251–267. https://doi.org/10.1002/arp.1810
52. Martens K, Walraevens K (2009) Tracing soil and groundwater pollution with electromagnetic profiling and geo-electrical investigations. In: Ritz K, Dawson L, Miller D (eds) Criminal and environmental soil forensics. Springer, Dordrecht. https://doi.org/10.1007/978-1-4020-9204-6_12
53. Yusuf (2016) Overview of effective geophysical methods used in the study of environmental pollutions by waste dumpsites. Afr Res Rev 10(2). https://doi.org/10.4314/afrrev.v10i2.8
54. Benabdelouahab S, Salhi A, Himi M et al (2019) Geoelectrical investigations for aquifer characterization and geoenvironmental assessment in northern Morocco. Environ Earth Sci 78: 209. https://doi.org/10.1007/s12665-019-8221-4
55. Awad A, Eldeeb H, El-Rawy M (2020) Assessment of surface and groundwater interaction using field measurements: a case study of Dairut City, Assuit Egypt. J Eng Sci Technol 15(1): 406–425
56. Dhakad NK, Deepak S, Choudhary P (2008) Water quality index of ground water (GWQI) of Jhabua Town, M. P. (India). J Environ Res Dev 2(3):443–446
57. Chatterji C, Raziuddin M (2002) Determination of water quality index (WQI) of a degraded river in Asansol Industrial area, P.O. Raniganj, District Burdwan, West Bengal. Nat Environ Pollut Technol 1:181–189
58. Ismail E, Abdelhalim A, Ali A, Ahmed M, Scholger R, Khalil M (2022) Isotopic, geophysical, and hydrogeochemical investigations of groundwater in West Middle Upper Egypt. ACS Omega 48:44000–44011. https://doi.org/10.1021/acsomega.2c05387
59. Fallatah O, Khattab MR (2023) Evaluation of groundwater quality and suitability for irrigation purposes and human consumption in Saudi Arabia. Water 15(13):2352. https://doi.org/10.3390/w15132352

60. Giao NT, Nhien HTH, Anh PK et al (2023) Groundwater quality assessment for drinking purposes: a case study in the Mekong Delta, Vietnam. Sci Rep 13:4380. https://doi.org/10.1038/s41598-023-31621-9
61. Barnett B, Townley LR, Post V, Evans RE, Hunt RJ, Peeters L, Richardson S, Werner AD, Knapton A, Boronkay A (2012) Australian groundwater modeling guidelines. Waterlines Report Series No. 82. National Water Commission, Canberra
62. Wang HF, Anderson MP (1982) Introduction to groundwater modeling: finite difference and finite element methods. Academic Press. Text Book
63. Sharma N, Jyoti PJ (2011) Experimental study of groundwater quality improvement by recharging with rainwater. https://api.semanticscholar.org/corpusid:130248912
64. Ohta T, Hattori S, Kikuchi Y, Shimofusa D (2019) Experimental and numerical study of the groundwater quality in altered Volcanic Rock Area. In: Shakoor A, Cato K (eds) IAEG/AEG annual meeting proceedings, San Francisco, California, 2018, vol 4. Springer, Cham. https://doi.org/10.1007/978-3-319-93133-3_12
65. Abdoulhalik A, Abdelgawad AM, Ahmed AA (2020) Impact of layered heterogeneity on transient saltwater upconing in coastal aquifers. J Hydrol 581:124393
66. Zhu S, Zhou Z, Werner AD, Chen Y (2023) Experimental analysis of intermittent pumping effects on seawater intrusion. Water Resour Res 59:e2022WR032269. https://doi.org/10.1029/2022WR032269
67. Kumar CP (2002) Groundwater flow models. National Institute of Hydrology, Roorkee – 247667 (Uttaranchal)
68. Wels C (2012) Guidelines for groundwater modelling to assess impacts of proposed natural resource development activities. Report no. 194001 British Columbia, Ministry of Environment
69. Ger M, Baran OU, Irfanoglu B (2004) Numerical simulation of groundwater contamination. Building partnerships, building partnerships, ASCE, https://doi.org/10.1061/40517(2000)373
70. Abd-Elaty I, Sallam GA, Strafacec S, Scozzari A (2019) Effects of climate change on the design of subsurface drainage systems in coastal aquifers in arid/semi-arid regions: case study of the Nile Delta. Sci Total Environ J 672:283–295. https://doi.org/10.1016/j.scitotenv.2019.03.483
71. Abd-Elaty I, Eldeeb H, Vranayova Z, Zelenakova M (2019) Stability of Irrigation Canal slopes considering the sea level rise and dynamic changes: case study El-Salam Canal, Egypt. Water 11(5):1046
72. Chen C-S, Tu C-H, Chen S-J, Chen C-C (2016) Simulation of groundwater contaminant transport at a decommissioned landfill site, a case study, Tainan City, Taiwan. Int J Environ Res Public Health 2016(13):467. https://doi.org/10.3390/ijerph13050467
73. Li L, Yin Z (2017) Numerical simulation of groundwater pollution problems based on convection diffusion equation. Am J Comput Math 7:350–370. https://doi.org/10.4236/ajcm.2017.73025
74. Abd-Elaty I, Pugliese L, Zelenakova M, Mesaros P, Shinawi AE (2020) Simulation-based solutions reducing soil and groundwater contamination from fertilizers in arid and semi-arid regions: case study the eastern Nile Delta, Egypt. Int J Environ Res Public Health 17(24):9373
75. Abd-Elaty I, Shahawy AEL, Santoro S, Curcio E, Straface S (2021) Effects of groundwater abstraction and desalination brine deep injection on a coastal aquifer. Sci Total Environ:148928., ISSN 0048-9697. https://doi.org/10.1016/j.scitotenv.2021.148928
76. Bahrami M, Zarei AR (2023) Assessment and modeling of groundwater quality for drinking, irrigation, and industrial purposes using water quality indices and GIS technique in fasarud aquifer (Iran). Model Earth Syst Environ. https://doi.org/10.1007/s40808-023-01725-2
77. Khiavi AN, Tavoosi M, Kuriqi A (2023) Conjunct application of machine learning and game theory in groundwater quality mapping. Environ Earth Sci 82:395. https://doi.org/10.1007/s12665-023-11059-y
78. Eid MH, Elbagory M, Tamma AA et al (2023) Evaluation of groundwater quality for irrigation in deep aquifers using multiple graphical and indexing approaches supported with machine

learning models and GIS techniques, Souf Valley, Algeria. Water 15. 182Return to ref 2023 in article

79. Khondaker AN, Al-Layla RI, Husain T (1990) Groundwater contamination studies the state of the art. Crit Rev Environ Control 20(4):231–256. https://doi.org/10.1080/10643389009388399
80. Abd-Elaty I, Straface S (2022) Mathematical models ensuring freshwater of coastal zones in arid and semi-arid regions. In: Furze JN, Eslamian S, Raafat SM, Swing K (eds) Earth systems protection and sustainability. Springer, Cham. https://doi.org/10.1007/978-3-030-98584-4_3
81. Ezzeldin R, El-Ghandour H, El-Aabd S (2022) Optimal management of coastal aquifers using artificial jellyfish search algorithm. J Hydrol Regional Stud 41:101058., ISSN 2214-5818. https://doi.org/10.1016/j.ejrh.2022.101058
82. Vahdat-Aboueshagh H, Tsai HF, Clement TP (2022) Multi-objective optimization of aquifer storage and recovery operations under uncertainty via machine learning surrogates. J Hydrol 612(Part C):128299., ISSN 0022-1694. https://doi.org/10.1016/j.jhydrol.2022.128299
83. NRMMC-EPHC-NHMRC (2009) Australian guidelines for water recycling. Managed Aquifer Recharge. National Water Quality Management Strategy, p 24
84. Vanderzalm J, Page D, Gonzalez D, Barry K, Dillon P, Taylor A, Dawes W, Cui T, Knapton K (2018) Assessment of managed aquifer recharge (MAR) opportunities in the Fitzroy, Darwin and Mitchell catchments. A Technical Report to the Australian Government from the CSIRO Northern Australia Water Resource Assessment, part of the National Water Infrastructure Development Fund: Water Resource Assessments, CSIRO, Australia
85. Abdelhalim A, Sefelnasr A, Ismail E (2019) Numerical modeling technique for groundwater management in Samalut city, Minia Governorate, Egypt. Arab J Geosci 12:124. https://doi.org/10.1007/s12517-019-4230-6
86. Ashu A, Lee S-I (2019) Reuse of agriculture drainage water in a mixed land-use watershed. Agronomy 9:6. https://doi.org/10.3390/agronomy9010006
87. Abd-Elaty I, Zelenakova M, Straface S, Vranayová Z, Abu-hashim M (2019) Integrated modelling for groundwater contamination from polluted streams using new protection process techniques. Water 11:2321
88. Dillon P (2005) Future management of aquifer recharge. Hydrgeol J 13:313–316. https://doi.org/10.1007/s10040-004-0413-6
89. Dillon P, Fernández Escalante E, Megdal SB, Massmann G (2020) Managed aquifer recharge for water resilience. Water 12:1846. https://doi.org/10.3390/w12071846
90. Casanova J, Devau N, Pettenati M (2016) Managed aquifer recharge: an overview of issues and options. In: Jakeman AJ, Barreteau O, Hunt RJ, Rinaudo JD, Ross A (eds) Integrated groundwater management. Springer, Cham. https://doi.org/10.1007/978-3-319-23576-9_16
91. Dillon P, Pavelic P, Toze S, Rinck-Pfeiffer S, Martin R, Knapton A, Pidsley D (2006) Role of aquifer storage in water reuse. Desalination 188:123–134
92. Ríos IH, Cruz-Pérez N, Chirivella-Guerra JI, García-Gil A, Rodríguez-Alcántara JS, Rodríguez-Martín J, Marazuela MÁ, Santamarta JC (2023) Proposed recharge of island aquifer by deep wells with regenerated water in Gran Canaria (Spain). Groundw Sustain Dev 22:100959. https://doi.org/10.1016/j.gsd.2023.100959. ISSN 2352-801X
93. Page D, Vanderzalm J, Gonzalez D, Bennett J, Castellazzi P (2023) Managed aquifer recharge for agriculture in Australia – history, success factors and future implementation. Agricultural water management, vol 285, p 108382, ISSN 0378-3774. https://doi.org/10.1016/j.agwat.2023.108382
94. Perzan Z, Osterman G, Maher K (2023) Controls on flood managed aquifer recharge through a heterogeneous vadose zone: hydrologic modeling at a site characterized with surface geophysics. Hydrol Earth Syst Sci 27:969–990. https://doi.org/10.5194/hess-27-969-2023
95. Pool M, Ramírez JC (2010) Dynamics of negative hydraulic barriers to prevent seawater intrusion. Hydrgeol J 18(1):95–105. https://doi.org/10.1007/s10040-009-0516-1

96. Ozaki S, Akl CA, Nagino T, Hiroshiro Y (2021) Investigating effect of pumping ratio on effectiveness of barrier wells for saltwater intrusion: lab-scale experiments and numerical modeling. Water 13:2100. https://doi.org/10.3390/w13152100
97. Saad S, Javadi AA, Chugh T, Farmani R (2022) Optimal management of mixed hydraulic barriers in coastal aquifers using multi-objective Bayesian optimization. J Hydrol 612(Part A):128021., ISSN 0022-1694. https://doi.org/10.1016/j.jhydrol.2022.128021
98. Shackelford CD, Sevick GW, Eykholt GR (2010) Hydraulic conductivity of geosynthetic clay liners to tailings impoundment solutions. Geotext Geomembr 28(2):149–162. https://doi.org/10.1016/j.geotexmem.2009.10.005
99. Sjöholm MH, Hämäläinen JH (2022) Performance of a needle punched geosynthetic clay liner in groundwater protection on roads in cold climate. In: Clay geosynthetic barriers. CRC Press, pp 357–363
100. Eithe AW, Koerner GR (1997) Assessment of HDPE geomembrane performance in a municipal waste landfill double liner system after eight years of service. Geotext Geomembr 15:277–287. https://doi.org/10.1016/S0266-1144(97)10010-3
101. Sun XC, Xu Y, Liu YQ, Nai CX, Dong L, Liu JC, Huang QF (2019) Evolution of geomembrane degradation and defects in a landfill: impacts on long term leachate leakage and groundwater quality. J Clean Prod 224:335–345. https://doi.org/10.1016/j.jclepro.2019.03.200
102. Lambert S, Touze-Foltz N (2000) A test for measuring the permeability of geomembranes. Proceedings Eurogeo 2, Second European Conference on Geosynthetics
103. Koerner GR, Koerner RM, Martin JP (1994) Geotextile filters used for leachate collection systems: testing, design and field behavior. J Geotech Eng 120(10):1792–1803
104. Rowe RK, Hsuan YG, Lake CB, Sangam P, Usher S (1998) Evaluation of a composite (Geomembrane/clay) liner for A lagoon after 14 years of use. In: Proceedings of the sixth international conference on geosynthetics, vol 1, March 1, Atlanta. Industrial Fabrics Association International, St. Paul, MN, pp 191–196
105. Elkamhawy E, Zelenakova M, Abd-Elaty I (2021) Numerical Canal seepage loss evaluation for different lining and crack techniques in arid and semi-arid regions: a case study of the River Nile, Egypt. Water 13(21):3135. https://doi.org/10.3390/w13213135
106. Moo-Young H, Johnson B, Johnson A, Carson D, Lew C, Liu S, Hancock K (2004) Characterization of infiltration rates from landfills: supporting groundwater modeling efforts. Environ Monit Assess 96:283–311
107. Dixit A, Singh D, Shukla SK (2023) Effect of expansive soils on swelling behavior of encapsulated sodium bentonite of geosynthetic clay liner (GCL). Mater Today Proc. https://doi.org/10.1016/j.matpr.2023.02.220
108. Gupt CB, Kushwaha A, Prakash A, Chandra A, Goswami L, Sekharan S (2021) Mitigation of groundwater pollution: heavy metal retention characteristics of Fly Ash based liner materials. Fate Transport Subsurf Pollut:79–104. https://doi.org/10.1007/978-981-15-6564-9_5
109. Touze-Foltz N, Xie H, Stoltz G (2020) Performance issues of barrier systems for Land Fills: a review. Geotext Geomembr 49(2):475–488. https://doi.org/10.1016/j.geotexmem.2020.10.016
110. Elkhawas NM (2014) Mitigating the effects of the rising sea and saltwater ingress along the Egyptian north coast. Zagazig University. M.Sc. Thesis, Fac. of Eng., Zagazig University
111. Pearlman L (1999) Subsurface containment and monitoring systems: barriers and beyond (overview report). National Network of Environmental Management Studies Fellow, U.S Environmental Protection Agency, Washington, DC, http://www.clu-in.org
112. Jefferis SA (1995) UK practice: materials and their properties. In: Rumer RR, Mitchell JK (eds) Assessment of barrier containment technologies: a comprehensive treatment for environmental remediation applications. Proceedings of the international containment
113. Schneider S, Mallants D, Jacques D (2012) Determining hydraulic properties of concrete and mortar by inverse modelling. Mater Res Soc Proc 1475:367–372. https://doi.org/10.1557/opl.2012.601

114. NAVFAC, Naval Facilities Engineering Command (2012) Permeable reactive barrier cost and performance report. Final technical report TR-NAVFAC-ESC-EV-1207, Battelle Memorial Institute, aviable at: Permeable Reactive Barrier Cost and Performance Report (frtr.gov)
115. Fang Y, Qian J, Zheng T, Wang H, Zheng X, Walther M (2023) Submarine groundwater discharge in response to the construction of subsurface physical barriers in coastal aquifers. J Hydrol 617:129010
116. Evans JC, Dawson NR (1999) Slurry walls for control of contaminant migration a comparison of UK and US Practices. The proceedings of the geo-engineering for underground facilities conference, University of Illinois, Urbana-Champaign, IL, June 13–17
117. EPA (1998) Evaluation of subsurface engineered barriers at waste sites. United States Environmental Protection Agency, EPA 542-R-98-005 EPA Office of Research and Development's (ORD), 5Twww.clu-in.com5T
118. Luca De DA, Lasagna M, Morelli di Popolo e Ticineto A (2007) Installation of a vertical slurry wall around an Italian quarry lake: complications arising and simulation of the effects on groundwater flow. Environ Geol 53:177–189. https://doi.org/10.1007/s00254-006-0632-3
119. LeBoeuf EJ, Thackston EL (2005) Design guidance for confined disposal facility lateral seepage control. DOER-R7, U.S. Army Engineer Research and Development Center, Vicksburg
120. Zheng T, Zheng X, Chang Q, Zhan H, Walther M (2020) Timescale and effectiveness of residual saltwater desalinization behind subsurface dams in an unconfined aquifer. Water Resour Res. https://doi.org/10.1029/2020WR028493
121. McMahon DR (1995) Vertical barriers: sheet piles. In: Rumer RR, Mitchell JK (eds) Assessment of barrier containment technologies. National Technical Information Service, Springfield, pp 77–93
122. Luyun R, Momii K, Nakagawa K (2009) Laboratory-scale saltwater behavior due to subsurface cutoff wall. J Hydrol 377(3–4):227–236. https://doi.org/10.1016/j.jhydrol.2009.08.019. ISSN 0022-1694
123. Chang Q, Zheng T, Zheng X, Gao C, Song X, Walther M (2023) Repulsion driven by groundwater level difference around cutoff walls on seawater intrusion in unconfined aquifers. Sci Total Environ 874:162535., ISSN 0048-9697. https://doi.org/10.1016/j.scitotenv.2023.162535
124. Abdoulhalik A, Ashraf Ahmed GA, Hamill (2017) A new physical barrier system for seawater intrusion control. J Hydrol 549:416–427., ISSN 0022-1694. https://doi.org/10.1016/j.jhydrol.2017.04.005
125. Qian X et al (2019) Applications of geomembrane cut-off walls in remediation of contaminated sites. In: Zhan L, Chen Y, Bouazza A (eds) Proceedings of the 8th international congress on environmental geotechnics volume 2. ICEG 2018. Environmental science and engineering. Springer, Singapore. https://doi.org/10.1007/978-981-13-2224-2_41
126. Gao M, Zheng T, Chang Q, Zheng X, Walther M (2021) Effects of mixed physical barrier on residual saltwater removal and groundwater discharge in coastal aquifers. Hydrol Process 35(7):e14263. https://doi.org/10.1002/hyp.14263
127. Zheng T, Gao M, Chang Q, Zheng X, Walther M (2022) Dynamic desalination of intruding seawater after construction of cut-off walls in a coastal unconfined aquifer. Front Mar Sci 9: 857807. https://doi.org/10.3389/fmars.2022.857807
128. Oude Essink GHP (2001) Improving fresh groundwater supply – problems and solutions. Ocean Coast Manag 44(5):429–449. https://doi.org/10.1016/S0964-5691(01)00057-6
129. Hu L, Jiao JJ (2010) Modeling the influences of land reclamation on groundwater systems: a case study in Shekou peninsula, Shenzhen, China. Eng Geol 114(3):144–153. https://doi.org/10.1016/j.enggeo.2010.04.011
130. Abd-Elaty I, Pugliese L, Straface S (2022) Inclined physical subsurface barriers for saltwater intrusion management in coastal aquifers. Water Resour Manag 36:2973–2987. https://doi.org/10.1007/s11269-022-03156-7

Risk Assessment of Potential Groundwater Contamination by Agricultural Drainage Water in the Central Valley Watershed, California, USA

Mohamed Galal Eltarabily and Mohamed Kamel Elshaarawy

Contents

1 Introduction 38
2 Methods 42
2.1 Study Area Description 42
2.2 Nitrate Contamination in the Central Valley 44
2.3 DRASTIC Method 46
2.4 Geodetector Method 48
2.5 Principal Component Analysis 52
2.6 Geographically Weighted Regression 53
3 Results and Discussions 53
3.1 DRASTIC Method 53
3.2 Geodetector Method 57
3.3 Principal Component Analysis Method 62
3.4 Geographically Weighted Regression Method 66
4 Conclusions 70
5 Recommendations 71
References 72

Abstract Risk assessment of potential groundwater contamination is essential for safeguarding public health, ensuring agricultural sustainability, and maintaining

M. G. Eltarabily (✉)
Department of Land, Air, and Water Resources, University of CA, Davis, CA, USA

Civil Engineering Department, Faculty of Engineering, Port Said University, Port Said, Egypt
e-mail: meltarabily@ucdavis.edu; eng_m_trabily@eng.psu.edu.eg

M. K. Elshaarawy
Civil Engineering Department, Faculty of Engineering, Horus University-Egypt, New Damietta, Egypt
e-mail: melshaarawy@horus.edu.eg

Shakir Ali and Abdelazim Negm (eds.), *Groundwater Quality and Geochemistry in Arid and Semi-Arid Regions*, Hdb Env Chem (2024) 126: 37–76, DOI 10.1007/698_2023_1051,

Published online: 9 December 2023

ecological balance. Groundwater, a primary source for drinking and irrigation, is threatened by pollutants. Proactive risk assessments prevent costly remediation and guide informed policy making and community actions, ensuring a clean and sustainable water source for current and future generations. The Central Valley aquifer in California had extensive groundwater nitrate (NO_3) contamination due to excessive nitrogen fertilizer seeping into the aquifer. Numerous hydrogeological conditions of the region affect the percolation of NO_3. Hydrogeologic conditions are used in groundwater contamination vulnerability mapping to identify susceptible regions. This chapter introduced DRASTIC and geodetector-based frequency ratio (GFR) methods to develop indices for NO_3 vulnerability. In this study, seven different variables with different weights and rating values were combined using the DRASTIC method. The vulnerability index was improved by the quantitatively derived rating values and weights used in the GFR methods compared to the DRASTIC method. These values showed superior concordance with the observed NO_3 contamination pattern to map the vulnerability to groundwater contamination, indicating that GFR was a superior method. Moreover, to create a NO_3 sensitivity index map, the geodetector method (GED) was combined with the frequency ratio (FR) method to derive rating values and the geodetector method to derive relative power of determinant (PD) values as weights. Furthermore, geodetector (GED), principal component analysis (PCA), and geographically weighted regression (GWR) were used in a comparison study to find the most effective technique for identifying environmental elements influencing GW-NO_3 concentration. The outcomes revealed that the San Joaquin and Tulare Lake Basin watersheds had notably higher proportions of wells with NO_3 concentrations exceeding 5 mg/L compared to the Sacramento Valley. Overall, the GED method performed better in identifying the impact of explanatory variables on GW-NO_3 contamination than the PCA and GWR methods.

Keywords Centre Valley California, DRASTIC, Frequency ratio, Geodetector, GIS, Groundwater contamination, Statistics

1 Introduction

The Central Valley aquifer is one of the USA's greatest agricultural regions, where fruits and vegetables are grown year-round. It is one-sixth of the country's irrigated area and one of the most intensively irrigated in the world [1]. The Central Valley contains 17% of all irrigated territory in the country and roughly 75% of all irrigated land in California [2]. However, for the past 50 years, this intensive level of agriculture, particularly the rising fertilizer application rate, has resulted in groundwater contamination with NO_3. Numerous studies have demonstrated the connection between Central Valley land use practices and groundwater nitrate (NO_3) contamination. A high GW-NO_3 vulnerable area in the nation had already been found in parts of Central Valley. According to statistics from 2007, 6.7×10^6 acres of irrigated lands in CA received 740,000 tons of nitrogen fertilizer [3]. Nearly 40% of shallow wells in the USA had GW-NO_3 concentrations higher than the maximum contamination safe limit (MCL = 10 mg/L) according to a national assessment of groundwater survey results by the US Environmental Protection Agency (US EPA). Shallow wells are particularly vulnerable since NO_3 needs time to seep into the deeper aquifer. It had been discovered that several wells in the Sacramento Valley, Tulare Lake Basin, and San Joaquin Basin were above the MCL.

Drinking water NO_3 could pose a health risk and, if consumed in quantities above the maximum contamination limit, might have serious adverse effects on one's health. The CA State government heavily prioritized NO_3 contamination in the Central Valley to ensure clean drinking water. The leading causes of NO_3 contamination in the groundwater of the Central Valley were also thought to be household animals, septic tanks, and fertilizer applications [4]. However, environmental variables may affect aquifer NO_3 concentrations, such as permeability, rainfall, soil quality, and the geochemical condition of the watershed. Although the Central Valley had seen many municipal-level studies, only a few for the entire aquifer had been conducted [3, 5, 6]. Conducting aquifer elevation studies was crucial in creating regional-scale policies to safeguard the viability of the Central Valley aquifer, considered one of the most important aquifers in the USA. The USGS National Water-Quality Assessment Program (NAWQA) examined groundwater quality and trends at the aquifer level for each of the 52 major aquifers in the USA. For instance, the groundwater center region of the Central Valley had an additional reducing environment that caused NO_3 to be converted to nitrogen gas. As a result, NO_3 levels were lower than in the Central Valley eastern region.

The eastern region had more rainfall and coarse-grained alluvial sediments, leading to an oxic environment. However, small-grained sediments and the prolonged groundwater residence time lessen the problem as the water flows toward the valley's axis [7, 8]. The northern SM Valley receives more rainfall than the southern San Joaquin Basin and Tulare Lake Basin, which can lower the NO_3 concentration or allow dissolved NO_3 to seep into the aquifer's deeper layers. Groundwater pumping in the Central Valley had already drastically altered the San Joaquin Valley's hydrogeology. Due to this, most of the groundwater has begun to change its directions into the groundwater depression cone [9]. This modification to the groundwater flow pattern may also affect how NO_3 dissolved in the groundwater is transported. The relation between GW-NO_3 contamination and predictor variables has been studied using statistical methods. For instance, numerous studies have employed multidimensional statistical techniques such as ordinary least squares regression (OLS) to examine the connection between land use and GW-NO_3 contamination [10–12].

OLS had the issue of overfitting the data when there were numerous variables by regressing random error rather than the correlation between the variables. The OLS made many assumptions about the data, including their linearity, homoscedasticity, normality, and multicollinearity. However, implementing these assumptions could be challenging because NO3 data were skewed in Tulare Lake depending on the researcher's expertise or the application of statistical techniques that minimized the total amount of variables. Logistic regression is another popular statistical method determining the likelihood that NO_3 concentration would rise above a particular level [13]. Nonetheless, its effectiveness might not have been immediately apparent to decision-makers due to its omission of instructions for Tulare Lake to analyze and compare NO_3 content. A statistically significant difference between the GW-NO_3 concentration during two different time intervals was determined by the Wilcoxon-Pratt signed rank test [12].

The DRASTIC method could create an index by combining seven variables with different weights: hydraulic conductivity, topography, aquifer media, recharge rate, and depth to water [14]. The pre-assignment of the model's determining variables increased the chance of overlooking other crucial variables that might be common in the study area. The drawback of this method is that each variable is subjectively weighted. Process-based methods required extensive data coverage and were only practical at the neighborhood level. This complex method used groundwater flow models like Darcy's Law to simulate the transport of contaminants and water movement in the environment [15]. It might be possible to identify the variables that contribute to the NO_3 contamination of groundwater by understanding aquifer vulnerability depending on the spatial distributions of the contributing variables (i.e., land use or fertilizer rate). For instance, multiple linear regressions could calculate a positive or negative coefficient for each independent variable and only establish a linear relationship with them, with higher coefficient values showing a more significant impact on the dependent variable [16]. Another drawback was that the R^2 (coefficient of variation) would always increase as the number of independent variables increased. Geodetector (GED) was a comparatively new statistical method assessing the significance of spatially stratified heterogeneity [17, 18]. It could identify significant contributing explanatory variables as well as contamination danger zones. It could also measure the significance of particular geographical strata of another suspected variable compared to other geographical strata (associated with a particular hypothesized explanatory variable). The method also examined whether two variables can combine to change the likelihood of aquifer contamination [17, 19]. The analysis of GW-NO_3 contamination was done using the GED method [20].

Numerous methodologies have been employed globally to assess groundwater vulnerability and contamination. These approaches encompass a range of techniques, including index-based methods, interpolation techniques, process-based models, and statistical models. Index methods involve assigning weights to parameters based on expert's knowledge, exemplified by the DRAV model [21], Susceptibility Index (SI) technique [22], GOD technique [23], ANIMO and EPIC models [24], and the DRASTIC model [25]. Interpolation methods utilize sensors and devices and rely on techniques like indicator kriging (IK) based on geostatistical procedures and the kriging interpolation algorithm [26–29]. Although these methods provide a convenient means of assessing groundwater pollution risk, they are hampered by the necessity for data-rich points and associated output uncertainties. Process-based models, such as agricultural management systems (GLEAMS) [30–33], the pesticide root zone model (PRZM-3) [30, 31], the water flow and nitrate transport global model (WNGM) [34, 35], and the groundwater flow model

(MODFLOW) [36], are sophisticated, yet their reliance on extensive input data hinders them. Moreover, these models often lack applicability at local scales due to their complexity [37–39]. In contrast, statistical models employ linear and nonlinear regression techniques to analyze the relationship between contaminant density and influencing parameters [40, 41]. These models require expert input for causal inference, which can challenge reaching a consensus.

Recently, the application of soft computing approaches, such as artificial intelligence (AI) and machine learning (ML) techniques, has gained prominence in predicting environmental risks and hazards [42–48]. While acknowledging the use of ML in groundwater assessment, this study emphasizes its broader application beyond the field. For example, ML has been utilized for modeling groundwater level changes in US agricultural regions [49], mapping groundwater contamination risk in multiple aquifers [50], and predicting groundwater nitrate concentrations from spatial data [31, 51, 52]. Researchers have employed advanced ML techniques such as K-nearest neighbor algorithms, random forests, and support-vector machines (SVM) to spatially model groundwater nitrate concentrations [53]. Moreover, ML has been used to assess environmental risks, engineering risks, and bioconcentration factors in various contexts [54–56].

Based on the aforementioned, the research gap identified revolves around the need to comprehensively understand and accurately assess the vulnerability and contamination of groundwater with nitrate (NO_3) in the Central Valley aquifer, a crucial agricultural region in the USA. Despite numerous studies demonstrating the connection between land use practices and groundwater NO_3 contamination, there is a lack of detailed and holistic studies that cover the entire aquifer. This gap is particularly evident in the Central Valley, a vital agricultural area with a significant groundwater NO_3 contamination issue due to intensive farming practices. While various statistical and process-based methods have been employed to assess groundwater vulnerability and contamination, there is room for improvement in accuracy and reliability, especially when considering the complex and varied factors affecting contamination.

This chapter aims to present an accurate GW-NO_3 vulnerability map using the GED and DRASTIC methods for the Central Valley, CA's most productive agricultural region. An evaluation of the methods is performed to identify which is better for mapping GW-NO_3 vulnerability. Additionally, comparisons were made between GED, PCA, and GWR results. This analysis aimed to ascertain the most effective approach for pinpointing environmental variables that have a greater impact on GW-NO3 concentrations. The goal was to enhance comprehension of the contamination process and to investigate Central Valley contamination. This endeavor contributes to a more profound insight into the contamination process and provides valuable perspectives for the development of efficient management strategies aimed at preserving the integrity of the aquifer.

2 Methods

The methodological approach used in this chapter can be described as shown in Fig. 1. Firstly, an introduction to DRASTIC and geodetector (GED) methods is presented. Then, indices for GW-NO_3 vulnerability are developed using the predictive variables for each method. Thirdly, the selection of the accurate method for mapping GW-NO_3 vulnerability. Moreover, to create a NO_3 sensitivity index map, the geodetector method (GED) was combined with the frequency ratio (FR) method to derive rating values and the geodetector method to derive relative power of determinant (PD) values as weights. Furthermore, geodetector (GED), principal component analysis (PCA), and geographically weighted regression (GWR) were used in a comparison study to find the most effective technique for identifying environmental elements influencing GW-NO_3 concentration.

2.1 Study Area Description

The Central Valley, covered by marine and continental sediments, was regarded as one of California's main structural troughs. The Central Valley was also known as

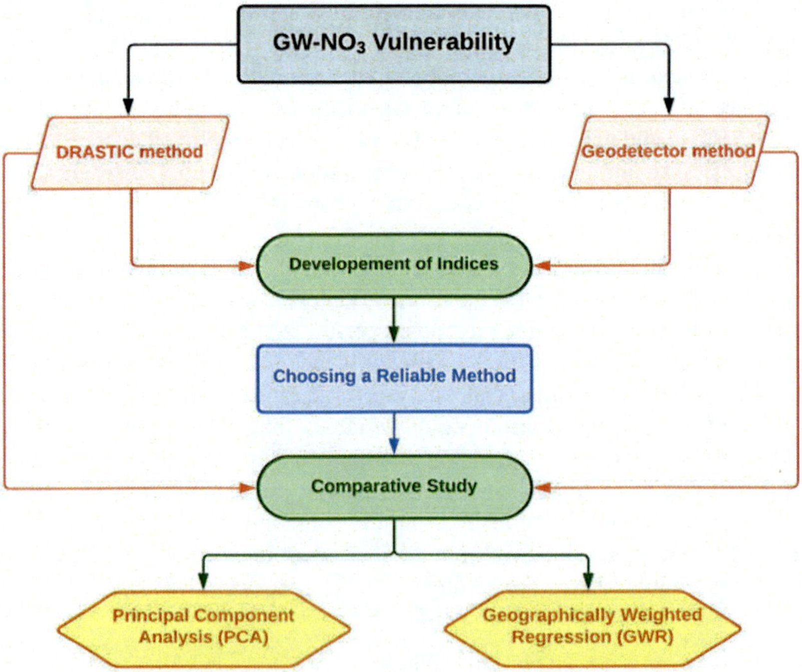

Fig. 1 Flow chart of the methodological approach adopted for this study

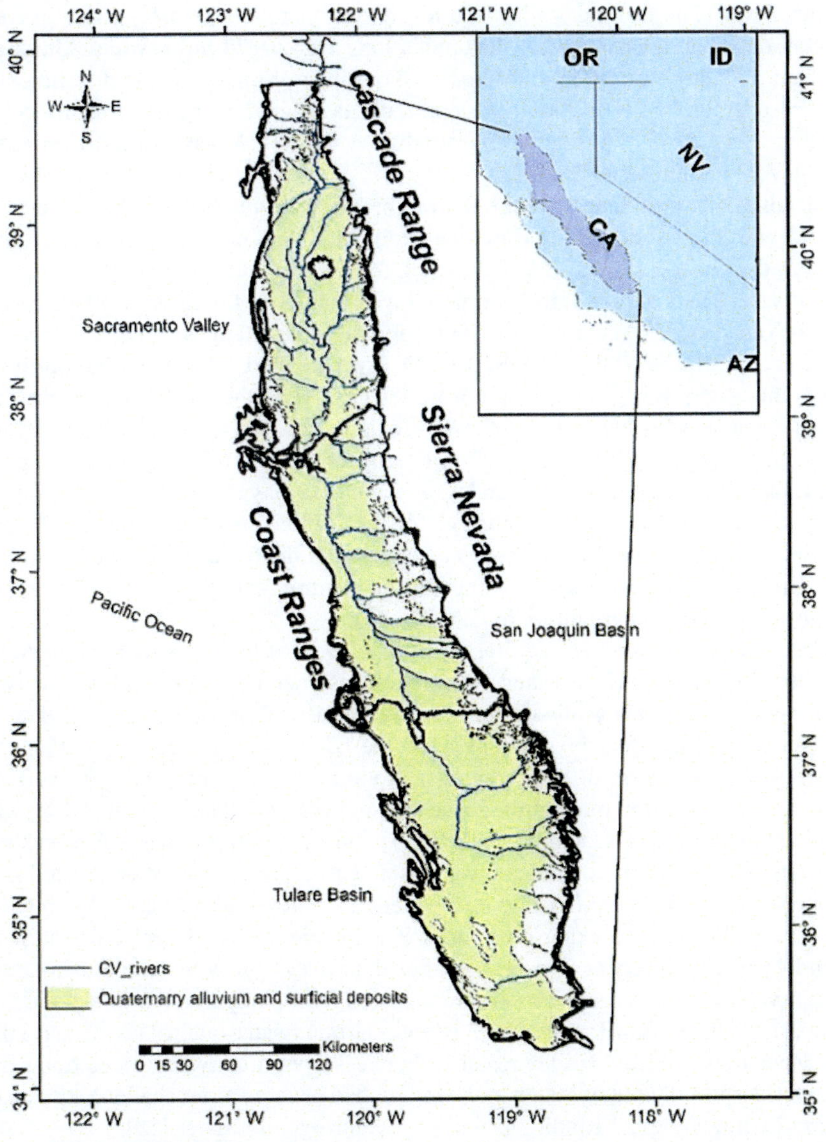

Fig. 2 Central Valley aquifer system [57]

the Great Valley of CA. It covered about 20,000 square miles, bounded by the Cascade Range to the north, the Sierra Nevada to the east, the Tehachapi Mountains to the south, and the Coast Ranges and San Francesco Bay to the west, as shown in Fig. 2. SM Valley, from which the SM River drained, occupied the northern part of the valley. The San Joaquin Valley comprised the southern two-thirds of the valley and was further split into the Tulare Lake Basin and San Joaquin Basin by the San

Joaquin River. The Central Valley aquifer system comprised mainly sand, gravel, and clay deposits, approximately 400 miles long and 20–70 miles wide. Although the Central Valley was filled with tens of thousands of unconsolidated sediments, most fresh groundwater is found at depths of less than 2,500 feet (762 m; [57]). Approximately, 83% of agricultural groundwater use in CA was extracted from the Central Valley's three sub-aquifers.

One of CA's significant structural troughs, the Central Valley, was primarily covered with the continental and marine sediments. Another name for the Central Valley is the Great Valley of CA. According to Fig. 2, it was roughly 20,000 square miles in size. The Cascade Range bordered it to the north, the Sierra Nevada to the east, the Tehachapi Mountains to the south, and the Coast Ranges and San Francesco Bay to the west. SM Valley, which the SM River drained, comprised the northern third of the watershed. The San Joaquin Valley comprised the southern two-thirds of the valley and was further split into the San Joaquin Basin and the Tulare Lake Basin by the San Joaquin River. The 400-mile-long and 20- to 70-mile-wide Central Valley aquifer system comprised sand, gravel, and clay layers. Although the Central Valley was covered in tens of thousands of feet of unconsolidated sediments, the majority of the fresh groundwater was discovered at depths of less than 2,500 feet [57]. The three sub-aquifers that comprise the Central Valley provide 83% of the groundwater used for agriculture in CA.

Most shallow, unconfined aquifers in the valley's central and southern regions were detached from surface water networks. These nearby creeks and rivers' seasonal runoff replenished groundwater networks. Seasonal precipitation caused shallow, unconfined aquifers in some parts of the SM Valley to reconnect with the surface water networks. In these regions, groundwater discharge into the surface water networks aided in maintaining base stream flow [58]. The Central Valley was assumed to be one diverse aquifer system [57]. Due to the groundwater flow system being altered, the hydraulic gradient was raised. Groundwater flow was directed into the deeper constrained aquifer due to the Central Valley's water table plummeting hundreds of feet due to extensive pumping [59]. Figure 3a, b depict the shift in groundwater flow direction before and after the valley's growth. The San Joaquin Basin's water level had dropped by more than 400 feet. Compared to wells in the SM Valley, those in the San Joaquin were deeper. Groundwater quality issues and land subsidence in the Central Valley result from excessive groundwater extraction. The fate and transport of contaminants in the region had become more challenging due to shifting hydrogeological conditions and excess nitrogen fertilizer [59].

2.2 *Nitrate Contamination in the Central Valley*

To explore the impacts of explanatory variables on GW-NO_3 contamination where the groundwater and surface water interacted in hydrologic processes at a watershed scale, considering the watershed as the fundamental analysis unit, to analyze the data, the 12-digit unit of Central Valley watersheds (Fig. 4) was chosen. The USGS

(a)

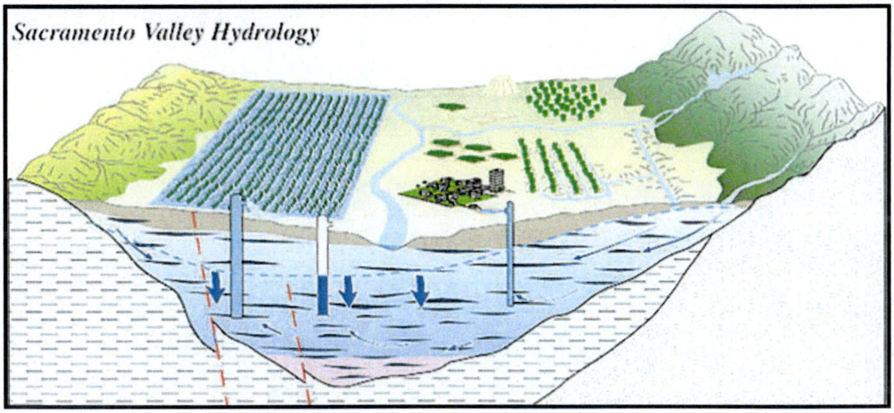

(b)

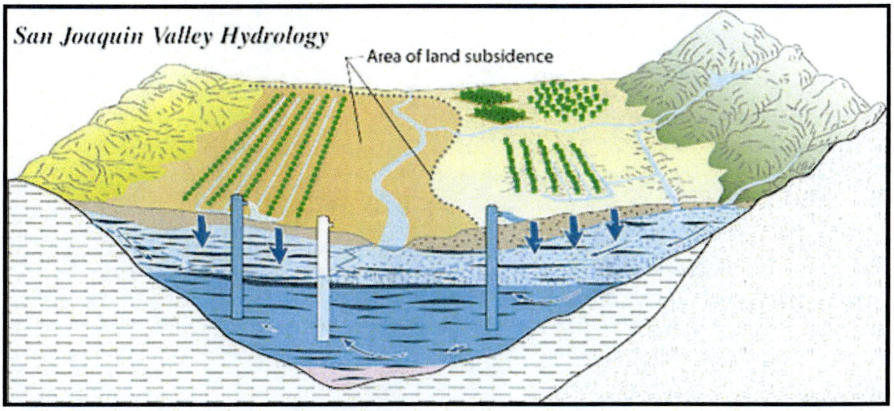

Fig. 3 San Joaquin Valley cross-section during (**a**) pre-development period; (**b**) post-development period [60]

Watershed Boundary Dataset and National Standard for Spatial Data Accuracy were used to obtain watershed data. The hierarchy of hydrologic units had no finer precision than this. The Central Valley contained a total of 656 watersheds.

To calculate the decadal average of GW-NO_3 contamination, the data on GW-NO_3 concentrations for 2018 were downloaded from the NAWQA, Groundwater Ambient Monitoring Assessment (GAMA), and National Water Information System (NWIS). One thousand and fourteen from 2,516 well samples had a mean content level (MCL) higher than 5 mg/L. NO_3 in potable water had an EPA MCL of 10 mg/L, and concentrations above that limit could have many adverse health effects [62]. To depict the NO_3 contamination's geographic variability with accuracy, a threshold of 5 mg/L was chosen [11]. Figure 5a depicts the overall number of wells and well samples with concentrations higher than 5 mg/L in the Central Valley's

Fig. 4 Central Valley watersheds [61]

groundwater watershed. The percentage of each watershed well with a mean concentration of NO_3 greater than 5 mg/L ($PW_{N>5}$) was determined in ArcGIS (Fig. 5b) using the formula:

$$PW_{N>5} = \frac{\text{Number of wells exceeding temporal average greater than 5 mg/L in a watershed}}{\text{Total number of wells in each watershed}} \quad (1)$$

2.3 DRASTIC Method

The DRASTIC index, as applied to Tulare Lake, was utilized to assess the susceptibility of the aquifer to potential groundwater contamination across various

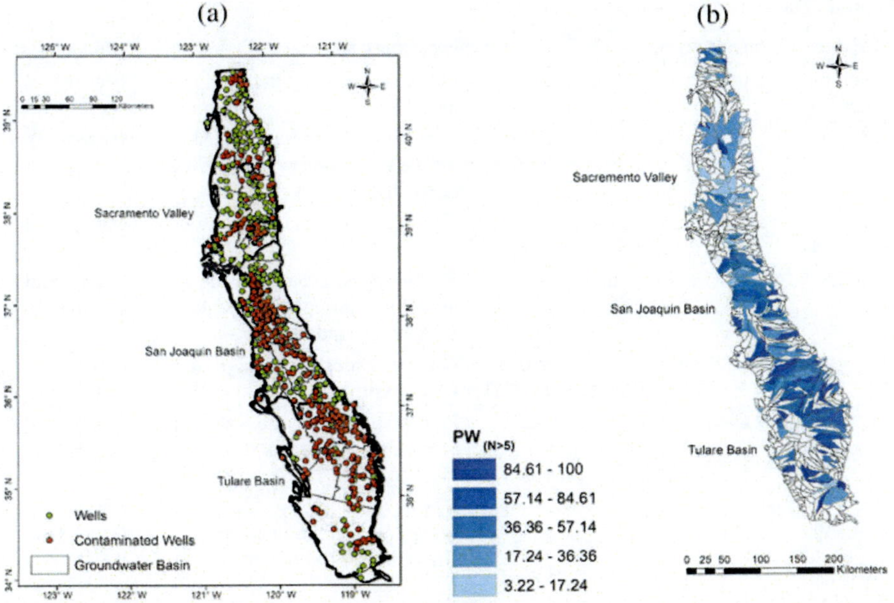

Fig. 5 (**a**) Groundwater basins and well samples in Central Valley [11]; (**b**) Watersheds with $PW_{N>5}$ mg/L [61]

hydrogeological conditions [63]. Neukum et al. [64] assumed that precipitation carries NO_3 from the Earth's surface into the groundwater. The effect of the vadose zone (*I*), topography (*T*), soil media (*S*), aquifer media (*A*), net recharge (*R*), hydraulic conductivity (*C*), and depth to water (*D*) were the seven variables influencing groundwater vulnerability that was used to calculate the DRASTIC index as follows:

$$\mathrm{DI} = D_r\, D_w + R_r\, R_w + A_r\, A_w + S_r\, S_w + T_r\, T_w + I_r\, I_w + C_r\, C_w \tag{2}$$

The subscripts r and w represent rates and weight, respectively. To capture the hydrogeological setting of the aquifer, natural breaks were used to separate each variable into five intervals and given ratings ranging from 1 to 10 [65]. Parameter weights, meanwhile, ranged from 1 to 5 based on the index assigned for the intensive agricultural activities in the study area. Table 1 depicts the data source for all DRASTIC variables. Determining regions more susceptible to groundwater contamination was possible after the DRASTIC index was assessed. Table 2 shows the grades and weights given to each DRASTIC variable. The groundwater vulnerability to NO_3 contamination increased with higher DRASTIC index values.

Table 1 DRASTIC variable sources [66]

Variable	Data source	Processing summary	Units
Depth to water (*D*)	[67]	Interpolation using the kriging method	Feet (ft)
Net recharge (*R*)	USGS Mean Annual Natural Groundwater Recharge [68]	Zonal statistics were calculated for the groundwater basin using raster data	Inches (in)
Aquifer media (*A*)	Surficial Geology of South-west Principal Aquifer [69]	Classified based on [70]	Categorical variable
Soil media (*S*)	USGS-NRCS data. Soil hydrologic group [71]	The weighted average of the soil hydrologic group was calculated for each groundwater basin	Categorical variable
Topography (*T*)	USGS Elevation Derivatives for National Application [72]	Zonal statistics for each ground-water basin	Percent slope
Impact of vadose zone (*I*)	California Department of Water Resources United States Geological Survey [73]	Processed depth to the water table Permeability data based on Piscopo [74]	Categorical variable
Hydraulic conductivity (*C*)	USGS [73]	Weighted average in each Groundwater basin	Gallons per day/ft^2

2.3.1 DRASTIC Variables Development

Table 2 shows the weights and ratings for the assigned DRASTIC variables to each variable and period to determine the DRASTIC index.

2.4 Geodetector Method

Using Geodetector Software 2007, the geodetector (GED) method was used [19]. According to GED, there was a relationship between the spatial distribution of the NO_3 contamination pattern and any probable environmental risk variables. The assumption was that if an environmental component caused NO_3 contamination, NO_3 contamination would display a geographical distribution similar to that of the environmental variable [19]. Figure 6a depicts the PWN > 5 overlays for each used watershed.

According to the amount of rainfall in the valley, the Central Valley was divided into various rainfall sub-regions (low to high) (Fig. 6b). In each different rainfall sub-region, the mean and variance values of $PW_{N>5}$ could be calculated by superimposing the watershed unit over the rainfall chart (Fig. 6c). Moreover, the $PW_{N>5}$ variance value (local variance) was then contrasted with the total Central Valley variance to determine the power of determinant (PD) from the following formula:

Table 2 Assigned ratings and weights to DRASTIC variables [66]

Depth to water (*D*; ft)	Recharge rate (*R*; mm)		
Range	Rating	Range	Rating
10.28–35.9	10	4.82–7.02	1
35.9–66.36	9	7.02–9.03	3
66.36–118.9	7	9.03–17.23	6
118.9–191.43	5	17.23–31	8
191.43–444.34	3	31–107.67	9
Aquifer media (*A*; Categorical variable)	**Soil media (*S*; Categorical variable)**		
Range	**Rating**	**Range**	**Rating**
Quaternary alluvium	8	Soil hydro group A (Sandy, loamy sand, sandy loam)	6
Sedimentary dominated the formation of all ages	6	Soil hydro group B (silt loam, loam)	5
Metamorphic or igneous units	3	Soil hydro group C (Sandy-clay-loam)	4
Soil hydro group D (clay loam, silty-clay-loam, sandy	3		
Clay, silty-clay, clay)			
Topography (*T*; %)	**Impact of vadose zone (*I*; Rating rate)**		
Range	**Rating**	**Range**	**Rating**
0.11–0.45	10	8–10	10
0.45–0.78	9	6–8	8
0.78–1.32	5	4–6	5
1.32–2.83	3	3–4	3
2.83–10.3	1	2–3	1
Hydraulic conductivity (*C*; Gal/day)			
Range	**Rating**		
5.23–13.82	1		
13.82–19.99	2		
19.99–29.87	4		
29.87–45.25	6		
25.25–62.64	8		

$$\text{Power of Determinant (PD)} = 1 - \frac{A_{C1} \cdot \text{Var}_{C1} + A_{C2} \cdot \text{Var}_{C2} + \cdots + A_{C5} \cdot \text{Var}_{C5}}{A \cdot \text{Var}E} \quad (3)$$

For each sub-region of rainfall class, A_{Ci} stood for the sub-region area, A stood for the size of the entire study area, Var_{Ci} ($i = 1, 2, ..., 5$) stood for the $PW_{N > 5}$ variances, and $\text{Var}E$ ($i = 1, 2, ..., 5$) stood for the entire Central Valley's watershed variances.

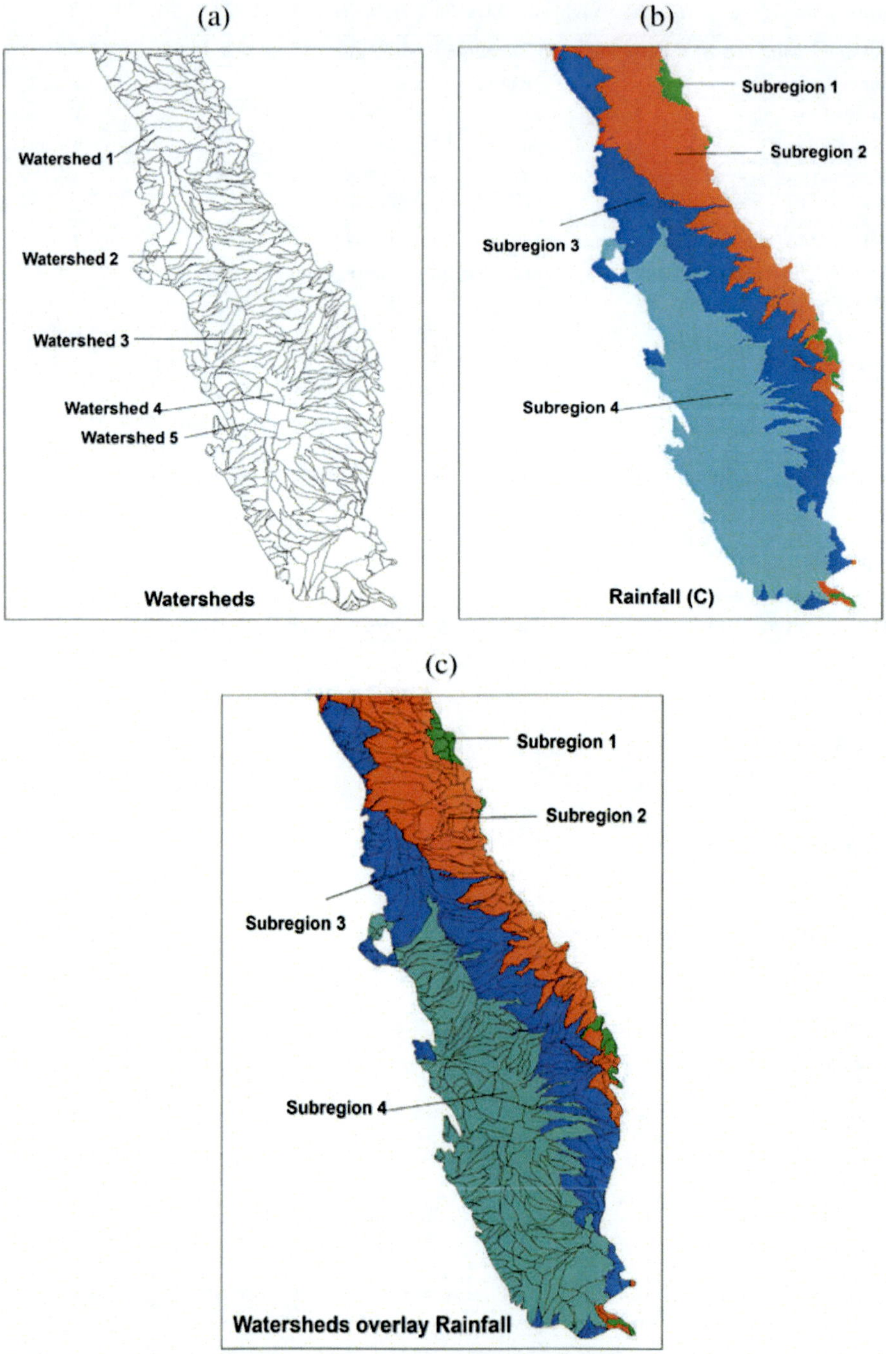

Fig. 6 Example of the geodetector method: (**a**) Basic analysis unit; (**b**) study area sub-regions based on one component (rainfall) [61]. (**c**) An overlay of the basic analysis unit and sub-regions for geodetector analysis [61]

PD had a number between 0 and 1. If the rainfall (*C*) variable controlled the GW-NO_3 contaminations, the PD value would be 1. It would be equal to 0 if it had no control at all values closer to 1, indicating that rainfall was a significant driver of GW-NO_3. The risk detector for each sub-region determined the average $PW_{N\,>\,5}$ value (M_{Ci}) where $i = 1, 2, \ldots, 5$. It determined whether the mean value deviated considerably from the average value for all other sub-regions. The risk detector identified the level with the highest mean NO_3 concentration compared to other levels, and its statistical significance was evaluated.

As the explanatory variable, rainfall as a continuous variable was categorized from Level 1 (low) to Level 5 (high). To ascertain whether the GW-NO_3 contamination was more heavily influenced by one geographic stratum of *C* (such as rainfall) than another. *C* would have a lower dispersion variance (σ^2) if it had more control over the contamination, such as through rainfall. The interaction detector determined the total impact of the two explanatory variables. To determine whether they strengthen, deteriorate, or are independent of one another, it was then compared to the sum of the individual PD values of the variables. For instance, nonlinear enhancement was offered if the PD value of the new component was higher than the total PD value of all the prior variables. Additional information on the method and the statistical analyses was found in Wang et al. [18, 19].

2.4.1 Geodetector Variables Development

Aquifer source variables, vulnerability variables, and geochemical conditions that could affect GW-NO_3 contamination of the Central Valley were the 12 different predictor variables subjected to the geodetector method. Shrestha and Luo [66] contained information about how the variables were processed in depth. Table 3 summarizes the data sources used to develop the predictor variables.

Table 3 Explanatory variables in the geodetector method [20]

Explanatory variables	Data source
Farm fertilizer (kg/ha)	United States Geological Survey (USGS) [75]
Manure (kg/ha)	United States Geological Survey (USGS) [76]
Cropland (%)	National Landcover Database (NLCD) [77]
Permeability (in/h)	STATSGO Soil Characteristics for the Conterminous United States [78]
Precipitation (mm)	PRISM Climate Data based long-term historical record (1981–2010) [79]
Slope (%)	Elevation derivatives for national applications [80]
Elevation (m)	National Elevation Dataset (NED) [81]
Clay (%)	United States Geological Survey (USGS) [82]
Recharge rate (mm/year)	United States Geological Survey (USGS) [68]
Dissolved oxygen (mg/L)	NAWQA and NWIS [83, 84]
Iron and manganese (mg/L)	NAWQA and NWIS [83, 84]

2.4.2 Frequency Ratio Method

The frequency ratio (FR) method was used to calculate the rating values for different intervals of significant predictor variables. Each number predictor variable was discretized into several intervals to depict stratified heterogeneity. Since groundwater basins were utilized as the fundamental unit in the study, the number of wells inside each period was considered while dividing the overall number of wells. The percentage of wells in each interval was the consequence of this. The frequency ratio, often known as the ratio of two percentages, was determined as follows:

$$\text{Frequency Ratio} = \frac{\text{Percent of contaminated wells in each interval}}{\text{Percent of wells of the related interval}} \tag{4}$$

The frequency ratio method was utilized in landslide investigation, specifically concerning Tulare Lake, to evaluate the frequency of landslides in regions susceptible to vulnerability. Greater than 1 demonstrated a strong correlation with the GW's NO_3 contamination, whereas less than 1 demonstrated a moderate correlation [85].

2.4.3 Index of Geodetector Frequency Ratio (GFR)

The weighted geometric mean method was used to determine the GFR index [86] using the relative weight of significant predictor variables and their frequency ratio values in each interval given by the formula below:

$$\text{GFR Index} = \prod_{k=1}^{n} (A_k)^{B_k} \tag{5}$$

where n is the total number of explanatory variables, B_k is the normalized PD values for the kth explanatory variable, and A_k is the kth explanatory variable frequency ratio.

2.5 Principal Component Analysis

Principal component analysis (PCA), a dimension-reduction technique, condensed the 12 explanatory variables into a small number of principal components (PCs), accounting for most of the data variance. Using SPSS, the data were examined for each of the 12 explanatory variables using PCA. The varimax method was used to rotate eigenvalues based on the data's standardized correlation matrix to enhance variety. To accomplish this, a new coordinate system was created by rotating the coordinate system of the initial variables. The PC-1 determined the highest variance in the data collection. The PC-2 determined a lower variance than the PC-1, the PC-3 determined a lower variance than the PC-2, and so on. The weight of each variable was its input to the newly created PC. A variable that described most of the variance in a PC had a higher weight in that PC. These new PCs were developed as uncorrelated variables from correlated or redundant explanatory variables. Before conducting the SPSS test, all assumptions of the PCA were confirmed. These encompassed the requirements for variables to be continuous and linearly correlated, as well as the necessity for an adequate sample size and outliers to be addressed, as indicated by previous work [87, 88].

2.6 Geographically Weighted Regression

Using ArcGIS 10.4, geographically weighted regression (GWR) was carried out on the data. Before performing the GWR analysis, an accurate ordinary least squares (OLS) model was fitted. Before conducting GWR, the OLS model's normality, multicollinearity, and heteroskedasticity assumptions were all satisfied. The GWR model calculated OLS-like regressions for each feature in the dataset. The OLS model was a straightforward linear model with presumed independent residuals. The association between the explanatory and the dependent variables was supposed to be constant over the study area. If residuals were autocorrelated and OLS was broken, spatial autocorrelation occurred. GWR performed well when spatial heterogeneity and variable relationships varied across the study area [89].

3 Results and Discussions

3.1 DRASTIC Method

3.1.1 DRASTIC Variables Maps

The Central Valley's depth to water (DTW) distribution is depicted in Fig. 7a. The DTW was deeper in the San Joaquin and Tulare Lake Basins than in the SM Valley, which had a very shallow DTW. Due to excessive groundwater pumping over the years, the hydrogeological features of the San Joaquin and Tulare Lake Basins had altered. As a result, the water level in this portion of the valley decreased [90]. Figure 7b depicts the net recharge rate in the Central Valley. The recharge rate was generally low throughout the valley, with a comparatively higher rate on the eastern flank, particularly in the northeastern Tulare Lake Basin and southeast San Joaquin Basin. The Coastal Range in the west acted as an orographic barrier to the saline onshore ocean breezes. Therefore, it was in the rain shadow. Higher elevation Sierra Nevada could be found in the eastern region of the valley. As a result of the orographic impact and rising recharge rate, the annual precipitation rate has risen.

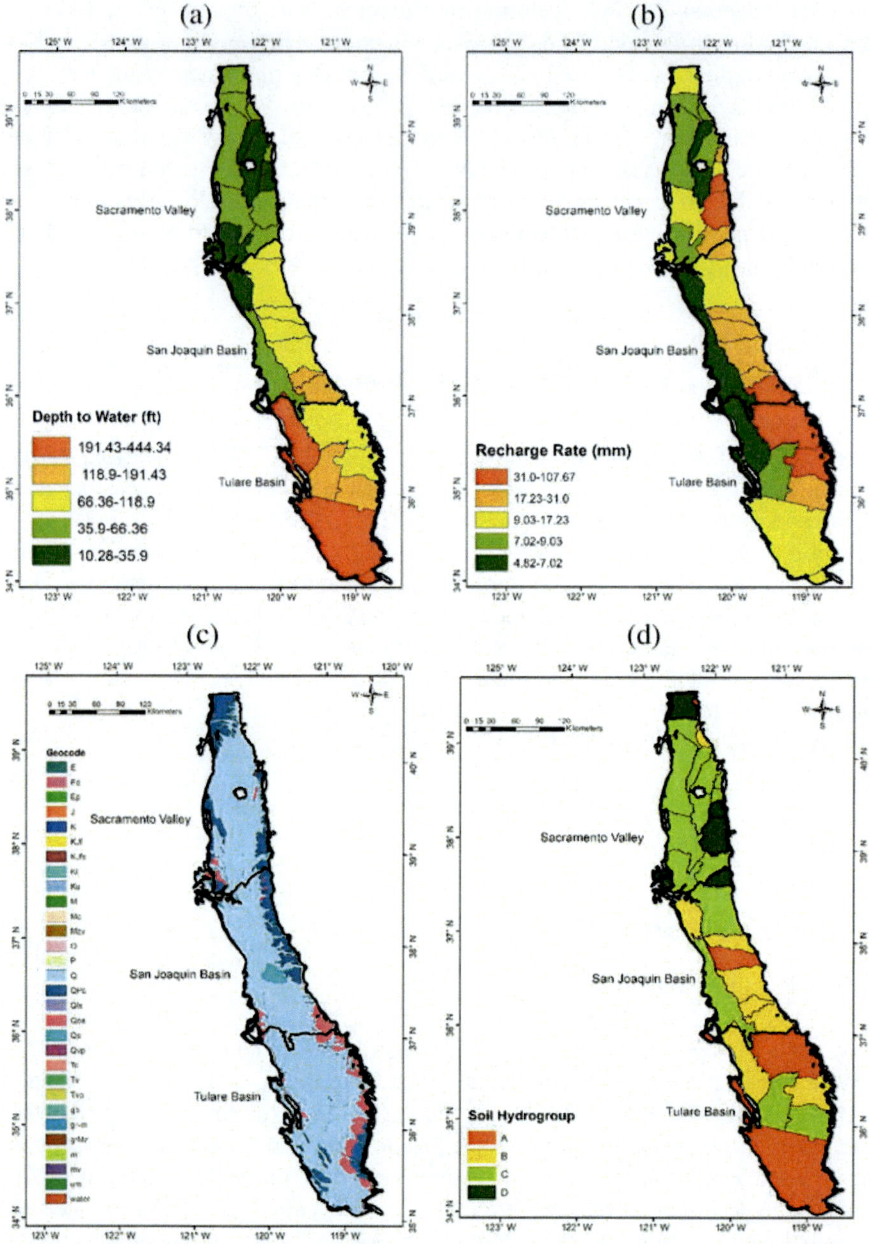

Fig. 7 DRASTIC variable maps in the Central Valley. (**a**) Depth to the water table; (**b**) recharge rate [66]. (**c**) Aquifer media; (**d**) soil media; (**e**) slope percentage; (**f**) impact of vadose zone [66]. (**g**) Hydraulic conductivity [66]

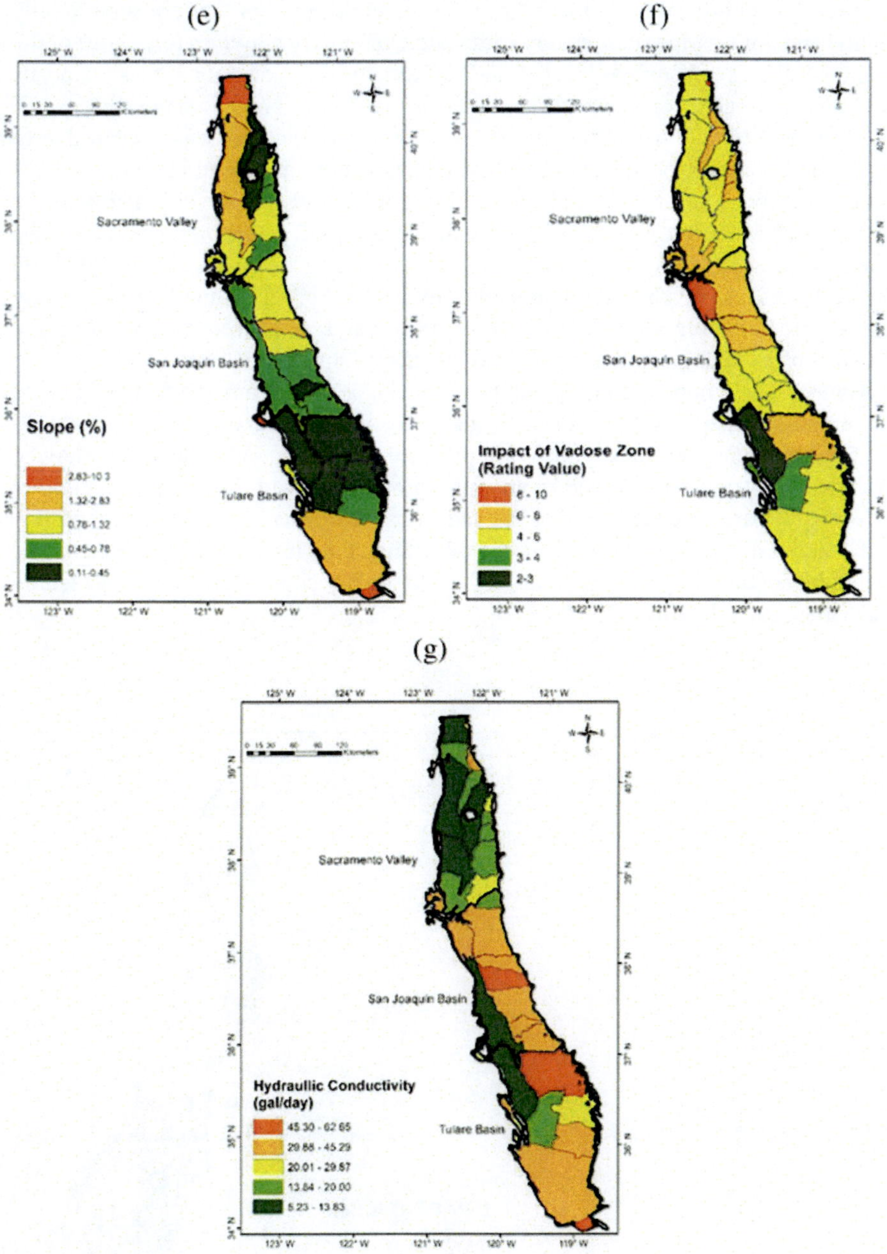

Fig. 7 (continued)

Rocks accumulated from the Jurassic to the Holocene era that are superficially marine and non-marine sedimentary rocks and Quaternary alluvium form most of the Central Valley's subsurface layer (Fig. 7c). Most of the materials in these deposits are unconsolidated or semi-unconsolidated, such as gravel, sandstone, shale deposits, and terrace deposits. The valley's freshwater aquifer was created due to the Central Valley's changing deposition environment, leading to continental sediments overlaying marine depositions [91]. Figure 7d depicts the soil media of the Central Valley aquifer. Soil from hydrologic categories C and D predominated in the SM Valley. Meteorological groups A and B were present in the San Joaquin and Tulare Lake Basins. This soil belonged to hydrologic group C and had a relatively fine texture, allowing for slow water infiltration. Sands or gravels were present, and the soils of hydrologic soil group A had significant infiltration rates.

Hydrologic group B soils typically range in density from moderately fine to moderately coarse, allowing for soils that were only moderately well drained. In the Central Valley, regions with 0–2% predominated (Fig. 7e). The figure showed that the Tulare Lake Basin had more valley floor areas with low slopes than other locations. Figure 7f illustrates that the vadose zone impact (IVZ) in the SM Valley and Tulare Lake Basin received lower IVZ ratings than the San Joaquin Basin. The hydraulic conductivity was minimal in the SM Valley compared to the San Joaquin and Tulare Lake Basins (Fig. 7g).

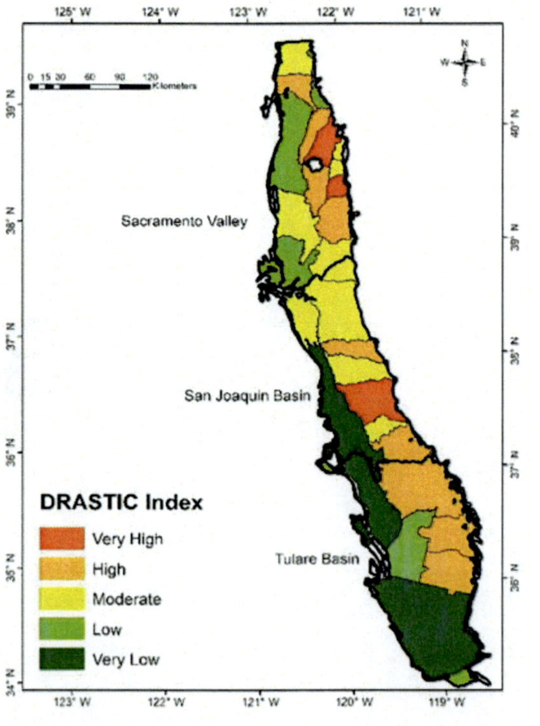

Fig. 8 DRASTIC index in the Central Valley [66]

3.1.2 Index of DRASTIC Method

The index of DRASTIC (DI) ranged from 103 to 193 m, as shown in Fig. 8. Categorizing natural breaks led to dividing the DRASTIC ratings into five intervals. Generally, the valley's eastern half showed higher index values than the western region. Index interpretations were high on the eastern side of the Tulare Lake Basin, while on the western side, they were low to very low. The San Joaquin Basin primarily had index values that were high to very high, except for a small portion of the region in the western region. Furthermore, the eastern part of the SM Valley exhibited notably higher index values than the valley's western region.

3.2 *Geodetector Method*

3.2.1 Geodetector PD Values

Table 4 depicts the GED method's results. The PD values for fertilizer and manure (source variables) and the precipitation, elevation, and %clay (groundwater vulnerability variables) had *p*-values of 0.05, making them statistically significant. This method revealed no statistically significant geochemical variables (dissolved oxygen, iron, or manganese).

3.2.2 Geodetector Variables Maps

Table 4 Explanatory variables PD values [61]

Variables	PD	*p*-Value
Precipitation	0.27	<0.01
Fertilizer	0.21	<0.01
Elevation	0.18	<0.01
Manure	0.16	0.01
Clay	0.10	0.03
Dissolved oxygen	0.09	0.09
Permeability	0.09	0.14
Iron	0.06	0.14
Slope	0.03	0.21
Cropland	0.07	0.22
Manganese	0.00	1.00
Recharge rate	0.02	0.86

According to the risk detector results, the San Joaquin and Tulare Lake Basin had higher $PW_{N>5}$ values (Fig. 9a). Level 1 (low fertilizer level) had an average

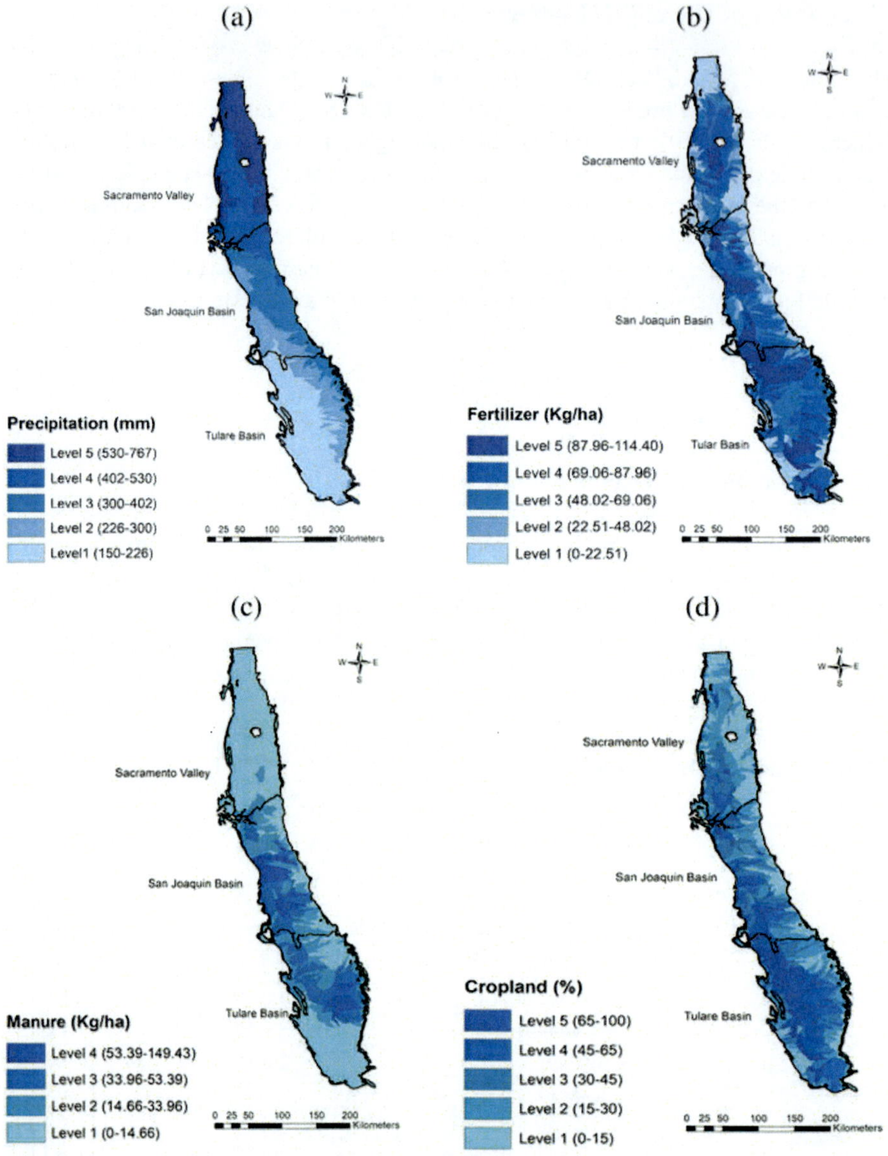

Fig. 9 Maps of geodetector variables in the Central Valley. (**a**) Rainfall; (**b**) fertilizer weight; (**c**) manure; (**d**) cropland percentage [66]. (**e**) Elevation; (**f**) clay percentage [66]

$PW_{N>5}$ of 21.87%. Levels 2 through 5 (55.97%) demonstrated an increasing tendency of $PW_{N>5}$ toward the highest level (i.e., Level 5). In Level 1, the difference between the average $PW_{N>5}$ value and all other higher fertilizer levels was statistically different (Fig. 9b), while for manure, the average $PW_{N>5}$ value was 29.46% with an increasing tendency of 52.90% in Level 4. Compared to the other

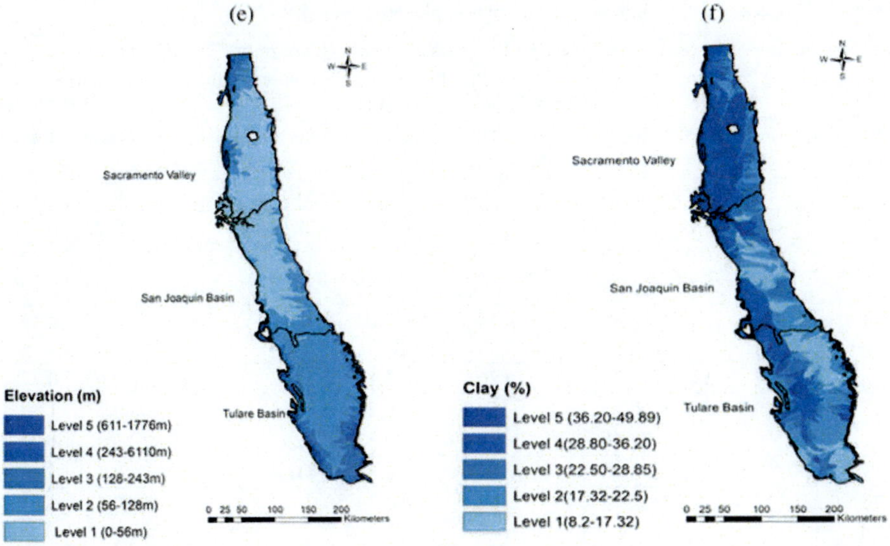

Fig. 9 (continued)

upper levels of manure, the average $PW_{N\,>\,5}$ was noticeably different (Fig. 9c). In Level 2, the average $PW_{N\,>\,5}$ was 54.08%, considerably higher than in Levels 1 and 3. Figure 9d represents the Central Valley, illustrating that most of the cropland (depicted in Fig. 9f) was situated within the Level 2 elevation region of the Tulare Lake Basin. Because clay particles prevent water from percolating into the aquifer, GW-NO_3 contamination dropped as the percentage of clay in the soil increased. At Level 1, the average value of $PW_{N\,>\,5}$ was 48.06%, which decreased to a statistically significant level of 26.97% at Level 3. Compared to Levels 1 and 2, Levels 4 and 5 had reduced average $PW_{N\,>\,5}$ values, but the difference was not statistically significant (Fig. 9f). Precipitation substantially interacted with permeability, elevation, and dissolved oxygen by 0.46, 0.41, and 0.43, respectively, according to the results of the interaction detector. Precipitation, permeability, and dung all had higher fertilizer interaction values of 0.38, 0.36, and 0.35, respectively.

3.2.3 Frequency Ratio Method

Table 5 shows the frequency ratio (FR) numbers for each level for the predictor variables. The bold highlighted values indicated the FR value greater than 1 for each level. When there was low precipitation. (1 and 2), the FR was higher than 1. The FR was more significant than 1 at higher levels of fertilizer (3, 4, and 5), lower elevations (2 and 3), and higher levels of manure. (3, 4, and 5). At levels 1, 2, 4, and 5, the clay (%) had an FR value greater than 1.

Table 5 Frequency ratio method of significant explanatory variables [66]

Factor	Range	Level	No. of wells	% Wells	No. of contaminated wells	Frequency ratio (FR)	
Precipitation (mm)	181.24–40.35	1	157	0.06	89	0.09	1.40
	240.35–338.4	2	998	0.40	519	0.52	1.29
	338.4–444.97	3	631	0.26	222	0.22	0.87
	444.97–557.87	4	469	0.19	126	0.13	0.67
	557.87–692.83	5	216	0.09	42	0.04	0.48
Fertilizer (kg/ha)	0–12.13	1	74	0.03	13	0.01	0.44
	12.13–41.19	2	261	0.11	34	0.03	0.32
	41.19–60.25	3	514	0.21	206	0.21	1.00
	60.25–74.68	4	798	0.32	329	0.33	1.02
	74.68–103.68	5	825	0.33	416	0.42	1.25
Elevation (m)	11.27–39.16	1	736	0.30	242	0.24	0.81
	39.16–71.86	2	1,142	0.46	479	0.48	1.04
	71.86–116.85	3	467	0.19	235	0.24	1.25
	116.85–237.71	4	126	0.05	41	0.04	0.81
	237.71–1180.33	5	1	0	0	0	0
Manure (kg/ha)	0–4.38	1	327	0.13	103	0.10	0.78
	4.38–12.1	2	457	0.18	111	0.11	0.60
	12.1–26.58	3	615	0.25	286	0.29	1.15
	26.58–40.47	4	236	0.10	133	0.13	1.40
	40.47–65.48	5	836	0.34	365	0.37	1.08
Clay (%)	13.31–18.55	1	785	0.32	348	0.35	1.10
	18.55–22.65	2	226	0.09	117	0.12	1.28
	22.65–28.30	3	678	0.27	209	0.21	0.76
	28.30–34.84	4	555	0.22	228	0.23	1.02
	34.84–39.54	5	228	0.09	96	0.10	1.04

3.2.4 Index of Geodetector Frequency Ratio Method

Figure 10 illustrates the results of calculating the geodetector frequency ratio (GFR) index using the geometric mean method (Eq. 5). The range of the GFR index was 3.70 to 5.17. To examine the distribution of the GFR index across the valley, natural breaks classification was applied to the GFR index. Compared to the SM Valley, the index values were most significant near the San Joaquin and Tulare Lake Basins.

The GFR method revealed that groundwater watersheds were more vulnerable to GW-NO_3 contamination than the DRASTIC method based on natural breaks classification intervals for both methods, as shown in Figs. 8 and 10. The highest interval range (4.96–5.17) of the GFR index had 29.27% of wells, compared to just 12.5%

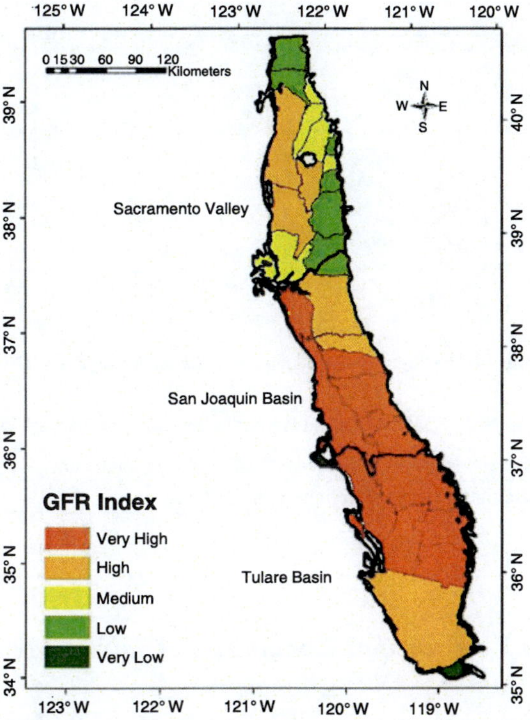

Fig. 10 Index of GFR method [66]

Table 6 Classification of the natural breaks for the DRASTIC and GFR indices [66]

		DRASTIC			GFR	
Index	DRASTIC index range	Number of basins	% Basin	GFR index range	Number of basins	% Basin
Very high	164–193	5	12.5	4.96–5.17	12	29.27
High	147–163	13	32	4.78–4.95	6	14.63
Moderate	129–146	10	25	4.66–4.77	7	17.07
Low	111–128	9	22.5	4.0–4.65	12	29.27
Very low	103–110	3	7.5	3.70–3.99	4	9.76

for the DRASTIC index (Table 6), which covered the highest interval range (164–193). In Fig. 11a, b, the DRASTIC and GFR indices data distribution is plotted as a histogram. The DI mean value was 141.87, while the GFR index was 4.70. Only a few groundwater watersheds were visible at the higher range of the positively skewed DRASTIC index values. The values of the GFR index were negatively skewed, indicating that more groundwater watersheds were in the top range.

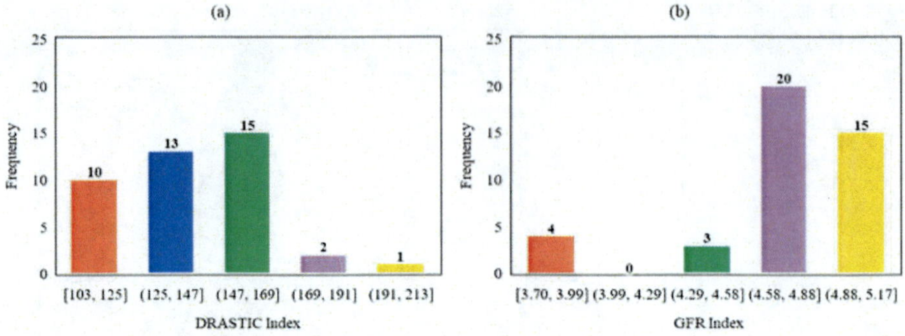

Fig. 11 Histogram of indices values at each range. **(a)** DRASTIC and **(b)** GFR [66]

Table 7 KMO value and Bartlett's test for sphericity of PCA analysis [61]

Kaiser–Meyer–Olkin measure of sampling adequacy	0.61
Bartlett's test of sphericity approx. Chi-square	547.63
Df	66
Sig	0.00

3.3 Principal Component Analysis Method

Kaiser–Meyer–Olkin (KMO) number was 0.63 based on principal component analysis (PCA) results (Table 7). The recommended minimal KMO number required to perform PCA analysis was 0.50. So, in terms of sampling adequacy, this study was valid. The assumption of variables being uncorrelated, as indicated by the identity matrix of the correlation matrix, was invalidated. This was supported by the Bartlett's test for sphericity, which yielded a significance level lower than 0.05. Cropland and manure had the most significant correlation of 0.65, followed by cropland and precipitation, clay and permeability, and manure and fertilizer of −0.58, −0.64, and 0.50, respectively, in correlation analysis, as shown in Table 8. Note a positive correlation between cropland, manure, and fertilizer. This illustrated how manure and fertilizer were used on farmland and how closely they were related.

The correlations between fertilizers, manure, and cropland with precipitation were negative, implying reduced rainfall in this area. On the other hand, a significant and unfavorable relationship existed between clay content and permeability. The significant but adverse connection between clay and permeability was due to the fact that a permeable region facilitated a higher rate of groundwater percolation. After all, water readily percolated through its pore spaces, unlike clay, which blocked water flow due to its small particle size. Dissolved oxygen and these elements negatively correlated since iron and manganese were found in low oxygen conditions. Since clay particles were small and had fewer pore spaces accessible for air, for the same reason, dissolved oxygen and clay had a negative correlation. Only 4 components were kept based on the screen plot and Eigenvalue greater than 1 (Table 9). The

Table 8 Explanatory variables correlation matrix [61]

	Slope	Recharge rate	Precipitation	Permeability	Manganese	Manure	Fertilizer	Iron	Elevation	Dissolved oxygen	Cropland	Clay
Slope	1.00											
Recharge rate	0.27[b]	1.00										
Precipitation	0.21[a]	0.11	1.00									
Permeability	−0.11	0.19[a]	−0.17	1.00								
Manganese	−0.15	−0.30[b]	−0.13	−0.23[a]	1.00							
Manure	−0.17	−0.14	−0.49[b]	0.17	0.10	1.00						
Fertilizer	−0.40[b]	−0.27[b]	−0.43[b]	0.34[b]	0.15	0.50[b]	1.00					
Iron	0.08	−0.19[a]	0.38[b]	−0.20[a]	0.03	−0.17	−0.05	1.00				
Elevation	0.36[b]	0.29[b]	−0.43[b]	−0.01	−0.07	−0.08	−0.08	−0.27[b]	1.00			
Dissolved oxygen	−0.10	0.55[b]	−0.26[b]	0.33[b]	−0.15	0.08	0.05	−0.42[b]	0.20[a]	1.00		
Cropland	−0.26[b]	−0.31[b]	−0.58[b]	0.00	0.28[b]	0.65[b]	0.40[b]	0.08	0.04	−0.09	1.00	
Clay	0.05	−0.30[b]	0.29[b]	−0.64[b]	0.45[b]	−0.14	−0.16	0.30[b]	−0.26[b]	−0.45[b]	−0.01	1.00

[a] Correlation is significant at the 0.05 level. (See Table 2 for full variable names)
[b] Correlation is significant at the 0.01 level

Table 9 Total variance explained by PCA [61]

Total variance explained									
	Initial eigenvalues			Extraction sums of squared loadings			Rotation sums of squared loadings		
Component	Total %	% of variance	Cumulative %	Total %	% of variance	Cumulative %	Total %	% of variance	Cumulative %
1	3.09	25.76	25.76	3.09	25.76	25.76	2.81	23.39	23.39
2	2.79	23.22	48.97	2.79	23.22	48.97	2.02	16.85	40.24
3	1.51	12.58	61.55	1.51	12.58	61.55	2.00	16.69	56.93
4	1.08	8.96	70.51	1.08	8.96	70.51	1.63	13.58	70.51
5	0.80	6.63	77.14						
6	0.68	5.70	82.84						
7	0.65	5.41	88.25						
8	0.48	4.03	92.28						
9	0.32	2.67	94.95						
10	0.25	2.11	97.06						
11	0.18	1.48	98.54						
12	0.18	1.46	100.0						

Table 10 Rotated component matrix [61]

	Component			
	1	2	3	4
Cropland	0.85			
Precipitation	−0.80			
Manure	0.78			
Fertilizer	0.67			
Clay		−0.80		
Permeability		0.76		
Manganese		−0.79		
Dissolved oxygen			0.80	
Iron			−0.79	
Recharge rate				
Elevation				0.79
Slope				0.78

See Table 5 for full variable names

variation was explained by PC-1, PC-2, PC-3, and PC-4 by 25.75%, 23.21%, 12.57%, and 8.96%, respectively. These four elements combined explained 70.50% of the variance. Table 10 displays variables that substantially affect each PC with a weight greater than 0.60. Manure, farmland, and fertilizer all have positive weights for PC-1, whereas precipitation has a negative weight.

In the San Joaquin Basin, precipitation was minimal and got even lower as it moved toward the Tulare Lake Basin. However, these regions had the most cropland, manure, and fertilizer. Despite the infrequent rainfall, irrigation using groundwater pumped from wells is used to maintain farmland in this area. Thus, the positive weight of crops, manure, and fertilizer is offset by the negative weight of precipitation. The correlation matrix demonstrating this opposite relationship showed a negative link between precipitation and cropland, fertilizer, and manure at −0.58, −0.43, and −0.49, respectively. In PC-2, the highest weight value was for clay at −0.80, followed by permeability and manganese at 0.76 and −0.73, respectively. In regions with high permeability, the intergranular area between the rocks might be easily penetrated by dissolved NO_3 in groundwater, allowing it to contaminate the aquifer. As a result, permeability had large positive weights. Conversely, clay had a negative weight because it was made up of tiny granules, which prevented water from flowing through it.

Figure 12a demonstrates that the Central Valley's highly permeable regions had a low clay content. The geochemical state of the groundwater in the reservoir was represented by the presence of manganese, iron, and dissolved oxygen. NO_3 could readily convert to nitrogen gas under anoxic conditions, lowering the amount of NO_3. Groundwater microbes preferentially used NO_3 as an electron acceptor in redox reactions to generate energy when oxygen was absent. As a result, the oxygen content was a good index of the aquifer's redox state. The redox process in groundwater also produces iron and manganese, and their concentration rises in anoxic environments. Manganese, which symbolizes the anoxic condition that can

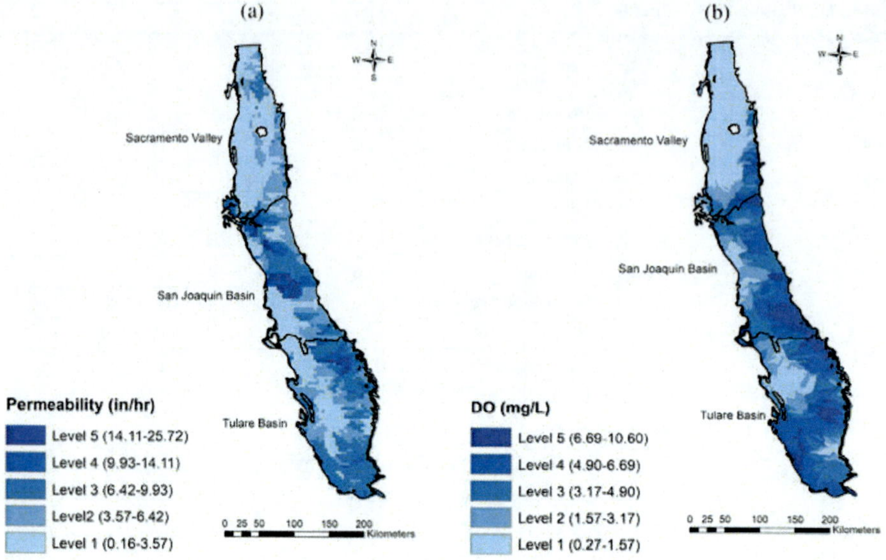

Fig. 12 Distribution in the Central Valley. (**a**) Permeability and (**b**) dissolved oxygen [61]

degrade GW-NO_3, has a negative weight. Greater dissolved oxygen levels were associated with higher NO_3 concentrations in groundwater, which were reduced when anoxic conditions occurred (Fig. 12b). This was also shown by PC-3, where dissolved oxygen had a positive weight. As anoxic conditions arise and denitrification occurs, the NO_3 content typically declines toward the deeper aquifer [11]. According to PC-4, slope and elevation have positive weights. Higher elevations result in steeper slopes. Groundwater contaminated with NO_3 is more prevalent at lower elevations due to a shallower water table at these altitudes.

3.4 Geographically Weighted Regression Method

Before using the geographically weighted regression (GWR) method, the OLS method was used. It revealed no multicollinearity; the variance inflation variable (VIF) was <7.5. Due to the non-significant result (p-value $= 0.16$) of the Jarque–Bera test, it can be concluded that the residuals are not regularly distributed. A residual with a normal distribution indicated that the model was biased and might lack a crucial explaining variable. The AIC number was 1611.28, and the adjusted-R^2 value was 0.21. The Akaike information criterion (AIC) gauges the model's effectiveness, while R^2 gauges the percent variation that the model measures. It was confirmed that there was no geographical clustering of the residuals by the insignificant Moran's I value of 0.026. The relationship between the dependent and explanatory variables was non-stationary, and the GWR model could enhance it, according

Table 11 Comparison between OLS and the GWR model [61]

Spatial autocorrelation					
	Adj R^2	AIC	Moran's I	Z Score	p-Value
OLS	0.21	1,611.28	0.026	0.49	0.62
GWR	0.26	1,606.96	−0.00028	0.05	0.96

to the combined F-statistics, which were significant (i.e., p-value = 0.05). The three OLS model variables that were statistically significant (i.e., p-value 0.05) were precipitation, fertilizer, and elevation. The elevation and manure coefficients were positive with $PW_{N\,>\,5}$, whereas the precipitation coefficient was negative.

Due to the inconsistent positioning observations throughout the study area, the geographic weighting in this GWR model was provided by utilizing the adaptive kernel. The GWR's output revealed that 119 closest neighbors were used to estimate each set of coefficients. It also calculated the proportion of data that falls under each kernel. The sum of the square residuals yielded a residual square value of 109,976. The residual square was lower, indicating that the GWR model adequately described the measured data. The adequate number gauged the model's intricacy, which was 16.63. The Sigma number was estimated as 16.63, which provided the expected standard deviation for the residuals, and lower values were desirable. The Akaike information criterion (AIC) was used as the bandwidth method because it automatically determined the bandwidth, minimized the AIC value, and provided the most accurate predictions. The AIC value dropped from 1611.28 in the OLS model to 1606.96 in the GWR model, indicating an improvement in the model's fit (Table 11).

Models were typically considered superior if their AIC values were lower and decreased by more than 3. Here, GWR's AIC value decreased by 5 when compared to OLS. The adjusted R^2 value of the GWR model increased from 0.21 to 0.26, outperforming the OLS model's total fit. The area R^2 values, however, were between 0.024 and 0.314. The model accurately predicted the SM Valley and San Joaquin Basin by considering localized R2 values. However, the Tulare Lake Basin exhibited lower local R^2 values. Only one watershed displayed a negative standardized residual on the GWR's standardized residual map.

In contrast, five watersheds had standardized residuals higher than 2. Moran's spatial autocorrelations were calculated to determine if the residuals were randomly generated. Moran's I had a number between −1 (complete dispersion) and +1 (complete clustering). Several zeros for Moran's I denoted absolute spatial irrationality. Moran's I was 0.00028 for the GWR, with a p-value of 0.96, showing that the existence of spatial autocorrelation cannot be ruled but must accept the random distribution of the residuals as the alternative. The GWR model eliminated any spatial correlations that might have existed in the OLS model (Fig. 13a, b). The GWR model's local condition (COND) value throughout the Central Valley varied between 24 and 27. There was no local multicollinearity because the COND value was less than 30.

The precipitation global coefficient was 5.6, showing a negative correlation between the precipitation and NO_3-contaminated wells, possibly due to the

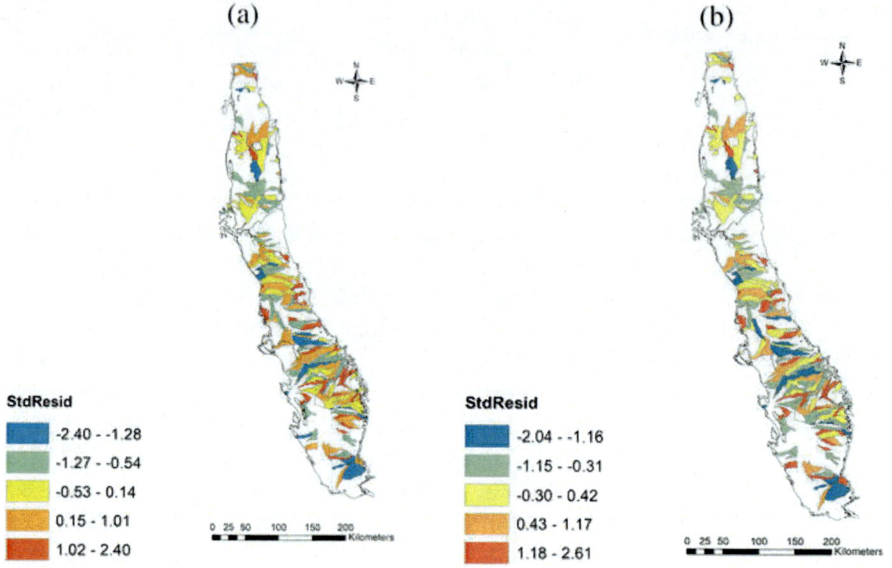

Fig. 13 Standardized residuals. (**a**) OLS model and (**b**) GWR model [61]

precipitation's diluting effect on the Central Valley. The spatial heterogeneity of the model was revealed by the local precipitation coefficient map (Fig. 14a), which indicated that the value varied from 13.63 (SM Valley) to 4.10 (Tulare Lake Basin). SM Valley got more rain than the southern regions of the Central Valley; as a result, NO_3 in GW might be diluted. Low rainfall caused irrigation in Tulare Lake and San Joaquin Basins to overly pump groundwater from wells. In this region, the level of the water table was situated at a lower position, causing groundwater to move in the direction of the groundwater depression cone.

Additionally, the application rates for manure and fertilizer were excellent here. These circumstances help NO_3 percolate into the groundwater. The local R^2 in this region also suggested that the model might benefit from adding another significant component. A positive correlation between NO_3 contamination and fertilizer application was shown by the fertilizer global coefficient, which was 4.05. There were greater fertilizer weights in the watersheds of all three hydrologic regions in the Central Valley. The local coefficient plot (Fig. 14b) shows that the model had some spatial heterogeneity. From 0 to 6.7 was the neighborhood coefficient range. Due to extensive irrigation and fertilizer application, fertilizer coefficients were generally more significant in the Tulare Lake Basin area, followed by the SM Valley and San Joaquin Basin. Elevation has a worldwide coefficient of 10.77 (Fig. 14c). From 0 to 22.4 was the neighborhood coefficient. The Tulare Lake and San Joaquin Basins had the lowest and greatest local coefficients, respectively. The San Joaquin Basin and

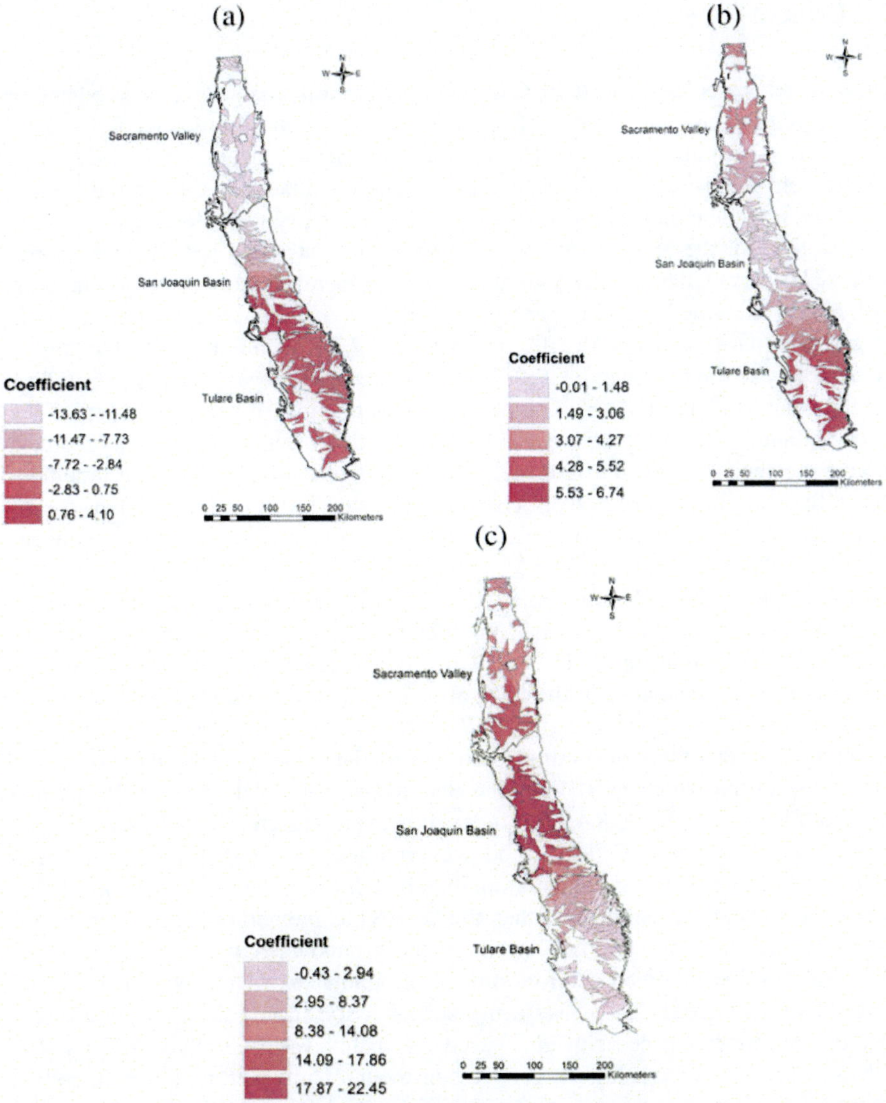

Fig. 14 Global coefficients for (**a**) precipitation, (**b**) fertilizer, (**c**) elevation [61]

SM Valley were located in areas with the lowest elevation, making them more susceptible to groundwater contamination. The most pronounced local coefficients were identified in line with Tulare Lake's characteristics. The GWR coefficient maps emphasized the spatial diversity of significant factors like precipitation, fertilizer usage, and elevation. This showcased that the localized impacts of groundwater contamination exerted a significant sway over the Central Valley.

4 Conclusions

This chapter presented an accurate GW-NO_3 vulnerability map using the geodetector method (GED) and the DRASTIC model for the Central Valley at CA's most productive agricultural region. Three statistical methods (i.e., GED, PCA, and GWR) were compared to analyze the GW-NO_3 contamination in the Central Valley. By examining the connections between the 12 explanatory variables and the NO_3 in each watershed, the analysis was conducted at the watershed level in the Central Valley. Moreover, the Central Valley aquifer's vulnerability to GW-NO_3 contamination was examined using the new GFR method. A comparison was also made between the GFR and DRASTIC methods. The GFR technique indicated higher vulnerability in the San Joaquin and Tulare Lake Basin compared to the SM Valley. On the other hand, the DRASTIC index values were higher in the eastern part of the Central Valley, known for frequent Tulare Lake activity. The GFR method predicted more watersheds falling into the higher index range than the DRASTIC method. These findings aligned with a prior study conducted in the Central Valley. Additionally, wells in the San Joaquin and Tulare Lake Basin displayed higher concentrations of NO_3 than those in the SM Valley [92].

Unlike the southern Central Valley, the SM Valley was made up of finer-grained sediments, which might lessen its vulnerability. Many studies added variables or changed intervals to adapt the DRASTIC method to site-specific conditions. Out of 41 groundwater watersheds in the Central Valley, 31 were found to have contaminated wells with NO_3 (greater than 5 mg/L). The groundwater vulnerability mapping advantage was that the index value could be calculated using the rating and weight values for all the watersheds [93]. The San Joaquin and Tulare Lake Basins had higher GFR indices. The DRASTIC index, however, was greater in the eastern part of the valley. The new GFR method was validated using Pearson's correlation coefficient and PD value, demonstrating that the GFR method was more compatible with observed contamination data than the DRASTIC method regarding vulnerability index and % of contaminated wells in the groundwater watersheds. Negative GFR index values showed disproportionate groundwater watersheds in the upper range. Different areas can measure the spatial association using the geodetector method, which was successful and based on spatial variance analysis. High PD values were discovered using the geodetector method; however, they were merely a statistical finding and did not imply causation. As a result, selecting the crucial elements required more significant consideration.

To check the accuracy of the GED method in identifying clay, manure, elevation, fertilizer, and precipitation as the key variables. More contaminated wells were found in regions with large percentages of cropland, fertilizer, and manure. Despite comparatively low annual precipitation, there were more NO_3-contaminated wells in the San Joaquin and Tulare Lake Basins. This could be ascribed to increased fertilizer application rates on agricultural land and evolving hydrologic conditions brought on by groundwater pumping over time that may have facilitated NO_3 percolation into aquifers downward. Precipitation, manure, fertilizer, and cropland were all grouped by the PCA results with high weights on PC-1, indicating a typical structure in the dataset that can cause issues with multicollinearity in the analysis. The GWR method successfully captured the regional precipitation, fertilizer, and elevation heterogeneity for $PW_{N > 5}$. The San Joaquin Basin and Tulare Lake Basin had only marginally positive local coefficients of precipitation, and the northern SM Valley had an increasing negative coefficient. In the Tulare Lake Basin, the coefficients of fertilizer were significant where more fertilizer was used. Due to their comparatively low elevation and increased vulnerability to aquifer contamination, the SM and San Joaquin Basins had the highest coefficient of elevation.

Overall, the GED method proved more helpful than PCA and GWR. PCA often made things simpler by using fewer details from Tulare Lake's data, but it sometimes made it hard to understand those simplified groups. The GWR model acted like a microscope for space, showing how things change across the region, but it had some issues with how variables were connected, which needed fixing with the OLS model. Unlike GWR, the GED method, guided by Tulare Lake, lets us compare places based on how much NO_3 was found on average. This way of comparing differed from GWR, which looked at how the effects changed throughout the research area. The GED method was great for studying how NO_3 contamination happened in the Central Valley and beyond because it gave us lots of information and could adapt to different data needs.

The comparison analysis demonstrated that the GED method gathered all the data disclosed by the four-detector PCA and GWR methods. Due to the following advantages, the GED method performed better than PCA and GWR: It worked effectively with both continuous and categorical data; it was more advantageous for policy makers because of its simplicity in understanding the effects of specific variables; it highlighted the most important variables, indicating the sections of each variable that were more vulnerable; and it investigated how different variables interact to simplify further and enhance the usefulness of data preparation.

5 Recommendations

The intricate dynamics of groundwater contamination from agricultural drainage water necessitate a multifaceted approach. Future studies should investigate localized assessments, accounting for varying soil types, drainage patterns, and agricultural practices. These studies should be bolstered by integrating meteorological, hydrological, and land use data, ensuring a comprehensive, data-driven foundation for subsequent strategies. Given the rapidly evolving nature of agricultural methodologies, continuous monitoring of emerging contaminants and updated assessment techniques are crucial to ensure relevance and accuracy in predictions.

For policy planners and decision-makers, a robust framework needs to be established. Proactive regulatory measures should be implemented, encompassing stricter guidelines on using and disposing of agricultural chemicals. Investments in cutting-edge drainage infrastructure prioritizing agricultural efficiency and environmental conservation are essential. For stakeholders at all levels, fostering a culture of collaboration is key. This includes educational outreach programs for farmers, community engagement initiatives, and public–private partnerships to develop innovative solutions. Together, a concerted effort across all tiers can greatly mitigate the risks of groundwater contamination from agricultural drainage.

Acknowledgments The authors would like to thank the editor and the anonymous reviewers for their valuable comments.

References

1. Lo M, Famiglietti JS (2013) Irrigation in California's Central Valley strengthens the southwestern US water cycle. Geophys Res Lett 40:301–306
2. USGS (2016) Contaminants found in groundwater, USGS Water Science School. US Geol Surv
3. Harter T (2009) Agricultural impacts on groundwater nitrate. Southwest Hydrol 8:22–23
4. Rosecrans CZ, Nolan BT, Gronberg JM (2017) Prediction and visualization of redox conditions in the groundwater of Central Valley, California. J Hydrol 546:341–356
5. Lockhart KM, King AM, Harter T (2013) Identifying sources of groundwater nitrate contamination in a large alluvial groundwater basin with highly diversified intensive agricultural production. J Contam Hydrol 151:140–154
6. van der Schans ML, Harter T, Leijnse A et al (2009) Characterizing sources of nitrate leaching from an irrigated dairy farm in Merced County, California. J Contam Hydrol 110:9–21
7. Javadi S, Kavehkar N, Mousavizadeh MH, Mohammadi K (2011) Modification of DRASTIC model to map groundwater vulnerability to pollution using nitrate measurements in agricultural areas. J Agric Sci Technol 13:239–249
8. Neukum C, Azzam R (2009) Quantitative assessment of intrinsic groundwater vulnerability to contamination using numerical simulations. Sci Total Environ 408:245–254
9. Harter T, Dzurella K, Kourakos G et al (2017) Nitrogen fertilizer loading to groundwater in the Central Valley. Final Rep to Fertil Res Educ Program, Proj, pp 11–301
10. Aller L, Bennet T, Lehr J, Petty R (1987) DRASTIC: a standardized system for evaluating ground water pollution potential using hydrogeologic settings. USEPA
11. Burow KR, Jurgens BC, Belitz K, Dubrovsky NM (2013) Assessment of regional change in nitrate concentrations in groundwater in the Central Valley, California, USA, 1950s–2000s. Environ Earth Sci 69:2609–2621
12. Lindsey BD, Rupert MG (2012) Methods for evaluating temporal groundwater quality data and results of decadal-scale changes in chloride, dissolved solids, and nitrate concentrations in groundwater in the United States, 1988–2010. US Department of the Interior, US Geological Survey
13. Nolan BT, Hitt KJ, Ruddy BC (2002) Probability of nitrate contamination of recently recharged groundwaters in the conterminous United States. Environ Sci Technol 36:2138–2145
14. Rupert MG (2001) Calibration of the DRASTIC ground water vulnerability mapping method. Groundwater 39:625–630
15. Focazio MJ (1984) Assessing ground-water vulnerability to contamination: providing scientifically defensible information for decision makers. US Government Printing Office
16. Minshew H, Selker J, Hemphill D, Dick RP (2002) NLEAP computer model and multiple linear regression prediction of nitrate leaching in vegetable systems. Hort Technol 12:250–256
17. Luo W, Jasiewicz J, Stepinski T et al (2016) Spatial association between dissection density and environmental factors over the entire conterminous United States. Geophys Res Lett 43:692–700

18. Wang J-F, Zhang T-L, Fu B-J (2016) A measure of spatial stratified heterogeneity. Ecol Indic 67:250–256
19. Wang J, Li X, Christakos G et al (2010) Geographical detectors-based health risk assessment and its application in the neural tube defects study of the Heshun region, China. Int J Geogr Inf Sci 24:107–127
20. Shrestha A, Luo W (2017) An assessment of groundwater contamination in Central Valley aquifer, California using geodetector method. Ann GIS 23:149–166. https://doi.org/10.1080/19475683.2017.1346707
21. Zhou J, Li G, Liu F et al (2010) DRAV model and its application in assessing groundwater vulnerability in arid area: a case study of pore phreatic water in Tarim Basin, Xinjiang, Northwest China. Environ Earth Sci 60:1055–1063
22. van Beynen PE, Niedzielski MA, Bialkowska-Jelinska E et al (2012) Comparative study of specific groundwater vulnerability of a karst aquifer in Central Florida. Appl Geogr 32:868–877
23. Foster SSD (1987) Fundamental concepts in aquifer vulnerability, pollution risk and protection strategy
24. De Filippis G, Ercoli L, Rossetto R (2021) A spatially distributed, physically-based modeling approach for estimating agricultural nitrate leaching to groundwater. Hydrology 8:8
25. Neshat A, Pradhan B, Pirasteh S, Shafri HZM (2014) Estimating groundwater vulnerability to pollution using a modified DRASTIC model in the Kerman agricultural area, Iran. Environ Earth Sci 71:3119–3131
26. Gong G, Mattevada S, O'Bryant SE (2014) Comparison of the accuracy of kriging and IDW interpolations in estimating groundwater arsenic concentrations in Texas. Environ Res 130:59–69
27. Kaown D, Hyun Y, Bae G, Lee K (2007) Factors affecting the spatial pattern of nitrate contamination in shallow groundwater. J Environ Qual 36:1479–1487
28. Narany TS, Ramli MF, Aris AZ et al (2014) Assessment of the potential contamination risk of nitrate in groundwater using indicator kriging (in Amol–Babol plain, Iran). In: From sources to solution: proceedings of the international conference on environmental forensics 2013. Springer, pp 273–277
29. Stigter TY, Ribeiro L, Dill AMMC (2006) Evaluation of an intrinsic and a specific vulnerability assessment method in comparison with groundwater salinisation and nitrate contamination levels in two agricultural regions in the south of Portugal. Hydrgeol J 14:79–99
30. Akbar TA, Lin H, DeGroote J (2011) Development and evaluation of GIS-based ArcPRZM-3 system for spatial modeling of groundwater vulnerability to pesticide contamination. Comput Geosci 37:822–830
31. Fontaine DD, Havens PL, Blau GE, Tillotson PM (1992) The role of sensitivity analysis in groundwater risk modeling for pesticides. Weed Technol 6:716–724
32. Leonard RA, Knisel WG, Still DA (1987) GLEAMS: groundwater loading effects of agricultural management systems. Trans ASAE 30:1403–1418
33. Leone A, Ripa MN, Uricchio V et al (2009) Vulnerability and risk evaluation of agricultural nitrogen pollution for Hungary's main aquifer using DRASTIC and GLEAMS models. J Environ Manage 90:2969–2978
34. Bonton A, Rouleau A, Bouchard C, Rodriguez MJ (2011) Nitrate transport modeling to evaluate source water protection scenarios for a municipal well in an agricultural area. Agr Syst 104:429–439
35. Qin R, Wu Y, Xu Z et al (2013) Assessing the impact of natural and anthropogenic activities on groundwater quality in coastal alluvial aquifers of the lower Liaohe River Plain, NE China. Appl Geochem 31:142–158
36. Nobre RCM, Rotunno Filho OC, Mansur WJ et al (2007) Groundwater vulnerability and risk mapping using GIS, modeling and a fuzzy logic tool. J Contam Hydrol 94:277–292
37. Anane M, Abidi B, Lachaal F et al (2013) GIS-based DRASTIC, pesticide DRASTIC and the susceptibility index (SI): comparative study for evaluation of pollution potential in the Nabeul-Hammamet shallow aquifer, Tunisia. Hydrogeol J 21:715

38. Garnier M, Lo Porto A, Marini R, Leone A (1998) Integrated use of GLEAMS and GIS to prevent groundwater pollution caused by agricultural disposal of animal waste. Environ Manag 22:747–756
39. Iqbal J, Gorai AK, Tirkey P, Pathak G (2012) Approaches to groundwater vulnerability to pollution: a literature review. Asian J Water Environ Pollut 9:105–115
40. Johnson TD, Belitz K (2009) Assigning land use to supply wells for the statistical characterization of regional groundwater quality: correlating urban land use and VOC occurrence. J Hydrol 370:100–108
41. McLay CDA, Dragten R, Sparling G, Selvarajah N (2001) Predicting groundwater nitrate concentrations in a region of mixed agricultural land use: a comparison of three approaches. Environ Pollut 115:191–204
42. Choubin B, Malekian A (2017) Combined gamma and M-test-based ANN and ARIMA models for groundwater fluctuation forecasting in semiarid regions. Environ Earth Sci 76:1–10
43. Choubin B, Malekian A, Samadi S et al (2017) An ensemble forecast of semi-arid rainfall using large-scale climate predictors. Meteorol Appl 24:376–386
44. Choubin B, Zehtabian G, Azareh A et al (2018) Precipitation forecasting using classification and regression trees (CART) model: a comparative study of different approaches. Environ Earth Sci 77:1–13
45. Ghorbani Nejad S, Falah F, Daneshfar M et al (2017) Delineation of groundwater potential zones using remote sensing and GIS-based data-driven models. Geocarto Int 32:167–187
46. Maqsoom A, Aslam B, Gul ME et al (2021) Using multivariate regression and ANN models to predict properties of concrete cured under hot weather. Sustainability 13:10164
47. Ullah F, Qayyum S, Thaheem MJ et al (2021) Risk management in sustainable smart cities governance: a TOE framework. Technol Forecast Soc Change 167:120743
48. Ullah F, Sepasgozar SME, Thaheem MJ, Al-Turjman F (2021) Barriers to the digitalisation and innovation of Australian smart real estate: a managerial perspective on the technology non-adoption. Environ Technol Innov 22:101527
49. Sahoo S, Russo TA, Elliott J, Foster I (2017) Machine learning algorithms for modeling groundwater level changes in agricultural regions of the US. Water Resour Res 53:3878–3895
50. Barzegar R, Moghaddam AA, Deo R et al (2018) Mapping groundwater contamination risk of multiple aquifers using multi-model ensemble of machine learning algorithms. Sci Total Environ 621:697–712. https://doi.org/10.1016/j.scitotenv.2017.11.185
51. Knoll L, Breuer L, Bach M (2019) Large scale prediction of groundwater nitrate concentrations from spatial data using machine learning. Sci Total Environ 668:1317–1327. https://doi.org/10.1016/J.SCITOTENV.2019.03.045
52. Ullah F, Sepasgozar SME, Thaheem MJ et al (2021) It's all about perceptions: a DEMATEL approach to exploring user perceptions of real estate online platforms. Ain Shams Eng J 12: 4297–4317
53. Rahmati O, Choubin B, Fathabadi A et al (2019) Predicting uncertainty of machine learning models for modelling nitrate pollution of groundwater using quantile regression and UNEEC methods. Sci Total Environ 688:855–866. https://doi.org/10.1016/j.scitotenv.2019.06.320
54. Hegde J, Rokseth B (2020) Applications of machine learning methods for engineering risk assessment – a review. Saf Sci 122:104492
55. Park SK, Zhao Z, Mukherjee B (2017) Construction of environmental risk score beyond standard linear models using machine learning methods: application to metal mixtures, oxidative stress and cardiovascular disease in NHANES. Environ Health 16:1–17
56. Ullah F, Al-Turjman F (2023) A conceptual framework for blockchain smart contract adoption to manage real estate deals in smart cities. Neural Comput Appl 35:5033–5054
57. Williamson AK, Prudic DE, Swain LA (1989) Ground-water flow in the Central Valley, California. US Government Printing Office
58. California Department of Water Resources (2018) California water plan. Public Review Draft December 2018. https://www.water.ca.gov/-/media/DWR-Website/Web-Pages/Programs/

California-Water-Plan/Docs/Update2018/PRD/California-Water-Plan-Update-2018-Public-Review-Draft.pdf
59. Planert M, Williams JS (1995) GroundWater Atlas of the United States: segment 1, California, Nevada. U.S. Geol Surv
60. Galloway D, Riley FS (1999) San Joaquin Valley, California. L Subsid United States US Geol Surv Circ 1182:23–34
61. Shrestha A, Luo W (2017) Analysis of groundwater nitrate contamination in the Central Valley: comparison of the geodetector method, principal component analysis and geographically weighted regression. ISPRS Int J Geo-Information 6. https://doi.org/10.3390/ijgi6100297
62. Ward MH, DeKok TM, Levallois P et al (2005) Workgroup report: drinking-water nitrate and health – recent findings and research needs. Environ Health Perspect 113:1607–1614
63. Croskrey A, Groves C (2008) Groundwater sensitivity mapping in Kentucky using GIS and digitally vectorized geologic quadrangles. Environ Geol 54:913–920
64. Neukum C, Hötzl H, Himmelsbach T (2008) Validation of vulnerability mapping methods by field investigations and numerical modelling. Hydrgeol J 16:641–658
65. Rupert MG (1999) Improvements to the DRASTIC ground-water vulnerability mapping method. US Geological Survey
66. Shrestha A, Luo W (2018) Assessment of groundwater nitrate pollution potential in Central Valley aquifer using Geodetector-based frequency ratio (GFR) and optimized-DRASTIC methods. ISPRS Int J Geo-Information 7. https://doi.org/10.3390/ijgi7060211
67. California Department of Water Resources (2017) The Department of Water Resources Groundwater Information Center Interactive Map Application (GICIMA). https://gis.water.ca.gov/app/gicima/%0A. Accessed 8 Aug 2017
68. Wolock DM (2003) Estimated mean annual natural ground-water recharge in the conterminous United States. US Geological Survey
69. McKinney TS, Anning DW (2009) Geospatial data to support analysis of water-quality conditions in basin-fill aquifers in the southwestern United States. US Department of the Interior, US Geological Survey
70. Anning DW, Paul AP, McKinney TS, Huntington JM (2012) Predicted nitrate and arsenic concentrations in basin-fill aquifers of the southwestern United States. Reston, VA, USA
71. Schwarz GE, Alexander RB (1995) State soil geographic (STATSGO) data base for the conterminous United States
72. EDNA (2005) Elevation derivatives for national applications. https://edna.usgs.gov/. Accessed 5 Mar 2017
73. Wieczorek M (2014) Area-and depth-weighted averages of selected SSURGO variables for the conterminous United States and District of Columbia. US Geological Survey
74. Piscopo G (2001) Groundwater vulnerability map explanatory map-Castelreagh catchment. Kingswood, Australia
75. Gronberg JM, Spahr NE (2012) County-level estimates of nitrogen and phosphorus from commercial fertilizer for the conterminous United States, 1987–2006. US Department of the Interior, US Geological Survey Reston, VA
76. Mueller DK, Gronberg JAM (2013) County-level estimates of nitrogen and phosphorus from animal manure for the conterminous United States, 2002. US Geological Survey
77. Homer C, Dewitz J, Yang L et al (2017) Completion of the 2011 National Land Cover Database for the conterminous United States – representing a decade of land cover change information. Photogram Eng Remote Sens
78. Wolock DM (1997) STATSGO soil characteristics for the conterminous United States. US Geological Survey
79. USDA Risk Management Agency (2016) PRISM climate data. http://www.prism.oregonstate.edu/. Accessed 10 May 2016
80. EDNA (2017) Elevation derivatives for national applications. https://edna.usgs.gov/. Accessed 5 Mar 2017

81. EDNA (2013) National elevation dataset (NED). http://edna.usgs.gov/Edna/edna.asp. Accessed 16 Mar 2013
82. Nolan BT, Hitt KJ (2006) Vulnerability of shallow groundwater and drinking-water wells to nitrate in the United States. Environ Sci Technol 40:7834–7840
83. USGS (2013) National Water Information System (NWIS). https://waterdata.usgs.gov/nwis. Accessed 25 Aug 2013
84. Water Resources Mission Area (2013) National Water-Quality Assessment (NAWQA). https://water.usgs.gov/nawqa/. Accessed 25 Aug 2013
85. Lee S, Pradhan B (2006) Probabilistic landslide hazards and risk mapping on Penang Island, Malaysia. J Earth Syst Sci 115:661–672
86. Luo W, Liu C-C (2018) Innovative landslide susceptibility mapping supported by geomorphon and geographical detector methods. Landslides 15:465–474
87. Kim H, Park S (2016) Hydrogeochemical characteristics of groundwater highly polluted with nitrate in an agricultural area of Hongseong, Korea. Water 8:345
88. Wu T-N, Su C-S (2008) Application of principal component analysis and clustering to spatial allocation of groundwater contamination. In: 2008 fifth international conference on fuzzy systems and knowledge discovery. IEEE, pp 236–240
89. Fotheringham AS, Charlton ME, Brunsdon C (2001) Spatial variations in school performance: a local analysis using geographically weighted regression. Geogr Environ Model 5:43–66
90. Bertoldi GL, Johnston RH, Evenson KD (1991) Ground water in the Central Valley, California: a summary report. US Geol Surv Prof Pap 1401-A, 44 pp
91. Faunt CC (2009) Groundwater availability of the Central Valley aquifer, California. US Geol Surv Prof Pap
92. Ransom KM, Nolan BT, Traum JA et al (2017) A hybrid machine learning model to predict and visualize nitrate concentration throughout the Central Valley aquifer, California, USA. Sci Total Environ 601:1160–1172
93. Osborn NI, Hardy RH (1999) Statewide groundwater vulnerability map of Oklahoma. Oklahoma Water Resources Board Oklahoma City, OK, USA

Impact of Bugun' Reservoir on Groundwater and Soil: A Case Study from South Kazakhstan

Vladimir M. Starodubtsev and Maryna M. Ladyka

Contents

1 Introduction 78
2 The Bugun' Reservoir, Its Design, Implementation, and Problems of Functioning 79
3 Controversial History of the Project Scientific Justification 82
4 Bugun' Reservoir 83
4.1 Features of Functioning 83
4.2 Impact on Groundwater and Soils 85
5 Conclusion 93
6 Recommendations 95
References 95

Abstract The influence of the Bugun' reservoir, created for irrigation in the semi-desert zone of South Kazakhstan, on groundwater and soils of the reservoir basin and its coast during long-term operation is analyzed. The rise in the level of groundwater on the banks of the reservoir and in downstream of the Bugun' and Karazhantak dams in the process of the reservoir filling in 1963–1965 and further use for irrigation of more than 93,000 ha of land for growing cotton, grain, fodder, and fruit and vegetable crops was estimated. The leaching of toxic salts for plants into the groundwater of adjacent territories by seepage water from soils and rocks of the reservoir basin was calculated. The seasonally water-salt regime of soils and rocks of the aeration zone, as well as the regime of the groundwater level (10 days) and their mineralization (monthly) under conditions of strong seasonal fluctuations in the level of surface waters in the reservoir, was studied. A delay in the phases of rise and fall of the surface water level and groundwater level on the coast, reaching 2 months, depending on the distance from the reservoir, was revealed. Desalination of

V. M. Starodubtsev and M. M. Ladyka (✉)
Department of Agrosphere Ecology and Environmental Control, National University of Life and Environmental Sciences of Ukraine, Kyiv, Ukraine
e-mail: vladimir_starodubtsev@nubip.edu.ua; mm.ladyka@nubip.edu.ua

Shakir Ali and Abdelazim Negm (eds.), *Groundwater Quality and Geochemistry in Arid and Semi-Arid Regions*, Hdb Env Chem (2024) 126: 77–96, DOI 10.1007/698_2023_1053,

Published online: 9 December 2023

groundwater from salts is observed only in the bowl of the reservoir and on its banks, and in relief depressions, water mineralization (salt content) increased to 35 g/l. A strong salinization of coastal soils was noted in relief depressions and in the downstream of dams (up to 300–500 t/ha in a soil layer of 2 m). The satellite images (Landsat and Sentinel satellite images) reveal that the process of soil flooding (waterlogging) with groundwater and their salinization (salt concentration increase) continues for 60 years even after the reservoir has been filled. Geomorphological conditions, lithology of soils and existing rocks of the coast, and the winds direction contribute significantly to the intensive soil erosion process on the reservoir's left bank. We found that the coastal destruction in the 60s–70s occurred at a rate of up to tens of meters per year, and later it began to slow down (averaging 7–11 m/year), due to the strengthening of the banks with stone filling and fragmentary overgrowth of wood-shrub vegetation. In the 2000s, it hardly exceeded 1–2 m/year. Nevertheless, in the last decade, climate change has caused changes in the reservoir filling regime in the winter-spring months and in the direction of winds, which has led to an increase in erosion processes. Monitoring erosion processes with satellite images and a quadrocopter (UAV) made it possible in 2020–2022 to timely identify the prominent erosion-prone areas and prevent possible reservoir destruction.

Keywords Bank protection, Dams, Geomorphological conditions, Groundwater, Lithology of soil, Reservoir, Salinization, Soil erosion, Waterlogging

1 Introduction

Reservoirs in the foothills of the arid and subarid zones are the most important (often the only) source of irrigation and water supply for millions of inhabitants of these territories. Accumulating water from small mountain rivers, including those that dry up in summer, they provide a relatively stable supply of water to the dissected foothill plains for the drinking needs of the population and livestock, municipal water consumption, irrigation of agricultural land, and the creation of comfortable (or at least acceptable) living conditions for local settlements. The centuries-old history of the peoples who lived and now live in such countries which receive high insolation confirms the necessity and expediency of such a redistribution of local water flow. At the same time, it is quite obvious that the reservoirs and irrigation systems built many years ago were based on the material resources of the peoples available at that time, the experience of accumulating and distributing water, the level of science and technology [1–6]. And, of course, these structures solved the urgent specific tasks that faced the local population at that time. Therefore, now such reservoirs and systems often need to reconstruct dams, reduce unproductive water losses, and optimize the structure of water consumption. The tasks of preserving the area's biodiversity, preserving the monuments of everyday life, culture and architecture of peoples, and even more so, their historical spiritual heritage are becoming increasingly acute on the agenda. Today, it is also becoming mandatory to clarify the role and tasks facing such reservoirs in terms of analyzing water consumption, economic development, and environmental protection of the entire river basin.

Irrigation reservoirs, mostly small and medium, are located in a continuous chain along all the mountain ranges of the arid and subarid zones at the places where mountain rivers flow to the foothill plains, and sometimes to intermountain depressions. For example, in South Kazakhstan and in adjacent regions [6], these are the Bartogai reservoir on the Chilik River, the Orto-Tokoi and the Tashutkul reservoirs on the Chu River, the Kirov reservoir on the Talas River, the Chardara reservoir on the Syrdarya (or Syr Darya) River, and many others. One of the authors of this publication had a chance to work in these regions for many years. The names of the reservoirs do not take into account the massive changes that have taken place in Kazakhstan in recent years. The Bugun' irrigation reservoir on the Bugun' River is the object of detailed analysis in this work.

2 The Bugun' Reservoir, Its Design, Implementation, and Problems of Functioning

After the end of the Second World War, the Republic of Kazakhstan, which was earlier part of the former Soviet Union, was tasked with increasing the cultivation of cotton, food, fodder, fruit, melons, grapes, etc. For this purpose (among other projects), the semi-desert vast Pre-Karatau plain ("Otrar steppe") was chosen, stretching on the slopes of the Karatau ridge from the Arys' River to the ancient city of Turkestan. The climate here is strongly continental, dry, with an abundance of insolation and thermal resources. The average annual rainfall is 178 mm, the average annual air temperature is 12.1°C, the absolute maximum temperature is $+49^0$, the minimum is -39^0, and the frost-free period is 6–7 months. To provide irrigation water to this arid territory, it was proposed to build the reservoir with a capacity of 370 million m^3, fill it with water from the Bugun' river, and feed it with water from the Arys' river.

Irrigation development of the foothill plain in South Kazakhstan, thanks to the construction of the Bugun' reservoir and the Arys'-Turkestan irrigation system, already in those years was of great importance for the socio-economic development of the region. It is no less important right now in connection with the creation of a new Turkestan region of Kazakhstan. Industrial and other crops grown here have become a strategic resource of this region, and the population has received employments, gained invaluable experience in irrigation and land reclamation in difficult natural conditions. The very view from space of the Arys'-Turkestan irrigation system with the Bugun' reservoir (Fig. 1) evokes an understanding of the importance and timeliness of providing freshwater to this vast semi-desert region [7]. We are also satisfied with our direct participation in those already quite distant years in the soil-reclamation substantiation of the project, the assessment of its positive and, unfortunately, some negative characteristics [6, 8–10].

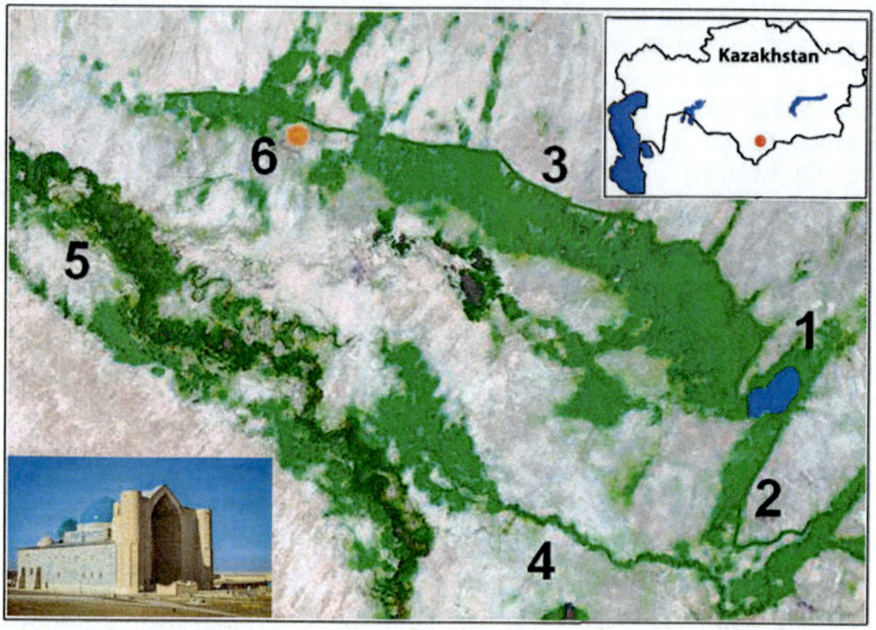

Fig. 1 The Bugun' reservoir and the Arys'-Turkestan irrigation system: *1 – reservoir; 2 – the Arys' main canal (AMC); 3 – the Turkestan main canal (TMC); 4 – the Arys' river; 5 – the Syrdarya (Syr Darya) river; 6 – the city of Turkestan with the Yasawi Mausoleum* (Landsat-8 satellite image (2019) [7] and figures from open sources were used for this picture [11])

The project's technical implementation was complex, lengthy, controversial, and yet generally successful. However, by now, problems have accumulated that require urgent solutions in connection with the region's development, the transformation of the population's rights to land and the creation of farms, and the aging of hydraulic and reclamation facilities. First of all, this is an aggravating lack of water for irrigation and water supply. It is caused by: (1) siltation of the Bugun' reservoir, (2) seepage losses from the main canals laid at a shallow occurrence of pebble deposits without sufficient impervious measures (the Turkestan Canal), and in loess deposits with noticeable filtration and subsidence properties (the Arys' Canal), (3) water losses in distribution canals of the first order; (4) unauthorized water withdrawals for irrigation of additional areas; (5) local water losses in certain sections of canals (main and distribution), leading to flooding and waterlogging of adjacent lands, (6) the use of increased irrigation and irrigation norms and other relevant reasons.

But first of all, we note changes in the regime of filling and drawdown of the Bugun' reservoir (Fig. 2) in recent years, apparently caused by climatic processes. These changes make some adjustments to the relationship between surface and groundwaters that have developed over half a century in the reservoir area

Fig. 2 Mode of filling and drawdown of the Bugun' reservoir in 2022 according to data from the Sentinel-2 and Landsat-9 satellites from NASA and ESA open funds

[12]. They especially affect the soil erosion processes on the reservoir's banks, which will be considered in more detail.

3 Controversial History of the Project Scientific Justification

The idea of building the Bugun' reservoir and the Arys'-Turkestan irrigation system was already expressed in a schematic design in 1948, and the project assignment was drawn up in 1953 by the "Kazgiprovodelectro" Institute (the Bugun' reservoir – 370 million m^3, canal length 194.5 km, regular irrigation – 70,500 ha, estuary irrigation – 53,800 ha, cotton area – 28,400 ha). Based on the material and technical resources of the state and the level of land reclamation science and practice of that time, the irrigation network was designed in earth channels, measures to combat filtration are not provided, drainage and wastewater networks are not designed. Soil and reclamation substantiation for the design task, completed in 1949–1950, and refined by research in 1952–1953 ("Sredazgiprovodkhlopok") and in 1956–1958 (SoyuzNIIKhI), was based on a reassessment of the draining role of pebble deposits of foothill fans and on an analysis of the soil cover structure that existed at that time, without taking into account the variability of soil-reclamation and hydrogeological conditions during irrigation. All this led to the construction of an irrigation system without the use of engineering melioration. The very first years of development of the massif revealed the shortcomings of such a technical solution and made it necessary to "clarify" the design task. The specified task provides for the development of 121,800 ha, of which 52,400 ha are regular irrigation, 53,000 ha of estuary irrigation and the land of the Darmina state farm (wormwood) – 16,400 ha. Estuary irrigation was subsequently excluded from the project. To combat the rise of groundwater and soil salinization, it is planned to build a collector-drainage network on an area of 22,000 ha and flush out the most saline soils. However, according to the reviewer S.F. Averyanov, "the drainage area planned in the project (22,000 ha) is the minimum, based on the current state of affairs, without considering the forecast of the water-salt regime of soils. This area can be conditionally taken as the volume of work of the first stage. In the near future (1-3 years), this area will increase."

Unfortunately, this forecast was justified in the process of irrigation development of the Arys'-Turkestan massif (as well as other similar systems [3, 4, 6] in the early 1960s). This is determined first of all by the presence of difficultly reclaimed saline (sometimes alkaline) soils on the massif [3, 6, 13, 14] and filtration water losses from the Turkestan canal, which passes through the territory with a shallow occurrence of gravel-pebble deposits with strong water permeability. Moreover, there were not always prompt construction of drainage (closed and open), often unjustified spatial distribution of drainage systems without taking into account the reclamation conditions of the massif, as well as insufficient experience and knowledge of the peculiarities of reclamation processes in the alluvial fans of the foothill Karatau plain. All this led to a significant increase of saline soil in this area. Salt survey of the massif, carried out on the instructions of the Institute "Soyuzgiproris" by a number of organizations, including the Dzhambul Institute of Water Management (Vyshpolsky F.F. [15] and others) and the Institute of Soil Science of the Academy of Sciences of the Kazakh SSR [6, 8–10] showed that the area of saline soils by that time had increased by 17,000 ha.

In subsequent years, the development of the massif continued, sometimes by trial and error, especially in the field of salinization and waterlogging control, construction and operation of open, closed, and vertical drainage. Irrigation methods, watering norms, and a set of crops were changed. Complex engineering and technical solution of this project, lack of experience in the development of large areas with saline soils that are difficult to reclaim, insufficient scientific justification, and sometimes just technical violations have created many problems on the massif. However, these problems are being gradually resolved.

4 Bugun' Reservoir

4.1 *Features of Functioning*

The reservoir was created in the valley of the Bugun' River by the construction of the Bugun' and the Karazhantak dams (Fig. 3).

The right (northern) bank of the reservoir is a rugged piedmont plain, the left (southern) bank is the watershed between the Bugun' River and the Karazhantaksai (Figs. 4 and 5). The reservoir's surface area is 6,300 ha, the volume is 370 million m^3, and the depth is 15–17 m. This reservoir accumulates water of the Bugun' River and partially – the Arys' River. It was filled annually from October to April, in

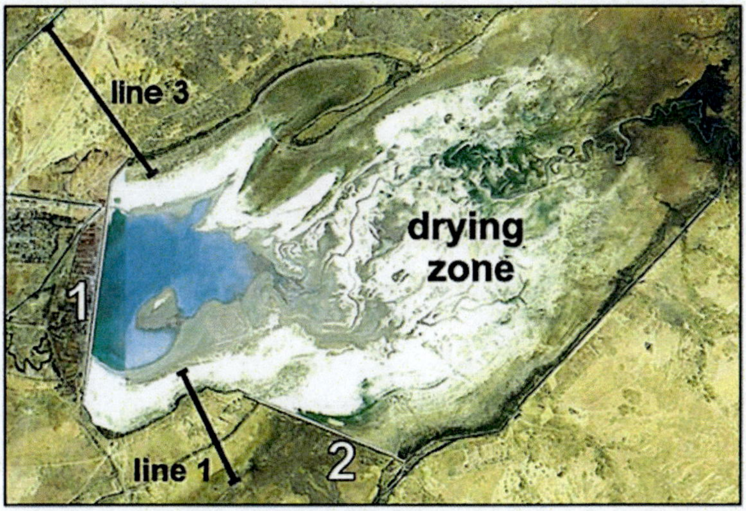

Fig. 3 The Bugun' reservoir with the Bugun' (*1*) and the Karazhantak (2) dams and research lines (transects) on the southern (*line 1*) and northern (*line 3*) banks (Landsat 8-9 satellite images (2022) from the NASA open fund were used for this figure)

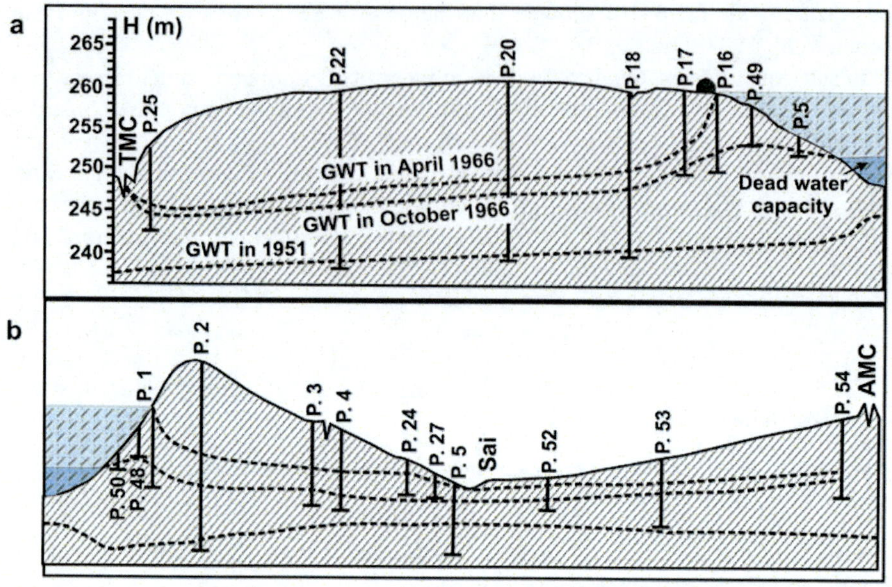

Fig. 4 Cross profile of the reservoir and the coast: (**a**) – *the northern part of the profile;* (**b**) – *the southern part.* Symbols: *GWT – groundwater table; AMC – the Arys' main canal; TMC – the Turkestan main canal; P. 1-54 – soil-lithological profiles; Sai – the Karazhantak Creek*

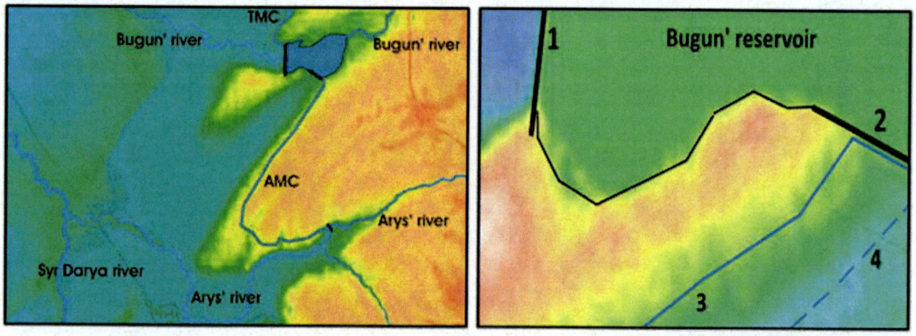

Fig. 5 Digital relief model of the Bugun' reservoir area (*left*) and an enlarged part of its southern coast (*right*) (this figure was prepared by authors with the use of the free topographic map [17])

April–May the maximum level was maintained, and from June to September there was a drawdown of water [6, 8–10] for irrigation almost to the "dead reserve" (Figs. 2 and 4). As already noted, in the last decade, due to climatic changes in the basins of the Bugun' and Arys' Rivers, the filling of the reservoir occurs much earlier (December–February), and the drawdown of water for irrigation occurs more strongly and to a lower level [7, 16].

A layer of loams represents the lithology of the territory underlain at a great depth (more than 20 m) by gravel-pebble deposits. Groundwater in 1951 (before the construction of the reservoir) in the bowl of the projected area lay at depths: in the floodplain of the river Bugun' – 2–5 m, and with a distance from it, they plunged up to 20 m. In the Karazhantaksai, groundwater was found to be at a depth of 5–10 m. Zonal soils here are light southern grayzem soils [3, 9], saline from a depth of 1.0–1.5 m to 20–30 m, and in the Karazhantaksai, hydromorphic solonchak soils were common. In the bowl of the created reservoir, the soil was subjected to desalination, and on the flooded coast – to secondary salinization [6, 8, 10]. In the bowl of the projected reservoir, the salt content in the 0–10 m layer was about 1,200 t/ha, and about 10 million tons of salts were contained in the entire bowl area.

As a result of the filling of the reservoir, which began in 1960–1966, the soils of the shores were flooded and secondary salinization occurred. And about 8 million tons of salts were washed into the groundwater and soils of the coast by seepage water from the bowl of the reservoir [10, 16].

4.2 Impact on Groundwater and Soils

The filling of the Bugun' reservoir, which began in 1960–1966, caused a sharp change in hydrogeological and soil-reclamation conditions in the territory adjacent to the reservoir, as well as an intensive redistribution of salt masses in soils. Because of a water filtration from the reservoir and from the main canals, as well as due to hydrostatic pressure, the level of groundwater by 1966–1968 rose on the coast by 7–10 m, and in the zone of periodic flooding – by 15 m. The groundwater level has reached the surface in the dams' lower pool (bief). In 1973, a further, albeit slow, rise in groundwater was noted throughout the entire study area and in 1976, the level had already relatively stabilized and is still preserved, reacting mainly to the mode of filling and drawdown of the reservoir. And only in the lower part of the southern slope, adjacent to the Karazhantaksai, a slow rise in the level of groundwater has been observed for 15 years. At the same time, the maximum and minimum levels are observed 1–3 months later than the water levels in the reservoir. With distance from the reservoir, the amplitude of level fluctuations decreases from 10–12 m to 1–2 m. Water filtration from the reservoir has led to the formation of a special hydrochemical regime of groundwater in the adjacent territory. For example, on the southern coast, mineralization varies from 1,000–3,000 mg/l at the reservoir to 20,000–35,000 mg/l at a distance of 5 km (Fig. 6), the amount of chlorides sharply increases in the ionic composition.

Further, 10-year observations showed (Table 1 and Fig. 6) that mineralized groundwater is gradually replaced by fresher water on the coast due to filtration from the reservoir, and highly mineralized waters are concentrated in relief depressions (pp. 4, 5, 24, 52). Salts also migrate in the same direction in soils, accumulating in waterlogged or flooded soils in relief depressions up to 300 t/ha in the 0–100 cm layer and 500–600 t/ha in the 0–200 cm layer. The waterlogged and flooded soils on

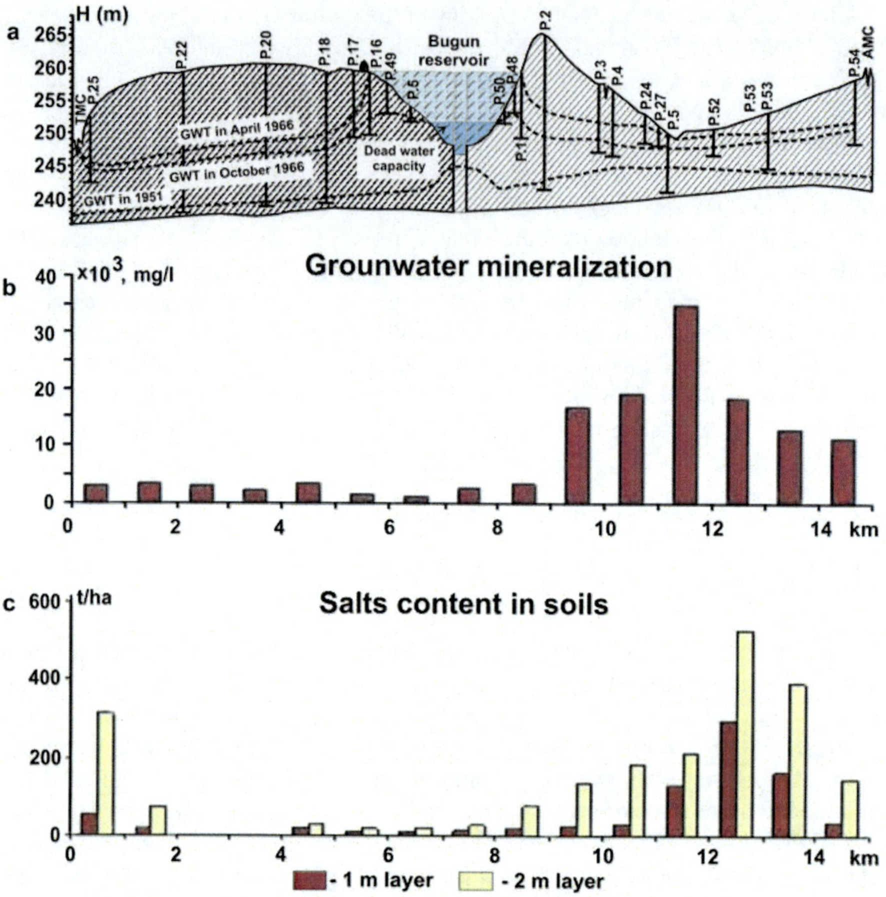

Fig. 6 Accumulation of salts in groundwater and soils of the reservoir coast (**a** – *cross profile of the reservoir and the coast;* **b** – *groundwater mineralization;* **c** – *salt content in soils*)

the territory of the dams downstream influence is still highly saline (Fig. 7), the area of these soils is increasing.

The hydrochemical features of the reservoir itself depend on the rate of water masses exchange. As already mentioned, the reservoir accumulates the runoff of the Bugun' and the Arys' Rivers with water salinity of 400–500 mg/l. Every year, more than 90% of the water is discharged for irrigation, and in June–July, water is supplied in transit through the reservoir for irrigation from the Arys' River. All this favors the low salinity of water in the reservoir. But in the first years of the reservoir filling, we observed an active exchange of salts between the soils of the banks and water masses. In 1966, the mineralization of water near the coast reached 1,800–2,000 mg/l, in 1967 it decreased to 700–900 mg/l, and in 1968–1976 – to 400–600 mg/l. In recent years, episodic pollution of river water has sometimes

Table 1 Groundwater mineralization (salts content) in the reservoir impact zone (mg/l)

Point of samples and dates	Ions content, mg/l						Sum of salts, mg/l
	HCO_3^-	SO_4^{2-}	Cl^-	Ca^{2+}	Mg^{2+}	Na^++K^+	
P.1, 30.07.1966	366	213	360	80	36	288	1,343
01.07.1967	145	28	193	28	19	97	510
28.07.1968	251	7	85	33	31	43	450
P.2, 01.07.1967	105	296	1833	127	88	797	3,246
30.07.1968	149	318	1,081	140	102	426	2,216
03.08.1973	390	229	599	10	31	557	1816
30.07.1976	176	369	816	10	49	592	2012
P.4, 30.07.1966	732	3,905	6,240	650	540	4,022	15,859
03.08.1973	235	1992	2,492	230	188	1953	7,090
30.07.1976	420	2,770	6,720	400	415	3,927	14,652
P.24, 10.05.1966	293	923	7,920	500	300	3,355	13,291
04.08.1973	683	4,216	16,596	384	1,286	8,066	31,231
P.5, 30.07.1966	659	10,011	12,480	600	1,020	10,085	34,855
01,07.1967	286	3,130	11,466	368	44	7,047	22,341
04.08.1973	471	5,412	12,458	672	994	7,000	27,007
Well in the downstream of the Bugun' dam (self-discharging)							
01.08.1967	287	32	975	127	97	266	1784
05.08.1973	322	21	415	86	73	97	1,014
31.07.1976	300	17	264	90	52	49	772

Fig. 7 Soil salinization downstream of dams. On the left – the Karazhantak dam, Sentinel-2 image (*1 – bog solonchak soils, 2 – highly saline meadow soils, 3 – solonchaks*. On the right – solonchaks in downstream and on the lower slope of the dam [14]

begun to occur when it flows into the reservoir after it flows through the Krasny Most village on the Shymkent-Turkestan highway (Fig. 8).

Based on the nature of changes in the soil cover and groundwater regime, the zone of direct impact of the Bugun' reservoir on reclamation processes was estimated by us in the upper pool (bief) at 5–6 km [3, 6]. When assessing soil changes, it was taken into account that soil regimes, processes, morphological features, and physicochemical properties of soils change at different rates. This creates difficulties in soil diagnostics. With a change in hydrogeological conditions on the waterlogged

Fig. 8 Water pollution of the Bugun' River in 2022: *1 – river water before the Krasny Most village, 2 – after flowing through the village, 3 – when it flows into the reservoir, 4 – "blooming" of water after the pollutants sedimentation* (Landsat 8-9 satellite images (2022) from the NASA open fund were used for this figure)

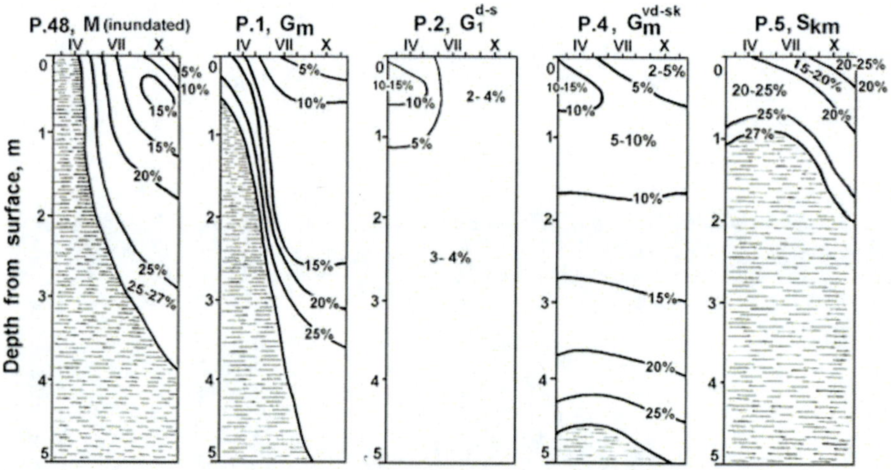

Fig. 9 Seasonal (over the growing season) dynamics of the groundwater level and soil moisture (%) in the soils of the southern coast. Soil names (according to [9]): *M – meadow soil, Gm – meadow greyzem,* $G1^{d\text{-}s}$ *– deeply saline light greyzem (modal soil),* $Gm^{vd\text{-}sk}$ *– saline meadow greyzem, Skm – meadow solonchak.* Locality of soil profiles is shown at Fig. 4

and flooded coast, soils' water and salt regimes change first of all (Fig. 9 and Table 2), which determine the further development of new features and properties. The main morphological features of waterlogged (flooded) soils changed more slowly than the groundwater regime (starting from 5–6 to 15 years). Vegetation cover corresponding to the changed water and salt regime was formed on these soils in 13–16 years. On flooded soils, halophytes (various saltworts, kermek), azhrek appear in the composition of vegetation. On the meadow solonchaks, thickets of the comber are developing.

Table 2 Changes in soil salinity and their nomenclature under waterlogging by groundwater (%)

	Ion content (%)						
Samples depth, cm	HCO_3^-	Cl^-	SO_4^{2-}	Ca^{2+}	Mg^{2+}	Na^++K^+	Sum of salts (%)
1	2	3	4	5	6	7	8
P.2, Greyzem light deeply saline (zonal soil), 1966							
0–7	0.022	0.007	0.042	0.008	0.002	0.020	0.101
25–44	0.019	0.003	0.056	0.008	0.002	0.023	0.111
81–111	0.050	0.007	0.024	0.006	0.003	0.022	0.112
144–176	0.026	0.088	0.269	0.062	0.031	0.065	0.542
196–217	0.026	0.085	0.341	0.128	0.038	0.009	0.627
450–500	0.029	0.127	0.540	0.130	0.040	0.126	0.992
700–750	0.024	0.090	0.708	0.145	0.042	0.160	1.169
750–1,200	0.015	0.109	0.504	0.085	0.022	0.178	0.913
P.1, Meadow greyzem (desalinated by surface water), 1966							
0–27	0.027	0.005	0.034	0.010	0.002	0.014	0.092
55–100	0.023	0.008	0.038	0.006	0.002	0.021	0.098
150–180	0.024	0.002	0.057	0.005	0.002	0.028	0.118
350–375	0.026	0.003	0.024	0.005	0.002	0.013	0.063
750–775	0.042	0.016	0.086	0.009	0.007	0.044	0.204
800–875	0.030	0.099	0.235	0.020	0.022	0.123	0.529
950–1,000	0.020	0.072	0.294	0.015	0.015	0.149	0.565
P.1, Meadow greyzem (desalinated by surface water), 1976							
0–30	0.034	0.001	0.001	0.006	0.002	0.004	0.048
30–49	0.032	0.001	0.001	0.006	0.002	0.003	0.045
49–84	0.029	0.003	0.001	0.006	0.002	0.003	0.044
100–125	0.041	0.003	0.003	0.007	0.003	0.006	0.063
150–175	0.032	0.003	0.001	0.006	0.002	0.004	0.048
175–200	0.037	0.001	0.001	0.007	0.002	0.004	0.052
P.3, Greyzem light deeply saline (secondary saline), 1966							
0–15	0.033	0.002	0.028	0.007	0.001	0.017	0.088
15–36	0.026	0.007	0.022	0.006	0.001	0.016	0.078
35–70	0.026	0.007	0.024	0.007	0.002	0.014	0.080
70–98	0.019	0.027	0.106	0.021	0.006	0.040	0.219
98–130	0.018	0.099	0.432	0.030	0.012	0.220	0.811
130–157	0.021	0.156	0.400	0.015	0.010	0.264	0.866
185–210	0.015	0.161	0.480	0.042	0.018	0.257	0.973
P.3, Greyzem light solonchakic (secondary saline), 1976							
0–10	0.049	0.001	0.014	0.012	0.002	0.012	0.090
30–50	0.027	0.080	0.050	0.028	0.006	0.045	0.236
50–70	0.020	0.165	0.429	0.098	0.044	0.131	0.887
70–100	0.020	0.157	0.331	0.029	0.007	0.222	0.766
125–150	0.015	0.153	0.479	0.051	0.009	0.259	0.966
175–200	0.015	0.143	1.034	0.253	0.017	0.272	1.734
P.4, Greyzem light deeply saline (secondary saline), 1966							

(continued)

Table 2 (continued)

Samples depth, cm	Ion content (%)						Sum of salts (%)
	HCO_3^-	Cl^-	SO_4^{2-}	Ca^{2+}	Mg^{2+}	Na^++K^+	
0–14	0.033	0.021	0.049	0.010	0.002	0.034	0.149
14–29	0.028	0.017	0.036	0.008	0.001	0.028	0.118
39–66	0.024	0.021	0.039	0.008	0.002	0.028	0.122
66–103	0.032	0.045	0.069	0.004	0.002	0.066	0.218
103–145	0.023	0.076	0.108	0.002	0.001	0.105	0.315
145–200	0.012	0.110	0.725	0.099	0.015	0.281	1.242
P.4, Greyzem light saline (secondary saline), 1976							
0–10	0.020	0.098	0.082	0.042	0.005	0.058	0.305
10–30	0.017	0.287	0.127	0.051	0.022	0.157	0.661
30–50	0.020	0.307	0.335	0.070	0.037	0.220	0.989
50–70	0.017	0.237	0.394	0.043	0.025	0.253	0.969
70–100	0.015	0.192	0.467	0.056	0.016	0.260	1.006
125–150	0.015	0.208	1.075	0.233	0.019	0.353	1.903
175–200	0.015	0.192	1.158	0.250	0.017	0.367	1.999
P.24, Meadow greyzem saline (secondary saline), 1966							
0–4	0.021	0.168	0.317	0.071	0.012	0.164	0.753
4–16	0.019	0.246	0.612	0.071	0.018	0.344	1.310
34–60	0.016	0.109	0.525	0.036	0.024	0.242	0.953
85–100	0.021	0.046	0.243	0.014	0.007	0.125	0.457
P.24, Meadow solonchak (secondary saline), 1976							
0–10	0.051	0.094	3.827	0.218	0.022	1.644	5.856
10–30	0.029	0.128	1.376	0.136	0.018	0.580	2.267
30–50	0.022	0.199	0.846	0.085	0.024	0.406	1.582
70–100	0.024	0.070	0.619	0.073	0.011	0.249	1.046
125–150	0.020	0.045	0.534	0.090	0.010	0.172	0.871
175–200	0.017	0.046	1.062	0.288	0.018	0.181	1.612
P.5, Meadow solonchak (secondary saline), 1966							
0–10	0.019	0.186	1.152	0.175	0.066	0.545	2.447
10–32	0.018	0.266	0.480	0.035	0.025	0.321	1.145
32–55	0.027	0.101	0.336	0.015	0.006	0.209	0.696
74–100	0.034	0.066	0.211	0.007	0.002	0.146	0.468
340–370	0.013	0.033	0.720	0.205	0.036	0.066	1.083
440–470	0.012	0.021	0.816	0.200	0.069	0.049	1.166
P.5, Meadow solonchak (strongly secondary saline), 1976							
0–10	0.063	0.417	4.054	0.244	0.050	1.885	6.713
10–30	0.027	0.261	1.163	0.154	0.028	0.518	2.151
30–50	0.015	0.237	0.807	0.143	0.021	0.348	1.571
50–70	0.017	0.180	0.608	0.096	0.016	0.278	1.195
70–100	0.017	0.139	0.44	0.055	0.012	0.226	0.893
125–150	0.012	0.111	0.951	0.270	0.018	0.190	1.552

Fig. 10 Erosion of loess banks after the first year of reservoir's filling (*left*) and destruction of buildings on the bank (*right*) in 1967 (*photo by V.M. Starodubtsev*)

Fig. 11 Loess rock first eroded in 1967 (*left*) and redeposited silt (*right*) (*photo by V.M. Starodubtsev*)

Erosion processes on the reservoir coast appeared already in the first years of its filling (1963–1968). The loess rocks of the southern (left) coast were most strongly eroded under the impact of wave activity during strong winds and currents. The unique susceptibility of light loams facilitated this to wetting, the loss of structure, and the acquisition of strong fluidity. Severe water erosion occurred on the southern steeper bank already in the first year of the reservoir's filling to the normal water level (NWL) (1966). The loess rock was quickly eroded: a cliff 1–2 m high was formed, moving toward the buildings constructed on the shore and destroying the strip of trees planted along the water's edge (Fig. 10).

For the first time, the loess rock eroded by surface waters acquired a nostril-like character, and then was subjected to blowing by strong winds (Fig. 11). And the resulting bottom sediments were re-formed as a result of waves and currents during uneven drawdown of the reservoir water for irrigation (Fig. 12) and partially overgrown with sparse shrubs and herbs. For the first time, the loess rock eroded by surface waters acquired a nostril-like character, and then was subjected to

Fig. 12 Reformation of the muddy bottom during uneven drawdown of the reservoir (*left*) and overgrowth of the bottom with shrubs in the drying zone (*right*) (*photo by V.M. Starodubtsev*)

blowing by strong winds (Fig. 11). And the resulting bottom sediments were re-formed as a result of waves and currents during uneven drawdown of the reservoir water for irrigation (Fig. 12) and partially overgrown with sparse shrubs and herbs.

Modern water erosion processes in reservoir are shown in a series of satellite images of Sentinel-2 and Landsat 8-9 (Fig. 13). The upper part of the figure shows the erosion processes along the entire coast of the reservoir with strong winds periodically in this area. The middle part of the figure shows the erosion of the protruding sections of the coast during the spring months and the dangerous section of the coast where there is a threat of water breakthrough and destruction of the reservoir (red arrow). The lower part of the figure shows the erosion of bottom sediments by alongshore currents.

The coast receded in the first decades by tens of meters, creating a threat to the safety of the Bugun' Dam in its western corner (Fig. 13, top). In recent years, the rate of coastal erosion has decreased. If the average for the period 1966–2019 coastal retreat amounted to 7–11 m per year, in the 2000s it did not exceed 1–2 m, including due to protective stone filling along the southern coast. The rate of coastal retreat due to erosion processes for the period 1966–2019 is shown in Fig. 14. And changes in the general configuration of the southern coast due to erosive processes over the entire period of the reservoir's operation are shown in Fig. 15.

However, in recent years, due to a change in the reservoir filling regime (Fig. 2) and wind currents, erosion has increased in the area of the Karazhantak dam (Fig. 13), creating a real threat of water breakthrough into the Karazhantaksai and potential destruction of the reservoir. At our suggestion, in 2021, surveys of the area were carried out here with a quadrocopter [7, 13], confirming satellite information. These materials showed the site of extreme danger of water breakthrough through the ridge of the southern coast and the need for urgent measures to protect the coast. And in the conclusion of the consideration of erosion processes, we note the significant process of siltation of the reservoir over half a century.

Fig. 13 Modern processes of water erosion in the reservoir (satellite images of Sentinel-2 and Landsat 8-9 from the NASA and ESA open funds)

5 Conclusion

In general, the most important environmental consequences of filling the reservoir are [7]: (1) water filtration and flooding of the coast (mainly the left bank); (2) waterlogging of soils in downstream of the Bugun' and the Karazhantak dams and in the upper reaches of the reservoir (at the confluence of the Bugun' river and

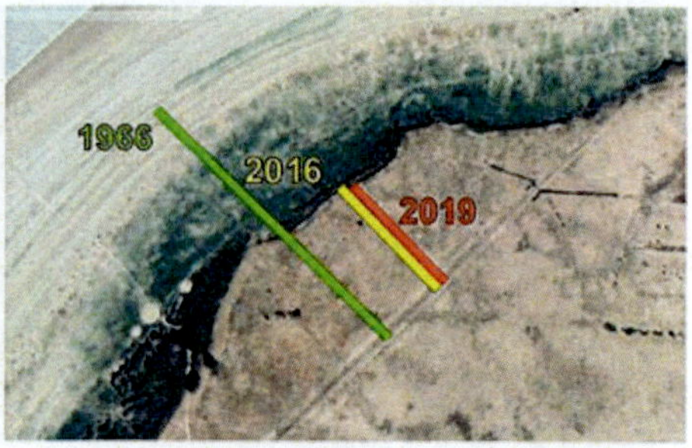

Fig. 14 Soil erosion of the southern coast in its southwestern (*left*) and southeastern (*right*) parts for the period 1966–2019 (satellite images of Landsat 8-9 (2019) from the NASA open fund were used for this figure)

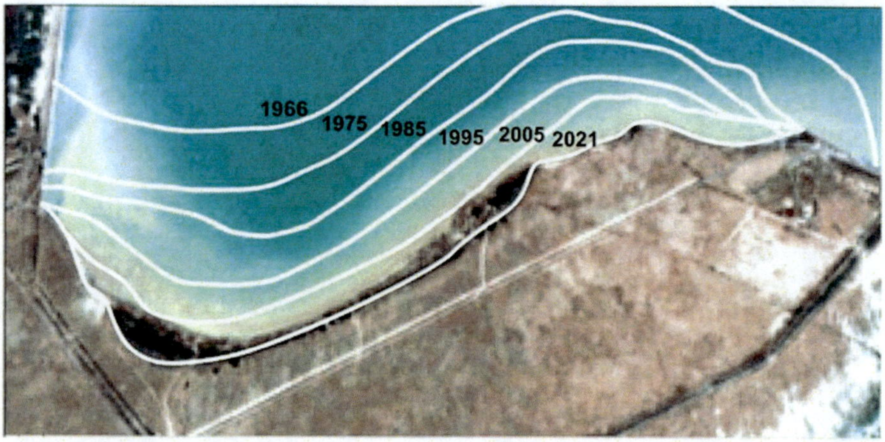

Fig. 15 South coastline change for the period 1966–2021 (satellite images of Landsat 8-9 (2019) from the NASA open fund were used for this figure)

the Arys' main canal (AMC); (3) salinization of soils during the flooding of the coast, downstream of the dams and the valley of the Karazhantaksai; (4) erosion of the banks with the threat of a breakthrough of water masses through the left bank to Karazhantaksai; (5) siltation of the reservoir basin; (6) inefficient design of the A-9 distribution channel on the slope of the Karazhantak dam, which does not provide irrigation of the designed areas; (7) overgrowth of large-grass coastal vegetation due to high air humidity above the banks of the reservoir (Figs. 4 and 5).

6 Recommendations

The presented investigation could be used for future studies of reservoirs and the environment, changes in arid ecosystems, policy planners, decision-makers, and other stakeholders.

Acknowledgments These investigations were carried out on the authors' personal initiative. The authors are sincerely grateful to the editors Dr. Shakir Ali Khan and Dr. Abdelazim Negm who provided useful advice during the preparation of the chapter for the edition.

References

1. World Commission on Dams (2000) Dams and development: a new framework for decision-making – the report of the world commission on dams.1st edn. Routledge, London
2. Dynesius M, Nilsson C (1994) Fragmentation and flow regulation of river systems in the northern third of the world. Science 266:753–762. https://doi.org/10.1126/science.266.5186.753
3. Starodubtsev VM, Fedorenko OL, Petrenko LR (2004) Dams and environment: effects on soils. Nora-Print, Kyiv
4. Voropaev GV, Avakyan AB (1986) Vodokhranilishcha i ikh vozdeystviye na okruzhayushchuyu sredu (reservoirs and their impact on the environment). Nauka, Moskva
5. Sun Y, Liu B, Yang G et al (2023) Analysis of spatiotemporal evolution patterns and driving forces of reservoirs on the northern slope of the Tianshan Mountains in Xinjiang. Sustainability 15:8824. https://doi.org/10.3390/su15118824
6. Starodubtsev VM (1977) Pochvenno-meliorativnyye protsessy v zone vliyaniya vodokhranilishch (soil-reclamation processes in the zone of influence of reservoirs). Problemy osvoyeniya pustyn 6:18–26
7. Starodubtsev VM, Beksultanov MK, Kalybekova AA, Nurimbetov RI (2019) Irrigatsionnoye osvoyeniye Arys'-Turkestanskogo massiva v Yuzhnom Kazakhstane (irrigation development of the Arys-Turkestan massif in South Kazakhstan). Modern Eng Innov 2:55–67. https://doi.org/10.30890/2567-5273.2019-10-02-040
8. Yegorichev GA, Starodubtsev VM (1970) Pochvenno-meliorativnyye usloviya v zone vliyaniya Bugunskogo vodokhranilishcha (soil-reclamation conditions in the zone of influence of the Bugun reservoir). Problemy melioratsii pochv Sredney Aziii Kazakhstana:280–285
9. Zhikhareva GA, Kurmangaliyev AB, Sokolov AA (1969) Pochvy Kazakhskoy SSR (soils of the Kazakh SSR), vypusk 12 (Chimkentskaya oblast'). NaukaKazSSR, Alma-Ata
10. Starodubtsev VM (1986) Vliyaniye vodokhranilishch na pochvy (influence of reservoirs on soils). Nauka, Alma-Ata
11. Mausoleum of Khoja Ahmed Yasawi. In: Wikipedia. The free encyclopedia. https://en.wikipedia.org/wiki/Mausoleum_of_Khoja_Ahmed_Yasawi. Accessed 20 Dec 2022
12. Zhang J, Shang Y (2023) Nexus of dams, reservoirs, climate, and the environment: a systematic perspective. Int J Environ Sci Technol. https://doi.org/10.1007/s13762-023-04765-4
13. Starodubtsev VM, Ladyka MN, Beksultanov MK (2022) Opasnost' erozii poberezh'ya Bugunskogo vodokhranilishcha (danger of erosion of the Bugun reservoir coast). Sovremennyye metody ekologicheskikh issledovaniy: Sb. nauch. trudov "TOO KAPE". pp 178–181
14. Starodubtsev VM, Ladyka MM, Naumovska OI, Beksultanov MK (2022) Soil erosion and salinization on the coast of the Bugun' irrigation reservoir. 47–51. https://doi.org/10.30888/2709-2267.2022-09-01-034

15. Vyshpol'skiy FF (1971) Razrabotka i obosnovaniye optimal'nykh norm dlya promyvki zasolennykh pochv na primere Arys'-Turkestanskogo massiva (development and substantiation of optimal norms for washing saline soils on the example of the Arys-Turkestan massif), Alma-Ata
16. Starodubtsev VM (2012) Impact of Bugun' water reservoir on coast for 50 years. Arid Ecosyst:132–138. https://doi.org/10.1134/S2079096112020102
17. Kazakhstan topographic map. In: Free topographic maps, elevation, terrain. https://en-gb.topographic-map.com/map-v24s/Kazakhstan. Accessed 20 Dec 2022

Impact of Climate Changes on Seawater Intrusion in the Nile Delta Aquifer (Egypt)

Asaad M. Armanuos, Mohamed Samir Taha, and Bakenaz A. Zeidan

Contents

1 Introduction 99
2 Groundwater Bearing Aquifers in Egypt 100
3 Intrusion of Saline Water into the Nile Delta Aquifer 101
4 Groundwater Recharge in Egypt 105
4.1 Recharge of Groundwater by Rainfall 106
4.2 Recharge of Groundwater from Excessive Irrigation Water 106
4.3 Groundwater Recharge from Irrigation Canals Network 108
5 Climate Change 108
5.1 Climate Change's Effects on Egypt 108
5.2 SLR Projections and Their Effects on Egypt's Coastal Zones 108
5.3 Climate Change Projections and Effects on the Nile Flows 109
5.4 Climate Change Projections and Their Effects on Water Demand 109
6 A Case Study: Nile Delta 111
6.1 Location of Nile Delta 111
7 Meteorological Aspects 112
7.1 Precipitation and Evaporation 112
7.2 Population 114
7.3 Climate Change 114
8 Hydro-Geological Settings 115
8.1 Surface Water System 116
8.2 Geometry of the Nile Delta Quaternary Aquifer 116
8.3 Hydraulic Parameters of the Nile Delta Quaternary Aquifer 117
8.4 Surface Water and Groundwater Relationship 119
8.5 Recharging the Nile Delta Quaternary Aquifer 119
8.6 Well Discharge Rate (Extraction from the Nile Delta Quaternary Aquifer) 120
8.7 Groundwater Level in the Nile Delta Aquifer 121

A. M. Armanuos (✉) and B. A. Zeidan
Irrigation and Hydraulics Engineering Department, Faculty of Engineering, Tanta University, Tanta, Egypt
e-mail: asaad.matter@f-eng.tanta.edu.eg; b.zeidan@f-eng.tanta.edu.eg

M. S. Taha
Faculty of Engineering, Delta University, Gamasa, Egypt
e-mail: Mohamed.Abdelshafy@deltauniv.edu.eg

Shakir Ali and Abdelazim Negm (eds.), *Groundwater Quality and Geochemistry in Arid and Semi-Arid Regions*, Hdb Env Chem (2024) 126: 97–164, DOI 10.1007/698_2023_1062,

Published online: 5 January 2024

8.8 Groundwater Salinity in the Nile Delta Aquifer ... 122
8.9 Solute Transport Parameters ... 124
9 Groundwater Flow and Solute Transport Simulation ... 125
10 Numerical Model Setup ... 126
10.1 Model Domain and External Boundaries ... 126
10.2 Model Layers Discretization ... 126
10.3 Model Boundary Conditions ... 128
10.4 Model Parameterization ... 130
10.5 Model Calibration ... 132
11 Advance of Shoreline Due to Climate Change ... 138
12 Saltwater Intrusion in the Nile Delta Aquifer (the Base Case Scenario) ... 140
13 Creep of Saltwater Interface in the Nile Delta Aquifer ... 145
14 Impact of Climate Change on Seawater Intrusion in the Nile Delta Aquifer ... 145
14.1 Saltwater Intrusion Distribution for Scenario No. 1 (0.25 m SLR + the Same Abstraction) ... 146
14.2 Saltwater Intrusion Distribution for Scenario No. 2 (0.25 m SLR + Half Abstraction) ... 147
14.3 Saltwater Intrusion Distribution for Scenario No. 3 (0.25 m SLR + Double Abstraction) ... 148
14.4 Saltwater Intrusion Distribution for Scenario No. 4 (0.5 m SLR + the Same Abstraction) ... 149
14.5 Saltwater Intrusion Distribution for Scenario No. 5 (0.5 m SLR + Half Abstraction) ... 151
14.6 Saltwater Intrusion Distribution for Scenario No. 6 (0.5 m SLR + Double Abstraction) ... 152
15 Conclusion ... 155
16 Recommendation ... 159
References ... 160

Abstract Climate change is one of the most significant natural processes and a serious threat to humankind, because it results in a rise in sea levels, a reduction in precipitation, and an increase in surface water evaporation. Coastal aquifers, which are a significant supply of freshwater in arid and semi-arid regions, are one of the areas most impacted by this phenomenon. Increased abstraction from coastal aquifers also reduces freshwater runoff into the ocean. As a result, there is an increase in inland seawater intrusion, and wells are contaminated by lowering water quality through increasing salinity. One of these aquifers is the Nile Delta aquifer (NDA) in Egypt, one of the world's largest groundwater aquifers with an area of about 22,000 km^2. This aquifer is exposed to severe seawater intrusion from the Mediterranean Sea.

The primary goals of this study are to apply the numerical models; Visual MODFLOW and SEAWAT to examine how climatic change would affect seawater intrusion in the NDA. In this study, Visual MODFLOW is used to simulate groundwater head. A groundwater model for the NDA was created using the SEAWAT program to simulate the intrusion of saltwater. In addition to the basic case, six other scenarios have been added considering a combination of sea level rise (SLR) and change in the withdrawal rate. A comparison between the Equi-concentration line

1,000 ppm which represents the freshwater line of the base case for the current study, and the Equi-concentration line for 1960, 1980, and 1992 is examined to study the creep of saltwater. The comparison revealed that the creep occurs in the west and middle of NDA but lagged toward the east. The results showed that the sixth scenario, which assumed sea level rise by 0.5 m and double the base case abstraction rate, is the worst scenario. Therefore, the withdrawal from wells must be reduced or at least maintained at the same rate, and the shore protection methods should be used to prevent the advancement of the shoreline.

Keywords Climate change, Egypt, Nile Delta aquifer, Sea level rise, SEAWAT, Seawater intrusion, Visual MODFLOW

1 Introduction

Groundwater (GW) in coastal aquifers is a potent freshwater supply globally, especially in arid and semi-arid regions with limited surface water resources and low rainfall. Development along the shore and urbanization are utilizing resources for GW in arid and semi-arid regions. One of these coastal aquifers is the Nile Delta aquifer (NDA), Egypt. In Egypt, substantial amounts of water are needed to support the population water demand and ongoing development. Securing water demand is quite difficult because the amount of water resources is almost constant. This necessitates defending against contamination, such as saltwater intrusion (SWI) and other pollutants, which reduce the available resources. GW storage has been increasingly extracted due to population concentrations along the coast and a rise in related activities. GW's salinity has grown due to seawater (SW) moving toward aquifers due to excessive withdrawal from coastal aquifers. Because of its high salinity, GW should not be used for drinking or agriculture except if it is desalinated or combined with less salty water. Under the situation of rising demands and depleting supplies, protecting GW resources becomes crucial because it causes the depletion of water resources. Population living along coastal areas are at higher risk due to GW contamination by seawater intrusion [1, 2]. Therefore, the management of GW resources is necessary. GW supplies, soil salinity, agricultural production, and quality in the seaside zone are all directly impacted by saltwater intrusion. Increased soil salinity decreased agricultural production, and socio-economic and health effects are all results of saltwater intrusion (SWI). Reduced abstraction rates, moved absorption wells, subsurface blocks, natural recharge, made-up recharge, and SW abstraction have all been employed to prevent SWI from polluting GW supplies [3].

2 Groundwater Bearing Aquifers in Egypt

In Egypt, there are several types of GW aquifers, including confined, unconfined, semi-confined, multi-layer, and perched aquifers. Aquifers of varied sizes make up the main Egyptian GW system. These include the Hard-Rock aquifer, the Fissured Carbonate Rock aquifer, the Moghra aquifer, the Fissured Carbonate Rock aquifer, and the NDA. A significant source of water for home use, industrial, and agricultural usage is GW. The flow of saltwater toward aquifers due to extreme pumping from coastal aquifers has raised GW salinity. Due to its high salinity, GW should not be used for drinking or agriculture except if it is desalinated or combined with less salty water. Under this situation of rising demands and limited supplies, protecting GW resources becomes crucial [4]. Despite some shallow GW sources being impacted by pollution, the Nile system's quality of GW is acceptable. Nearly 20% of the NDA's groundwater does not fulfill requirements for water supply, particularly in areas along the edges where there is limited or no protective clay layer. The quality of the GW in the ND is often higher than that of the Nile Valley, while that in Sinai and the eastern desert has high salinity and that in the western deserts is often extremely good (TDS). In general, the carbonate aquifer includes brackish water, however, some freshwater is present in recharge zones [4].

Due to the overlying clay layer, the NDA is thought to be semi-confined and considers a shallow aquifer mostly recharged by surplus irrigation. On the other hand, it contributes around 85% of Egypt's overall abstractions of GW. It is composed of a substantial layer of gravel and sand that is topped with a 50-meter-deep clay cap. The aquifer of Nubian Sandstone extends into Libya, Chad, and Sudan and occupies around 2 million km^2 from 200 m in east Owinat to 3,500 m in the northwest of El-Farafra Oasis, the fresh layer's thickness is estimated. Although this aquifer has a theoretical storage capacity of more than $150{,}000 \times 10^9$ m^3, much of it is very deep and is currently not economical [5]. The Fissured Carbonate Rock Aquifer, which sits atop the Nubian Sandstone Aquifer and acts as a limiting layer, covers over 50% of Egypt. It includes various natural springs and stretches from Sinai to Libya. Aquifer recharge is uncertain, and its potential is not understood with certainty. GW from the Moghra aquifer, which is in the northwestern desert, flows into the Qattara Depression. It receives the Nile aquifer's lateral inflow and rains to recharge and contains fresh GW, just as the salinity rises toward the north and west. Rainfall replenishes the coastal aquifer system, and the presence of SW beneath the freshwater lenses restricts the amount of water that can be extracted from it. Similarly, limited amounts of entering precipitation fill the aquifer of Fissured Hard Rock in the eastern desert and southern Sinai [5].

3 Intrusion of Saline Water into the Nile Delta Aquifer

Visual MODFLOW, a numerical GW flow model, and a solute transport model (MT3D) were employed by [6] to evaluate the qualitative and quantitative long-term effects of the nation's projected water strategy on the NDA systems. GIS (Geographic Information System) is extensively used by different researchers in Egypt to assess groundwater vulnerability and selection of suitable landfill sites in the ND [7, 8]. The specific goals are: (1) evaluating how development has affected GW in the ND between 1992 and 2008 is the first goal; (2) to develop and categorize qualitative risk maps for GW pollution; (3) to create a dynamic GIS-metadata bank that is integrated with the ND region's numerical solution. Due to the growth of water requirements and the scarcity in supply owing to increased agriculture, residential, and industrial usage, the environmental impact managing of GW resources in the ND is extremely critical. The analysis of the model results also confirmed that the salinity level of the counter line of 1,000 ppm intruded further north in the central part of the NDA (see Fig. 1). Monem [9] used the visual MODFLOW USGS code to investigate the environmental influences on the NDA GW system; overall environmental impact evaluation with mitigation of an environmentally useful existing strategy and substantial benefits is highly likely to affect 32%, whereas mitigation of an environmentally feasible new project with large advantages is highly likely to affect 75%. Mabrouk et al. [10] studied the effects

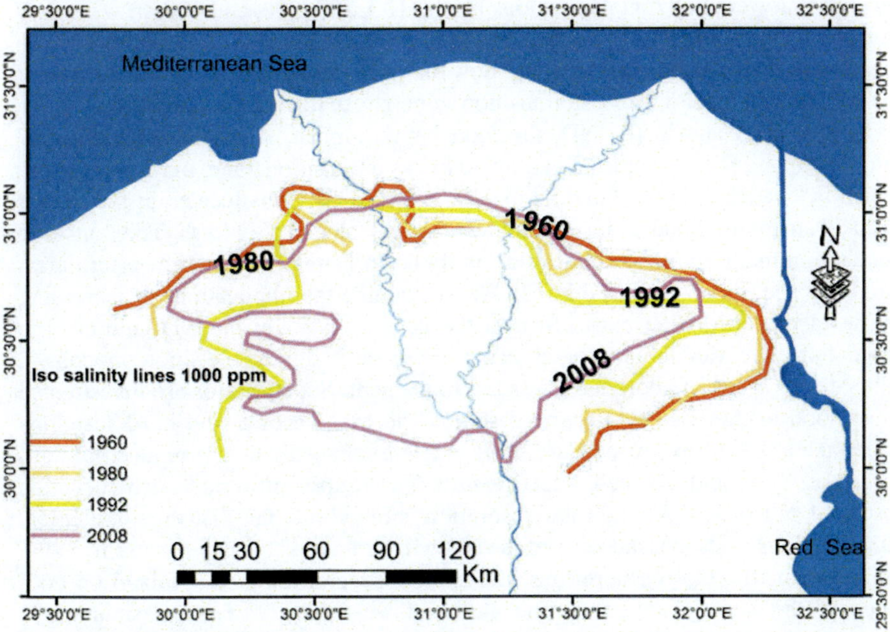

Fig. 1 The advancement of the contour line of 1,000 mg/l in the NDA, modified by [6]

of the rise in sea level in Egypt's northeastern ND and found that the brine water is anticipated to be pumped into deep wells to various depths, and changes in salinity and flow directions were observed during a 30-year period. A combined model incorporates brackish water absorption rates, brine water injection into the aquifer, and SWI while considering rising sea level. The model outcomes will provide eventually simulated perspective of the ND aquifer in the vulnerable area. In addition, a strategy plans for the aquifer's long-term management, to minimize the climate change impacts, rising of sea level, and shortage of water resources in this vital area.

Gaamea [11] developed the code of the SUTRA model for simulating the behavior of the ND's transition zone at various levels of abstraction. Gaamea (2000) observed that the middle Delta's northern section is more salinized than its southern section. The influence of pumping both fresh and brackish water at the same time was investigated using this model. A single saline well may be used to manage four or more freshwater abstraction wells at a set distance to achieve the equilibrium status of mixing zone. El Didy and Darwish [12] investigated how the NDA's SWI was affected by freshwater conservation in the northern lakes of Manzala and Burullus. For simulating the system, the scientists used the SUTRA and lake models, which they named lake. They guaranteed that SWI exists in their zone of effect's northern region, whereas lake freshwater mitigates infiltration. Sherif et al. [13] examined the idea of the equivalent head of freshwater in a series of horizontal SWI simulations in the ND using FEFLOW. However, due to a lack of data, the simulations were run as sequences in two dimensions (vertical layers). Four sections that were horizontal at various levels (100, 200, 300, and 400 m) were used in the horizontal modeling; every horizontal segment included a pressure head to measure its entire depth. Their results show that as we move down toward the base of the aquifer, the position of the transition zone shifts toward the land side. At the mean SW level and in the ND, the model was applied. At shallow depths, SW intruded inland for 40 km, followed by 30 km as a transition zone, in Mansoura city which is located across the middle of ND, Ismailia, which is located in the east of Delta, Damanhur, which is located in the west of Delta, GW is fresh. The transition zone's minimum width of about 6 km is found in Damanhur's northwestern area. Sefelnasr and Sherif [14] used FEFLOW to examine the implications of increasing sea level in the Mediterranean Sea on the issue of SWI in the ND aquifer. The simulations are run in horizontal perspective, with digital elevation models to simulate the effect of shoreline landward movement. Six scenarios are investigated in addition to the base case (current state). The first, second, and third scenarios undertake a 0.5 m rise in seawater level, while total abstraction is reduced to half, preserved as is, and doubled, subsequently. The fourth, fifth, and sixth scenarios assume 1 m rise in SW, and total absorption is taken like the first three scenarios. Huge portions of the coastal zone of the ND will be flooded by saline water from the sea; additionally, the seashore line will migrate several kilometers inland on both sides of the Delta that are on the east and west. The worst-case situation is represented by scenario no.6 in which the amount of freshwater is decreased to around 513 km^3 (billion m^3). Armanuos et al. [15] used the MODFLOW code to estimate the potential influence of abstraction scenarios and the Grand Ethiopian Renaissance Dam (GERD) on the variations of the level of GW in the NDA. The entire irrigation system canal system in the ND area was modeled using an integrated 3-D GW. Three distinct scenarios were seen for the GW model: (1) a reduction in the depth of the water canal, (2) an increase in the rate of abstraction from the NDA, and (3) a combining of the first two scenarios. The findings showed that raising the pumping rate had a significantly greater effect on the variations of level of GW in the ND than lowering the canal's water depth. The worst scenario was shown in the previous scenario, whereas the average GW level drop increased to 1.26 m in the western, central, and eastern regions of the ND, 1.70 m, and 1.35 m, accordingly. Armanuos et al. [16] investigated the possible effects of higher abstraction rates on the variations of GW levels in the NDA by building a 3D GW model using MODFLOW. The model was verified using the observed data of the GW level as the base case. The outcomes from 10 scenarios showed that lower GW levels in the central and southern regions of NDA were significantly impacted by higher pumping rates. The tenth scenario was named the most catastrophic scenario since it raised the rate of extraction compared to the remaining scenarios, which saw GW levels drop to 1.32, 1.59, and 2.41 m in the southern boundary. Armanuos and Negm [17] used SEAWAT and MODFLOW programs and the ND region's actual irrigation canal system, the NDA's comprehensive 3D GW model was constructed to predict SW infiltration across different climate change scenarios. Bank levels as well as the irrigation canals' top width in the ND region were estimated using Google Earth Pro software. The WetSpass hydrological model was utilized to evaluate the temporal and spatial fluctuations in groundwater replenishment from rainfall within the NDA. The land use classification maps were created using ENVI software using the Nile Delta's land cover images for the years 1972, 1984, 1990, 2000, and 2009. ENVI software was utilized to construct a categorization of land use. By comparing the simulated and estimated GW recharges, the WetSpass model was calibrated using the water balance equation technique. The outcomes confirmed that the outputs of the two approaches were remarkably similar in terms of GW recharge [18]. To study the saltwater intrusion in the NDA, the SEAWAT program was employed to produce an integrated GW model. There are three possibilities presented: rising of sea level, a drawdown in the groundwater head in the south owing to enhanced GW abstraction, and a mixture of the two examples previously described. The third case (a combination of rising sea levels and reduced GW head owing to increased pumping) is the most dangerous scenario [18]. Abdelaty et al. [19] used a 3D model (SEAWAT) to study the SWI in the ND aquifer. RIGW claims that the salinity concentration data from several wells in 2008 was utilized. The findings demonstrated that the 1,000 ppm Equi concentration line moved inland into the aquifer to 93.75 km from the shoreline of the Mediterranean Sea, while the 35,000 ppm Equi concentration line migrated inside by 63.75 km in the cross section at the middle part of the ND.

GW salinity contour maps were created by [20] for the NDA between 1960 and 2000. Based on the maps, RIGW found that the GW salinity was impacted by development activities in the NDA. Between 1980 and 1990, the salinity of GW

Table 1 The Nile Delta aquifer's intrusion length for the saltwater and the freshwater lines

The study	Year	The used code	The length of intrusion for 35,000 mg/l	The length of intrusion for 1,000 mg/l
[20]	2002	–	–	Intruded further south into the aquifer of the Nile Delta
[14]	2014	FEFLOW	43 km	75 km
[19]	2014	SEAWAT	63.75 km	93.75 km
[23]	2015	MODFLOW, MT3DMS	43.5 km	68 km
[21]	2016	2D-FEST	64 km	112 km
[17]	2018	SEAWAT, MODFLOW	56 km	105 km

increased, and the Equi concentration line 1,000 ppm migrated more toward the south, indicating more intrusion of SW. Abd-Elhamid et al. [21] used a coupled transient (2D-FEST) FEM to simulate solute transport and fluid flow across saturated and unsaturated zones to observe the impacts of climate change on SWI in the ND aquifer. According to the model's results, the iso-saline line 35,000 ppm migrated inland into the aquifer of ND to a point 6 km from the coastline, while the iso-saline line 1,000 ppm moved inland to a distance of 112 km at cross section in the Nile Delta's central region. Nofal et al. [22] studied the SWI in the NDA utilizing the 3D model finite element model (SEAWAT) which considered the existing heterogeneity information and the variation in GW density value in lately drilled boreholes. The appreciable agreement connected to the piezometric head and flow fluxes was revealed by the model results. The built model explained the location of the NDA's SWI. Elshinnawy et al. [23] investigated the Nile Delta Aquifer's hydraulic conductivity field. Due to a shortage of hydraulic conductivity data, Geostatistical Earth Modeling Software (GEMS) was employed to determine the best interpolation approach and perform a Geo statistics analysis to investigate the values of direct parameters measurements of hydraulic conductivity (employing pumping test data). The outputs of the Geo statistics study are employed to determine the NDA system's hydraulic conductivity heterogeneity. The aquifer system was calibrated using the visual MODFLOW, and MT3DMS (Model of 3-D Finite Differences with Constant Density), and the results of the Geo statistics study. The location of the SWI interface is determined after calibrating against the field observations. Previous investigations on the penetration length of SWI toe for the freshwater and the saltwater lines in the NDA are listed in Table 1.

Wassef and Schüttrumpf [24] employed FEFLOW software to build a model of a 3D finite element in the western ND to analyze the SWI considering various scenarios of climate change. The findings indicate that by the end of 2100, the SW-freshwater interface is anticipated to achieve a maximum of around 43 km under the RCP 2.6 scenario, and 57 km under the RCP 8.5 scenario. Van Engelen et al. [25] used a GW salinity paleo hydrogeological reconstruction for the last 32 ka to test the physical validity of the Holocene-transgression theory to analyze measured values of salinity, employing a modernized version of the SEAWAT code that enables for sequential computing with a complicated 3-D variable-density GW flow model.

Many situations with different lithology and provenances of hypersaline GW are explored and five were chosen as having the best situation with the data. Total freshwater quantities varied among these selections, varying between 1,526 and 2,659 km^3, owing to lithology uncertainty offshore and at greater depths. In all five cases, the overall quantity of hypersaline GW was greater than that of SW. Results also reveal that, due to rising sea levels, total freshwater amounts decreased dramatically over the previous 32 ka, by a factor ranging from 2 to 5. Elshinnawy and Almaliki [26] investigated the susceptibility of coastal GW systems to the effects of sea level rising (SLR) in the Gamasa (Ras El Bar) area, which is among the ND's most susceptible coastal regions. To examine, measure, and collect data, a field campaign was conducted. This available information was utilized with historical data to determine the future inundation zones, erosion and accumulation rates, coastal variations, wave climate and the SWI, and efficiency of drainage infrastructure. The estimates of a 0.73 m SLR in the research region, predicted up to the end of the present century, show the following. The projected variance in the GW heads because of rising of sea level will result in an enhancement in GW heads of 0 to 0.5 m over existing levels. The predicted variation in GW will result in a 1 km landward SWI. The results showed that rising GW will have a negative influence on around 271 km^2 (60%) of the research region. Armanuos et al. [27] verified that researchers had investigated SWI in the NDA in detail over the course of 30 years, utilizing various 2-D and 3-D model codes for the NDA. The MODFLOW and SEAWAT software packages are frequently used for investigating the magnitude of the SWI and the salinity distributions in the NDA Region. Most of the research on SWI in the ND has established that the ND's additional SW advancement and the degree of SWI have been accelerated by rising sea levels and increased pumping flow. The aquifer system's freshwater volume decreased and because of that the quality of the GW deteriorated. As a result, the interface between fresh and saltwater has moved from the coastline shoreline to the center of the ND, extending between 84 and 112 km. The GW aquifer was invaded by the 35,000 mg/l saltwater line up to 41–63.9 km. A fundamental information on the GW in the NDA, Egypt, and its assessment, modeling, and management has been presented by [27, 28]. The previous literature on GW vulnerability studies, GW modeling, SWI modeling, and GW management published studies in the NDA is summarized and extensively analyzed [27].

4 Groundwater Recharge in Egypt

The Nile River recharge, seepage from drainage systems and irrigation canals, extra irrigation water, and GW recharging from rainfall are the main resources of the recharge of GW in the NDA.

4.1 Recharge of Groundwater by Rainfall

For appropriate representation and simulation, an integrated GW model of the aquifer beneath ND requires input data relevant to the long-term recharging of the GW. For determining how climate change will affect the ND, integrated GW modeling is necessary (the rising sea level and hydrological circumstances), as well as the associated effects of growing improvement on the management of the NDA's GW resources to assess the variations in GW level, its water budget, and conditions for salinity in GW [29]. The NDA is replenished by rainwater that penetrates the top layer of clay, excess irrigation water, and infiltration from the system of irrigation canals. Only the winter months get GW recharging from rainfall in the ND. It is a crucial mechanism in coastal areas, especially where coastal dunes cover the coastal aquifer in the Mediterranean Sea [6]. A GW model of the ND was constructed by [14, 19] to explore SWI and the impacts of climate changes on the NDA's water resources. According to [10] most GW modeling studies ignore the rainfall recharging in the ND. Armanuos et al. [18] utilized WetSpass hydrological model for estimating the values of GW recharge from rainfall in the ND region for the following years 1970, 1980, 1990, and 2010. The outcomes of the hydrological model for the year 1991 showed that the GW recharge has a minimal value of 0.0 and a maximum amount of 304 mm each winter season, and a minimal value of 134 mm every winter season for the year 2000.

4.2 Recharge of Groundwater from Excessive Irrigation Water

The primary sources of recharge in the NDA are return flow from irrigation and infiltration from drainage systems and canals. Recharge varies based on the hydrogeological circumstances. Depending on the season, the infiltration/drainage pattern may be reversed [30]. Agricultural recharge to the Quaternary aquifer fluctuated between 0.8 and 1.1 mm per day for historic lands as well as 1.9 and 2.1 mm per day for reclaimed regions [31]. The Quaternary aquifer in the Nile Delta gets a mean percolation rate of about 0.8 mm per day, as reported by [32]. The level of infiltration in the central regions and southern regions of the ND is between 0.25 and 0.8 mm/day, according to Warner et al. [33]. The rate of percolation is influenced by soil type, irrigation schedules, and drainage strategies. In the western desert areas, this figure increases with furrow irrigation from 1.0 to 1.5 mm/day to values between 0.1 and 0.5 mm/day with drip and sprinkle irrigation. In the large reclamation operations, the infiltration rates in the eastern region of the ND vary from 0.2 mm/day to 5 mm/day [10]. According to El Ramly [34], the primary irrigation canals that intersect the ND region's GW aquifer system refresh it by deep subsurface drainage percolation and seepage. The extra irrigation water recharges the GW at an average rate of 0.25 to 1.1 mm/day. The quantity of excess

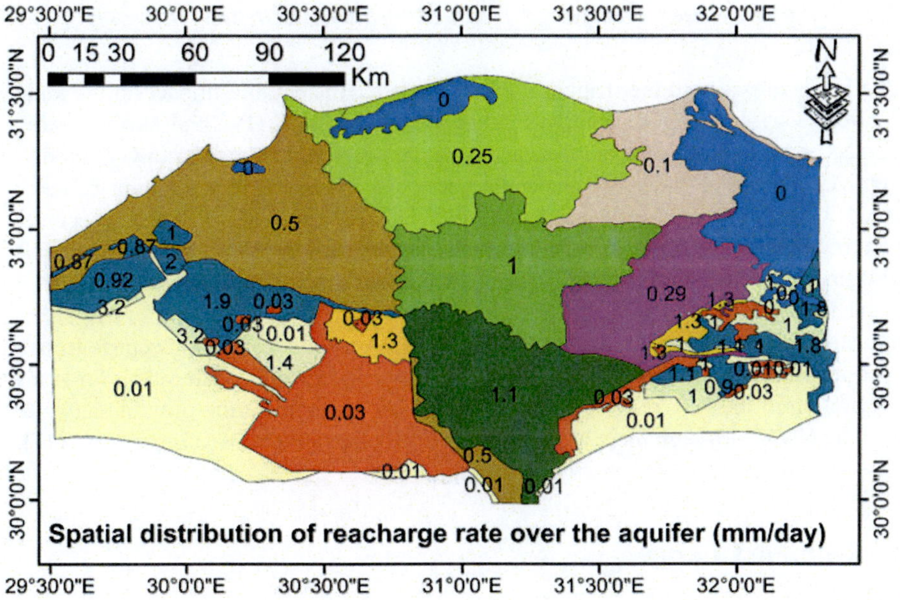

Fig. 2 Nile Delta region drainage surplus distribution of 2008 [6, 35]

Table 2 Groundwater recharging in the Nile Delta from excessive irrigation

Study	Recharge (mm/day)
[31]	0.8–1.1 (old lands) 1.9–2.1 (reclaimed areas)
[32]	0.8 (average)
[30]	0.25–0.8 (southern and central parts)
[10]	1–1.5 (furrow irrigation) 0.1–0.5 (sprinkler and drip irrigation) 0.2–5 in the large reclamation projects
[34]	0.25–1.1 (average)
[6]	0.25 (northern parts) 1.0 (middle parts) 1.1 (southern parts)

drainage in the ND is 0.25 mm/day in the northern area, 1.0 mm/day in the middle area, and 1.1 mm/day in the southern area, in accordance with [6]. These values decline in the elder lands, eventually reaching 0.29 and 0.5 mm/day in the ND's eastern and western regions. As shown in Fig. 2, the drainage surplus rate in the recently reclaimed portions of the western deserts ranges from 0.87 to 3.2 mm/day. The values of GW recharging from excessive irrigation water are summarized by mm/day in Table 2.

4.3 *Groundwater Recharge from Irrigation Canals Network*

The levels of water in the irrigation canals have a considerable impact on the main aquifer recharge as well. Due to their influence on how GW and surface water interact, moreover, these water levels are an important consideration in GW modeling. Mabrouk et al. [10] state that most modeling studies used a constant value of mean water level across the irrigation canals [36]. However, for a more accurate picture of the interactions between the surface water and the NDA, the canals' levels of water differ from month to month and throughout several canal segments. Surface water from the Nile is sent to the Delta of 35×10^9 m^3/year for use in irrigation, industry, and water supply. In such a way, the NDA is replenished by seepage from a vast network of canals and drains carrying excess agricultural water [37]. Previous research used the GW recharge from a few canals to simulate the flow of saltwater into the NDA, although there are more than 200 canals in the irrigation system [17].

5 Climate Change

5.1 *Climate Change's Effects on Egypt*

Water needs and resources must be balanced as part of the adaptation of water management to climate change in an unstable and dynamic environment. Resolving this ambiguity by supplying data on potential futures depends on political decisions. These forecasts are based on information currently available and contribute to vulnerability assessments.

5.2 *SLR Projections and Their Effects on Egypt's Coastal Zones*

The direct impacts of sea level rise (SLR), which submerges low-lying areas, are felt most acutely by coastal communities. Indirect effects on coastal areas include SWI and GW resource contamination, which exacerbate soil salinity and threaten food security. Egypt is extremely susceptible to SLR [38]. The effects of a 1.0 m SLR for 84 developing nations were considered. Egypt was classified as the second most severely damaged countries in terms of the coastline inhabitants affected and the fifth highest when it comes to the percentage of affected metropolitan regions. Currently, the ND area is undergoing changes, such as coastline alterations brought on by subsidence, erosion, and SLR due to climate change. The ND seaside zone is extremely sensitive to the effects of SLR due to direct floods and saltwater intrusion [38]. Due to sea level rise, coastal regions with low elevation are considered substantial risk areas (SLR). El-Raey [39] evaluated Alexandria City's projected to anticipated SLR, in this century, it is assumed that the SLR will be between 0.5 and 1.0 m. In respect to the 2nd Communication National Report [40] the IPCC's fourth assessment report's findings showed that, based on the most likely scenario, a worldwide SLR of 18 to 59 cm is anticipated by century's end. CORI (Coastal Research Institute) [41] calculated the susceptible areas that will be impacted by sea level rising under two situations of lake boundaries using these figures and considering the rates of land subsidence in the west, middle, and east of Delta. Lake borders were considered zero levels in the first case and protected in the second. Three scenarios were taken into account for each model, including the IPCC scenarios B1 and A1FI in addition to, the new CoRI scenario, which considers a linear growth in sea level until 2100. Table 3 displays the Nile Delta's susceptible areas under the worst scenario A1FI for the situation involving the borders of protected lakes. The majority of the lakes' borders are higher than zero (on average between 0.50 m and 2.50 m), while the Mohamed Ali seawall protects the lowlands of Abu-Quir Bay.

5.3 *Climate Change Projections and Effects on the Nile Flows*

Egypt is the country located furthest down the Nile, is influenced by how climate change is occurring not only within its boundaries but also because of changes in the basin, which it shares with 10 other nations. Pressure on Egypt's water resources is projected to increase due to economic advances in upstream nations and climate change adaptation measures taken by those nations. One obvious illustration of these effects of development is the quick development of the dams in Ethiopia. A 10% reduction in precipitation across the Nile's resources might result in a 31% drop in the river's flow at Khartoum, whereas a 10% increase in precipitation is expected to result in a 36% increase in the flow at the river's estuary at that exact place [40]. Due to the Nile's vulnerability to variations in precipitation and temperature, various researches on climate change's effects have revealed the possible for very huge variations in the Nile flow [42].

5.4 *Climate Change Projections and Their Effects on Water Demand*

Climate change's predicted temperature increases would certainly result in higher water demands for urban and agricultural use. Effects may include variations in average temperatures, the dates of crop seasons, and the amount of cooling that happens in the evening. Seasonal patterns of precipitation, including intensity, timing, and kind of precipitation, are anticipated to change similarly to temperature patterns. Temperature is different from precipitation in that the former has higher regional fluctuation and is harder to forecast [43]. Precipitation and temperature

Table 3 Nile Delta vulnerable zones, the situation of protected lake boundaries, and the A1FI scenario

	2025			2050			2075			2100		
Year	W[a]	M[a]	E[a]	W	M	E	W	M	E	W	M	E
SLR (cm)	13.00	14.80	27.90	34.00	37.50	68.80	55.00	60.30	109.60	72.00	79.00	144.00
Affected area in km^2	29.70	63.70	59.50	38.70	140.70	76.90	80.10	284.00	85.70	104.50	565.80	91.00
% of Nile Delta area	0.12	0.25	0.24	0.16	0.56	0.31	0.32	1.14	0.34	0.42	2.26	0.36
Total affected area in km^2	152.85			256.29			449.80			761.30		
% of Nile Delta area	0.61			1.03			1.80			3.04		

[a] W: West of Delta (Alex), M: Middle of Delta (AL-Burullus), and E: East of Delta (Port Said)

Table 4 Climate change's projected mean direct and indirect effects on water demand [5]

Year	2025	2050	2075	2100
Population (million)	104	146	191	237
Increase in average air temperature (°C)	1.00	1.70	2.50	3.50
Water requirements for ET_o and irrigation %	2.00	4.50	8.00	12.00
Urban water (billion m^3/year)	9.60	12.55	14.75	17.20
Industrial water (billion m^3/year)	2.20	3.40	4.00	4.90

variations typically interact, due to increased evaporation brought on by higher temperatures, the environment may become drier. GCM (General Circulation Models) models' predictions of temperature and precipitation variations will have an impact on water availability, resource management, critically influencing future agricultural production patterns. The following list could be applied to sum up the anticipated impacts of climate change on agriculture [43].

- Crop yield will decrease as temperatures rise and extreme occurrences occur more frequently.
- Changes in average temperature will lead to changes in how crops are distributed.
- A rise in temperature will harm marginal land and push farmers to abandon it.
- Marginal lands will also be impacted by a lack of water supplies, which will accelerate desertification.
- Social and political instability may result from the socio-economic effects of job loss, unemployment, and income loss.

According to Egyptian water policy, municipal water supply comes first. Plans for municipal water supplies in the future are based on anticipated population increase, and this is being done at the cost of the water allocations for farming. According to the water holding company's predictions, projections for the future growth and water supply output up until 2050 are shown in Table 4. Using the anticipated population, the projected numbers for 2075 and 2100 are generally estimated [5].

6 A Case Study: Nile Delta

6.1 Location of Nile Delta

The Nile Delta (ND) represents one of the largest river Deltas in the world. The Ismailia Canal borders it to the east, the Nubaria Canal borders it to the west, the Mediterranean Sea borders it to the north, and the Nile River borders it to the south. It is situated as presented in Fig. 3 in between Latitudes 30° N and 31°45′ N and Longitudes 29° E and 32° E. Flat, low-lying lands make up the ND, and most of them are used for agriculture. It has a shoreline that is about 240 km long, 170 km long from south to north, and covers an area of 25,000 km^2. It is said to be among the

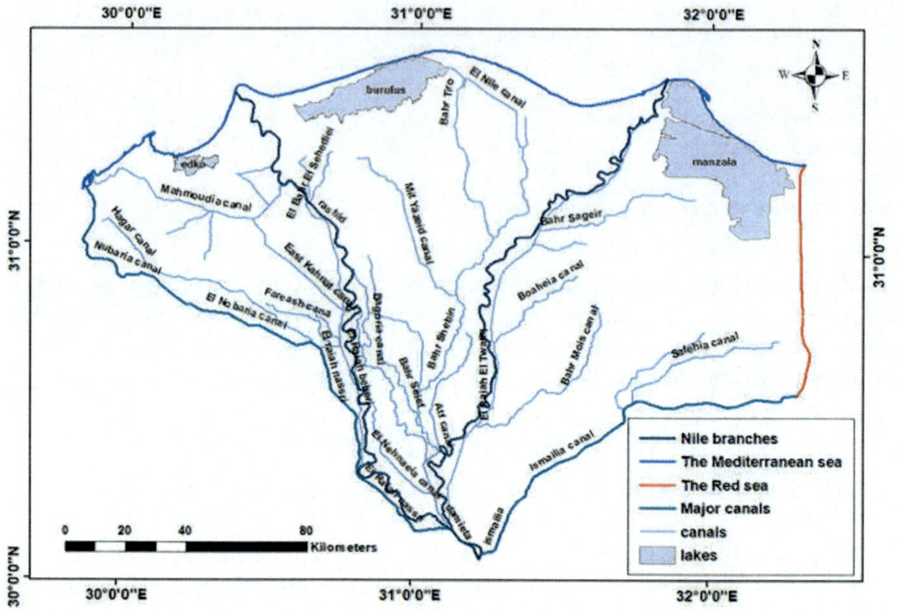

Fig. 3 The boundary of Nile Delta [5]

most populous agricultural areas in the world. It has 4 brackish lakes, 10 governorates, and approximately 25 big cities. One of Egypt's most essential water resources is the NDA system, it primarily reflects the need to satisfy the increasing water requirements. This aquifer is primarily Quaternary in age and was created from fluviomarine and deltaic deposits. Unconsolidated sand and gravel make up most of the aquifer, with the amount of clay lenses increasing toward the north [5].

7 Meteorological Aspects

The Mediterranean climate has a considerable impact on the ND, which can be found in the dry zone region of Southwest Asia and Northeast Africa. Consequently, this area may be described by having a Mediterranean climate close to the coastline and a semi-arid to dry climate in the south [5].

7.1 Precipitation and Evaporation

The annual average rainfall in the ND is quite low, approximately 25 mm/year in the middle and south of Delta, and 200 mm/year in the north [44]. Precipitation trends

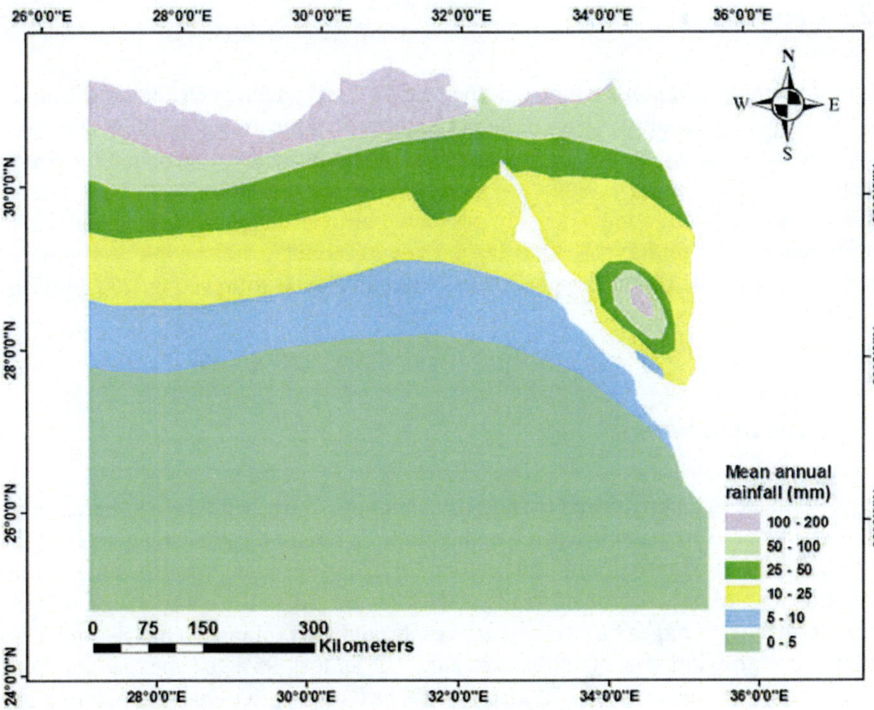

Fig. 4 Egyptian annual precipitation on average (mm/year) [46]

for Egypt are not available, according to the certain studies, the Mediterranean's wet season precipitation has usually declined since the mid-1960s [45]. The average yearly precipitation in Egypt (mm/year) is shown in Fig. 4. The Met-Office of Hadley Centre predicted a drop in precipitation, as did the great bulk of the Middle East and the wider Mediterranean. Strong ensemble agreement predicts decreases of more than 20% in the country's west. The southeast is expected to have less significant changes. Egypt's current evaporation rates range from around 4 mm per day at the North Mediterranean coast and 7 mm per day in the upper regions of Egypt [45]. It is expected that rising temperatures, shifting winds, and variations in humidity will impact these statistics. According to Eid [47] a temperature increase of 1°C may result in a 4 to 5% increase in evapotranspiration, whereas a rise of 3°C may result in a 15% increase. Accordingly, if Egypt's agricultural sector uses 41 billion cubic meters now, then a rise of 1°C requires an extra 2.0 billion cubic meters to keep output at the present level. Additionally, a 10% increase in yearly evapotranspiration was shown to cause a 6% decrease in GW recharging [48].

7.2 Population

Egypt is the most populous country in the Middle East, which is also the third-most populous country in all of Africa. The population is concentrated in three principal areas: the banks of the Nile valley and the ND, Alexandria and Cairo, and the Suez Canal. These three areas account for over 97% of the Egypt's population. These districts are some of the most densely populated all over the world and account for approximately 4% of Egypt's total territory, with about 1,500 people per square kilometer. By 2050, 146 million people will live in Egypt, followed by 237 million in 2100 [5].

7.3 Climate Change

Egypt's climate is distinguished by hot, dry summers, warm winters, and extremely little rain due to its location in a semi-arid region. To satisfy the growing needs of the domestic, industrial, and agricultural sectors, Egypt is dependent upon the Nile River as its primary and only source of freshwater. A severe risk to the entire nation would result from any changes in water supplies brought by climate change and the certainty of growing demographic pressure, as almost 95% of the people live along the ND. Sea level rise (SLR) also poses a hazard to agriculture and habitations in the ND and the Red Sea. In addition, rising temperatures by themselves would cause more water to evaporate, raise the demand for water resources, and deter visitors [5]. In two cases of lake boundaries in the West, Middle, and East Nile Delta [45], estimated the vlunerbale areas that will be impacted by the expected sea level rising in those areas. Lake borders were considered zero levels in the first case and protected in the second. Three scenarios, the two IPCC scenarios B1 and A1FI, plus a new CORI scenario that assumes sea levels rising at a constant rate until 2100 were considered for each model. The projected mean air temperatures and estimated sea level rise are displayed in Table 5, at 2000–2100, for the west, middle, and east ND

Table 5 Sea level rise and temperature changes under the B1 and A1FI scenarios

	Temperature and sea level rise projections for the years 2025, 2050, 2075, and 2100				
Year	Statistic	2025	2050	2075	2100
Scenario B1	Temperature (°C)	0.9	1.3	1.8	1.8
	SLR west of Delta (cm)	7	16	27	28
	SLR middle of Delta(cm)	8.75	29.5	32.25	35
	SLR east of Delta (cm)	18.12	39.5	64.3	72.5
Scenario A1F1	Temperature (°C)	1.2	2.2	3.2	4
	SLR west of Delta (cm)	13	34	55	72
	SLR middle of Delta(cm)	14.75	37.5	60.3	79
	SLR east of Delta (cm)	27.9	68.8	109.6	144

Table 6 Temperature and SLR impact under various IPCC scenarios (RCP scenario) [50]

Scenario	Description	Global warming (o_C) Mean and likely range		Global SLR (cm) Mean and likely range	
		2046–2065	2081–2100	2046–2065	2081–2100
RCP 8.5	Rising radiative forcing pathway leading to 8.5 W/m^2 (~1,370 ppm co_2 equivalent) by 2100	2 (1.4–2.6)	3.7 (2.6–4.8)	30 (22–38)	63 (45–82)
RCP 6.0	Stabilization without overshot pathway 6.0 W/m^2 (~850 ppm co_2 equivalent) at stabilization after 2100	1.3 (0.8–1.8)	2.2 (1.4–3.1)	25 (18–32)	48 (33–63)
RCP 4.5	Stabilization without overshot pathway 4.5 W/m^2 (~650 ppm co_2 equivalent) at stabilization after 2100	1.4 (0.9–2.0)	1.8 (1.1–2.6)	26 (19–33)	47 (32–63)
RCP 2.6	Peak in radiative forcing at ~3 W/m^2 (~650 ppm co_2 equivalent) before 2100 and then decline (the selected pathway declines to 2.6 W/m^2 by 2100)	1.0 (0.4–1.6)	1.0 (0.3–1.7)	24 (17–32)	40 (26–55)

areas under the low scenario (B1) and the high scenario A1FI of SRES (Signed Response), as stated by the First IPCC First Assessment Report as stated by [49].

According to the estimations, the earth is warming by 0.13° due to the increase in greenhouse gas concentrations in the atmosphere every 10 years [45]; in addition, by the end of this century is forecasted to possess a whole array of temperatures of 1.1° to 6.4°. This range of temperature rise is anticipated to cause the polar ice caps to melt and deep ocean water to expand, raising sea levels in the process. A Representative Concentration Pathway (RCP) is a course that the IPCC has approved for greenhouse gas concentrations rather than emissions. Four studies and modeling approaches were used for the 2014 IPCC Fifth Assessment Report (AR5). The pathways show various possible climatic futures, dependent on the amount of greenhouse gases (GHG) released in the upcoming years, all of which are thought to be possible. The RCPs (Representative Concentration Pathways), which were originally designated as RCP 2.6, RCP 4.5, RCP 6, and RCP 8.5, are titled for a variety of radiative forcing intensity in the year 2100 (2.6, 4.5, 6, and 8.5 W/m^2, respectively) as shown in Table 6 [50].

8 Hydro-Geological Settings

Owing to the more homogeneous aquifer type and absence of pollutant transport difficulties, the hydro-geologic environment for GW projects is frequently substantially different from that for hard rock or aggregate mining operations. It is also frequently relatively less complex. However, using GW extraction from huge aquifers might be difficult due to GW-surface water interactions [24]. Two hydrological systems are compressed by the ND: the Nile River's surface water systems and the ND Quaternary aquifer system.

8.1 Surface Water System

The overall effectiveness of the Egyptian irrigation system is greater than 70%. In accordance with volumetric quotas established according to the serviced area, cropping patterns, and soil and climate conditions, water is distributed through the irrigation system among primary canals and areas. The drainage system, pumping stations for irrigation and drainage, and hydropower generating are other significant infrastructures. The drainage system prevents salt from accumulating in the irrigation soils and allows irrigation water to be recycled. It comprises of a massive drainage system that includes main drains, field drains, and sub-collectors, which can replenish the Nile with the drainage water, discharge it into interior or coastal lakes, or deliver it directly to the sea [5]. In Egypt, precipitation is an insignificant source of water compared to other water suppliers. Forecasts for climate change suggested that its tendencies would decline and that there would be a rise in the chance of flash floods. As a result, rainfall is not expected to have a substantial effect on the water strategy's development, but there may be opportunities for flash floods to recharge groundwater aquifers. However, modern technology might assist in maintaining existing levels of usage and, ideally, even slightly enhance them to meet the increasing demands for every drop of water. Rainfall gathered annually as of 2010 is estimated to be 1.30 billion m^3 [5]. Water is provided via an intricate network of irrigation canals to the extensive agricultural practiced in the Delta. Figure 3 depicts a general picture of the Nile valley and Delta, displaying the irrigation network and major infrastructure. A significant amount of reusing the farm drainage water to compensate for a lack of freshwater, particularly in the low sections of the canals, in which the irrigation and drainage system is complicated. In the downstream, the irrigation and drainage canals' water quality is getting worse because of the growing pollution load from intensive agricultural operations, activities, and dense population [5].

8.2 Geometry of the Nile Delta Quaternary Aquifer

All the ND region's topography depends on information from topographic maps created by [51]. As illustrated in Figs. 5, and 6, it is also the base level of the semi-previous layer covering the aquifer, which includes the clay thickness and the bottom of the Quaternary and Moghra aquifer system level map in (m) relative to mean sea level.

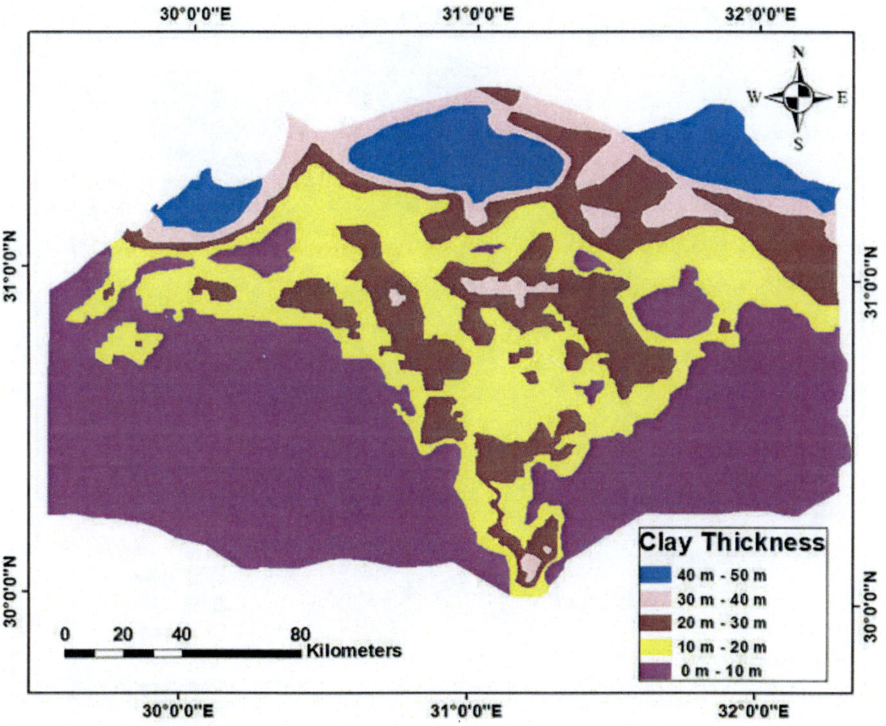

Fig. 5 Arial distribution of clay cap thickness [46]

8.3 Hydraulic Parameters of the Nile Delta Quaternary Aquifer

The hydraulic characteristics, including vertical (K_v) and horizontal (K_h) hydraulic conductivity, storage coefficient, specific yield, transmissivity, total and effective porosity, are the most crucial factor in influencing flow characteristics (Table 7 and Table 8). One of the most crucial techniques for assessing the aquifer system and determining its capacity for water extraction and storage is by examining its hydraulic properties.

A massive portion of the ND is capped with clay, which keeps the aquifer there in a semi-confined state as indicated in Table 9. The Nile Delta's clay layer ranges in thickness from 5 to 20 m in the south and middle, and it reaches 50 m in the north [55].

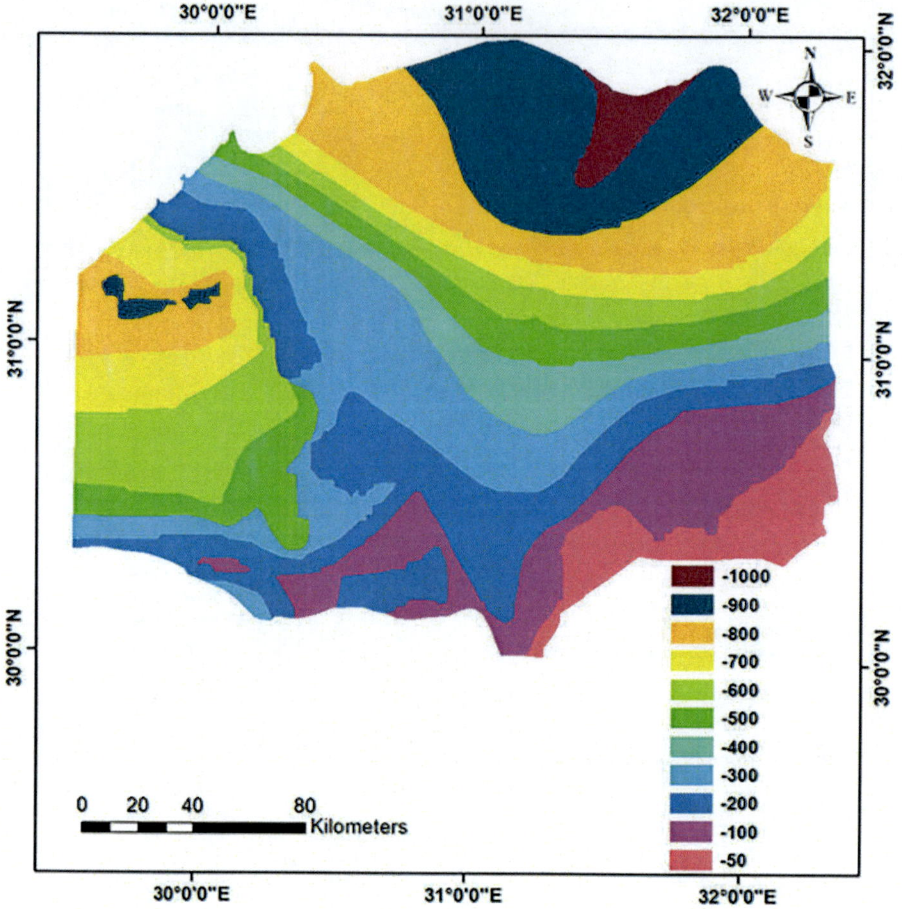

Fig. 6 Arial distribution of base of Quaternary and Mogra aquifer system [46]

Table 7 Hydraulic features of the Nile Delta's Quaternary aquifer as reported

Main hydraulic units	Hydraulic conductivity	Transmissivity m^2/day	Storage coefficient	Specific yield	Porosity	Effective porosity
	K (m/day)	T (m^2/day)	[–] (m/day)	S_s (1/m)	n (%)	n_{eff} (%)
[36]	50	–	–	–	–	–
[52]	75	15,000–75,000	10^{-4}–10^{-3}	–	25–40	–
[14]	70–100	–	–	–	–	–
[53]	119	–	10^{-4}–10^{-3}	0.15	30	–

Table 8 A semi-confined layer's hydraulic conductivity value

Study	k_h (m/day)	k_v (m/day)
[30]	0.216	0.0073
[54]	0.1–0.25	0.01–0.025
[13]	–	0.0005–0.005
[22]	0.05–0.5	0.0025
[52]	0.05–0.5	0.0025

Table 9 Hydraulic properties' input values for the model [54]

Main hydraulics units	Layers no	Hydraulic conductivity K_h (m/day)	Hydraulic conductivity K_v (m/day)	Storage coefficient S [−]	Specific yield S_s (1/m)	Effective porosity n (%)
Clay	1	0.1–0.25	0.01–0.025	0.001	0.1	50–60
Fine sand with lenses of clay	2,3,4, and 5	5–20	0.5–2	0.005	0.15	30
Coarse sand quaternary	6,7,8, and 9	20–75	2–7.5	0.0025	0.18	25
Graded sand and gravel	10 and 11	75–100	7.5–10	0.0005	0.2	20

8.4 Surface Water and Groundwater Relationship

There is a clear hydrogeological interaction between surface water and GW in the research area, which is divided among SW bodies in the north and east and numerous surface freshwater canals and drains on the west, east, and south. The relationship between groundwater and surface water in the research area changes depending on the kind of sediments, soil hydraulic conductivity, infiltration rate, GW water table in relation to surface water level, and variations in the hydraulic properties of GW and surface water [56].

8.5 Recharging the Nile Delta Quaternary Aquifer

The main NDA is replenished by water levels in irrigation canals, seepage from the Damietta and Rosetta branches in the Middle Delta, and large canals such as the Ismailia Canal toward east and the Nubaria Canal toward west. Excess irrigation water from percolating irrigated land acts also as recharge. Figure 2 illustrates the 2008 drainage surplus distribution in the ND region. Three steps are involved in the recharge through the research area: Only the winter months, with an average annual rainfall of 25 mm, act as the first source of recharge by rainwater to the study area aquifer. For the years 1991 and 2010 in the ND region, GW recharging from rainfall

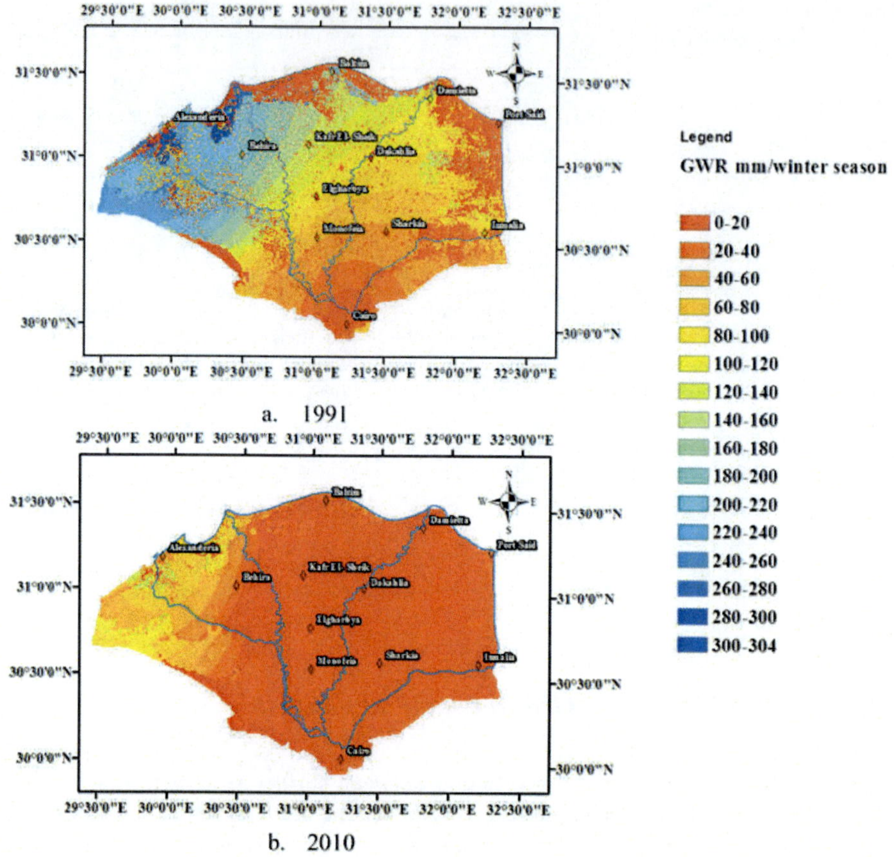

Fig. 7 Rainfall-induced groundwater recharge in the ND region between 1991 and 2010 (**a**) 1991 (**b**) 2010 [17, 18]

is illustrated in Fig. 7a, b [17, 18]. Based on the kind of soil, irrigation, and drainage methods, the second type of recharge occurs when surplus water from the Nile River is employed for irrigation and seeps into the aquifer through canals; this leakage occurs 0.25 to 0.80 mm/day. The third is GW inter-aquifer flow.

8.6 Well Discharge Rate (Extraction from the Nile Delta Quaternary Aquifer)

The ND region's abstraction inventory was managed to be completed by RIGW in 1992, 1995, 1997, and 2002 with the goal of estimating GW abstraction. Information sheets were then uploaded to the database to fit its needs, and new functionality was added in 2008. In the ND region, the extraction rate from GW wells between 1992

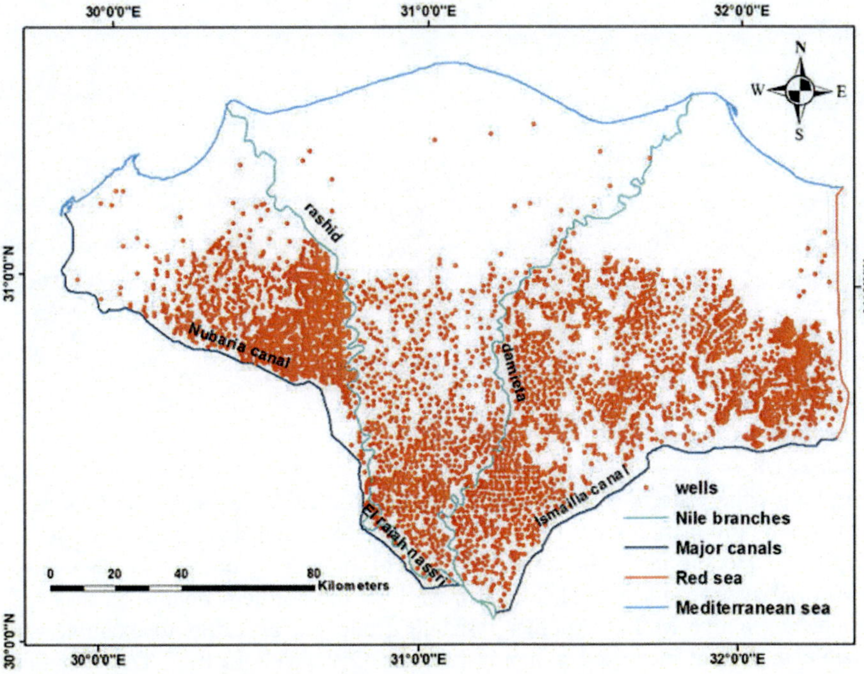

Fig. 8 Extraction wells location map in the Nile Delta region [3, 19, 57]

and 2008 was 3.03 and 4.90 (Bm^3)/year for various governorates [6]. Figure 8 illustrates extraction wells location map [19]. Additionally, SWI in coastal areas could result in GW seeping upward into the clay cap aquitard layer from the aquifer, this leads to deterioration of the soil and groundwater in both upward and downward GW movement [58].

Over the past 30 years, abstraction rates increased between 1980 and 2010. Significant increasing trends in abstraction were confirmed by the RIGW report [59], which reached in 1991 at about 2.6×10^9 m^3/year. The total annual abstraction in 2003 was 3.5×10^9 m^3/year [36]. In 2010, it was approximately 4.6×10^9 m^3/year. As illustrated in Fig. 9, the Nile Delta's abstraction is rising gradually at a rate of 0.10×10^9 m^3/year, except for the years 2003 to 2010, when it grew drastically at a rate of 0.20×10^9 m^3/year [29].

8.7 *Groundwater Level in the Nile Delta Aquifer*

The GW table ranges between 1–2 m deep in the north, 3–4 m deep in the middle, and 5 m deep in the south. The GW table's various estimated depths have been published by [6, 20]. (RIGW) observation wells, which feature well screens 20 to

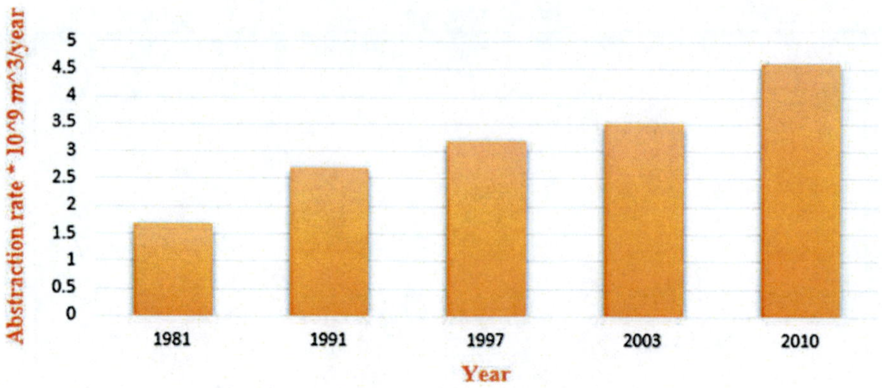

Fig. 9 Time-based abstraction rates in the Nile Delta [29]

75 m below the water's surface, are used to track the GW levels in the ND area. Furthermore, production wells for (RIGW) often draw water from the ND aquifer at a depth of 20 to 150 m. At the southern border of the ND region, the GW levels are 16.96 m above mean sea level (M.S.L.), while at the northern boundary, they are zero meters above M.S.L. Another research group put an effort to examine the possible impact of increased pumping rates on GW levels in the NDA. For this purpose, Armanuos et al. [15, 16] used MODFLOW to create a 3D GW model. As the basis case, the model was validated on the measurement of GW level, See Fig. 10).

8.8 Groundwater Salinity in the Nile Delta Aquifer

Although the Nile system's GW quality is acceptable, contamination has harmed some shallow GW basins. 20% of the Nile aquifer's SWI does not fulfill requirements for potable water, particularly in areas along the edges where the protective clay layer is either absent or very thin. Compared to the Nile Valley, the GW in the ND is frequently of superior quality. While the Sinai and eastern deserts have high salinity, the western deserts are often extremely good (TDS). Brackish water makes up most of the carbonate aquifer, but some freshwater is added during recharge. Indirect effects on coastal areas include saltwater intrusion and SWI contamination, which increase soil salinity and threaten food security. Additionally, storm surge severity and frequency will both increase, which will undoubtedly influence coastal infrastructure [5]. The ND SWI was evaluated and simulated using a variety of numerical techniques. In earlier research, the primary objective was to determine the freshwater thickness of the NDA using semi-analytical models that relied on the sharp interface modeling method. This research' examples can be found in [4, 13, 60–63]. Because there were not adequate records of the aquifer's salinity, the

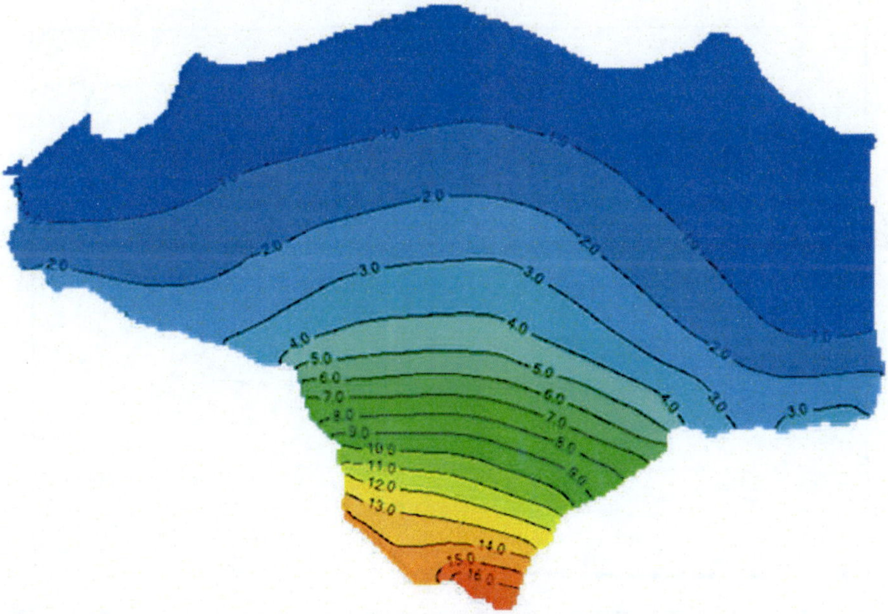

Fig. 10 Quaternary aquifer groundwater contour map taken in 2008 [15, 16]

majority of this research was more theoretical in nature. As already mentioned, the Nile Delta's transition zone is quite vast and is distinguished by the dynamic interaction between freshwater and ocean. Since freshwater and SW interact in the aquifer, variable density numerical models are better adapted to simulate these interactions. Such models have been created in past years as horizontal models in two dimensions for some of the NDA's cross sections or vertical models in two dimensions for some of the Delta's cross sections. Since the projected impacts of SLR on the NDA's salinization have only recently become known, the majority of previously developed variable density models were primarily concerned with assessing how increased GW abstraction rates would affect the aquifer's salinization. The main factor for the rise in GW salinity, especially in the northern portions, was SWI. The range of GW salinity was discovered to be between 227 ppm and 15,264 ppm after researchers examined additional sources, such as salinization caused by soil forms. GW salinity fluctuates when canal water levels change. The water salinity records from 1957 to 1984 revealed that it increased, and freshwater was dominating and defeating SW encroachment. GW salinity rose after 1984 due to widespread Nile flow reduction and abstraction. In 1990, as the Nile water flow increased, the GW's salinity decreased once more to its pre-increased levels. But in 2000, new reclamation operations and significant abstraction caused the salinity of GW to rise once more [4]. In accordance with RIGW's database on iso-salinity 1,000 ppm comparisons for the years 1960, 1980, 1992, and 2008. In GW wells mostly located at latitude 31° N in 1960, the GW salinity is around 1,000 ppm. The

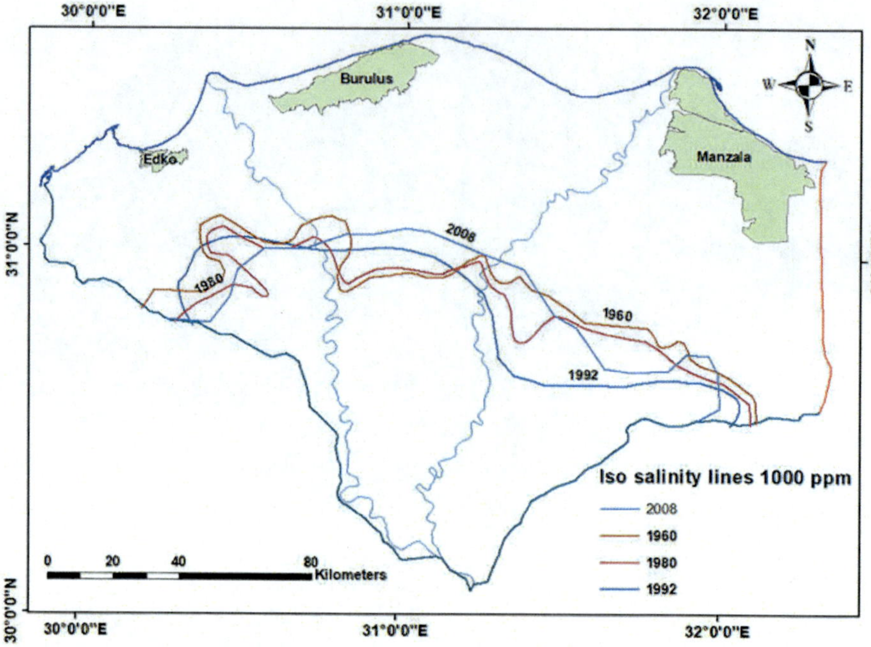

Fig. 11 Groundwater salinity (TDS) in the years 1960, 1980, 1992, and 2008 [6]

1,000 ppm contour line shifted somewhat toward the north, especially in the middle Delta and surrounding both Rosetta and Damietta branches, due to increased recharge from two branch and main canal, and increased salinity in the east and west Delta due to limited of recharge, but there was little change between 1960 and 1980 where the freshwater pushed the SW northward due to increased GW recharge. Between 1980 and 1992, there was an increase in salinity in the east and west Delta as a result of increased GW extraction, Moreover, during the dry years of 1983 to 1996, the salinity of the Delta increased in some areas. Due to increased recharging from Nubaria and Ismailia canal, the 1,000 ppm contour line in the central Celta moved farther north between 1992 and 2008, and salinity decreased particularly in the eastern and northwestern ND edges, as presented in Fig. 11.

8.9 Solute Transport Parameters

he same physical transport mechanisms, namely dispersion and advection, have an impact on all solutes. In contrast, the solute of interest and the aquifer's geochemical conditions determine the geochemical transport processes (sorption, precipitation/dissolution, and degradation). Effective porosity and dispersion parameters make up most of the aquifer's physical transport characteristics [64]. According to the field data from pumping tests, the effective porosity used for the advective process had an average value of 0.25. On the other hand, longitudinal dispersion (α_l) and lateral dispersion (α_T) are the dispersion parameters for the dispersion mechanism. These parameters are often derived from field studies employing tracer techniques and pumping experiments or laboratory scale utilizing column experiments [6]. According to Sherif et al. [63] the NDA has longitudinal and lateral dispersivity values of 100 m and 10 m, consequently. Additionally, the vertical to horizontal conductivity ratio was adjusted at 1:10.

9 Groundwater Flow and Solute Transport Simulation

A 3D model was built for the NDA by utilizing MODFLOW software. Scientific, technical, and water resource management issues are being addressed using numerical models that take fluid density impacts on GW flow into consideration [65]. To simulate saltwater intrusion and forecast solute concentrations in seaside aquifers, researchers employed the coupled fluid flow and the solute transport models. Overdraft, mining, depression, water stress, seawater intrusion, and groundwater pollution are the major concerns with groundwater [66]. To anticipate the migration of seawater inland under several situations of recharge in the coast plain [67] studied the consequences of dewatering. The current study was developed to use Visual MODFLOW and SEAWAT for simulating GW flow. Visual MODFLOW is a highly complete and approachable modeling solution accessible to real-world studies in 3D GW flow and pollutant transport simulation. Powerful analytical capabilities and a reasonable menu structure are included in this fully integrated solution. Graphical tools that are simple to use enable you to:

- Dimension the model domain and choose units quickly
- Specifying the boundary conditions and the model properties in a convenient way
- Run flow and pollutant transport model simulations
- Using manual or automated methods for calibrating the model
- Improve the rates and locations of the abstraction and remediation wells
- Apply 2D or 3D graphics to see the results

The most recent version of the SEAWAT computer program, which simulates GW flow in porous media of 3D, variable-density, and transient properties, was utilized for the simulation of the solute transport of SWI in the NDA. Modified versions of MT3DMS and MODFLOW-2000 were combined to create SEAWAT-2000, a single computer software. In addition to the processes supplied with MODFLOW-2000, SEAWAT-2000 also includes the integrated MT3DMS transport process and the variable-density Flow Process (as a substitute to the constant-density GW Flow Process) [68].

10 Numerical Model Setup

The process of transforming a conceptual, qualitative model into a numerical model, or a complex collection of mathematical equations that can be numerically solved, is known as numerical model setup. It is necessary to formulate the mathematical problem correctly to solve the numerical model. These needs defining the following [24]:

1. Domain and boundary conditions modeling
2. Discretization, model layers
3. Internal sinks and sources, boundary conditions
4. Parameterization of a model
5. Time stepping and initial conditions for transient simulations
6. Modeling convergence

10.1 Model Domain and External Boundaries

The specification of a suitable model domain, comprising the region where GW flow occurs and contaminant transport is to be researched, is the initial step in the creation of a model. The numerical model's domain often matches the conceptual model's domain, but it occasionally is smaller. The geological volume of interest that the mathematical modeling of GW flow and transport is determined within is enclosed by the model domain. The aquifer of ND is covered in the current study's GW flow and solute transport model, which has a total active area of more than 23,000 km^2. Conceptualizing the impact of the model boundaries on heads and flows is necessary, the model chooses the best or most suitable mathematical illustration of this impact (alternatively known as a "boundary condition"). The Mediterranean Sea lies to the north, the Suez Canal toward the east, the Ismailia Canal to the southeast, and the Nubaria Canal to the southwest, which form the study's exterior boundaries.

10.2 Model Layers Discretization

The modeling of the hydrogeological system with discrete constituent volumes is a fundamental component of numerical GW models. Either the "finite-difference approach" or the "finite-element method" represents this discretization within the model domain using an orthogonal grid of the model cells.

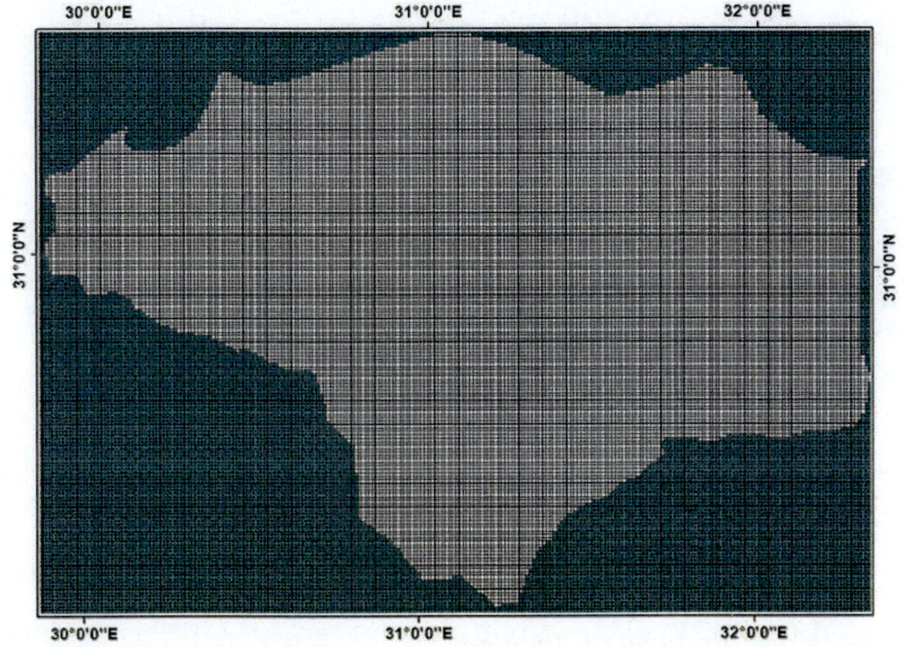

Fig. 12 Grids and model extension

10.2.1 Horizontal Discretization

Visual MODFLOW classic interface software was used to apply the numerical model, by using 170 rows and 240 columns for both active and inactive cells with 1.00 km × 1.00 km-sized cells as presented in Fig. 12. For achieving consistent modeling results that satisfy the necessary accuracy, it is best to begin with a reasonably coarse model grid and after that refine the model discretization.

10.2.2 Vertical Discretization

The simulated model consisted of 11 layers, with the second to eleventh levels representing the Quaternary aquifer and the first layer representing the clay top. The Quaternary aquifer is still present in the first layer, which is a clay cap with a depth range of 20 m in the south to 50 m in the north. Most of the aquifer is composed of Quaternary layers, which have silty and sandy clay and range in depth from south to north, with an average depth of 200 m near Cairo to 1,000 m near the coast. Two vertical sections were made in the central region of the domain, as illustrated in Fig. 13a, b, correspondingly. The first section was taken in the X-direction, through the Suez Canal toward the Nubaria Canal in the region to the east, whereas the

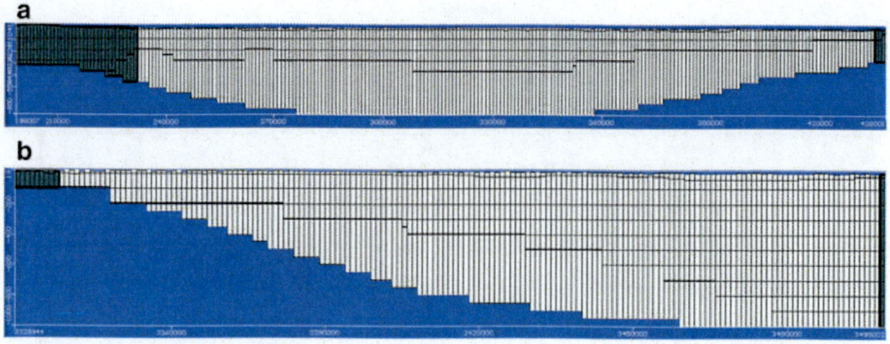

Fig. 13 (**a**) Vertical section at X (east to west) directions. (**b**) Vertical section at Y (north to south) directions

second part was obtained in the y direction, from the Mediterranean Sea coast in the north toward Cairo in the south.

10.3 Model Boundary Conditions

The boundary conditions are critical in determining if a GW flow model is adequate since they specify wherein the system's water intake and output occur. The model will provide a poor simulation of the real system for GW flow if the boundary conditions are inaccurate. The selection of the suitable boundary conditions is also influenced by the modeling objectives and the number of stresses to be simulated. When GW systems are under a lot of stress, the physical characteristics that govern the system may change as a result. In general, there are two conceptual forms of boundaries: hydraulic boundaries and physical boundaries. Hydraulic boundaries are dependent on the GW flow system and are susceptible to change because of the GW flow. Physical boundaries are not affected by GW flow or GW system stress, or the modification is negligibly low [64]. Hydraulic boundaries (artificial boundaries) include groundwater dividing lines and streamlines (flow lines) in addition to arbitrary contours on a regional map of the water table or distant head limits representing remote surface water bodies. For calibration purposes, a steady state flow field can be created using hydraulic boundaries, although for steady state or transient prediction models of applied stresses, these bounds might not be suitable. To establish whether using a hydraulic boundary might result in unacceptable model errors, it should be carefully considered. This study employed two different boundary types as:

10.3.1 Head Boundary Condition

At the north boundary, where a constant head of zero was utilized as the boundary, the model's boundary and hydrological settings had been constructed. Constant head with Cairo's southern edge at 16.96 m above mean sea level (MSL). The GW direction perpendicular to the water's contour line was left unrestricted at the east boundary. Ismailia Canal extended from south to east, with the water level in the field starting from 18.5 m in the south to 7.6 m in the east above mean sea level. The field's water level varied from 16.00 m in the south to 0.50 m in the north above mean sea level (MSL), with El Rayah El Behery and El Nubaria Canal forming a border from south to west. The boundary conditions for the first layer are revealed in Fig. 14. The primary lakes in the study area were designated with a constant head border. Idku, Burullus, and Manzala are three brackish lakes that have direct access to the Mediterranean Sea.

10.3.2 Salt Concentration Boundary Condition

The SEAWAT model's head and hydrological settings were set to the same parameters as those used in the MODFLOW model. A concentration of 40,000 mg/l is used to represent saltwater TDS when inland sea flow occurs along the Mediterranean Sea's coastal zone, while a concentration of 35,000 mg/l is used to represent SW

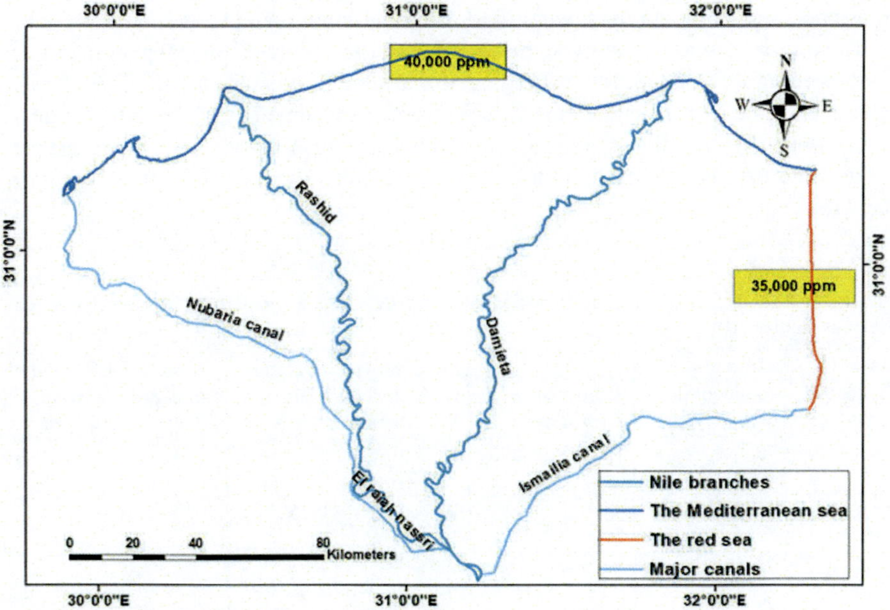

Fig. 14 Model boundary conditions for concentration

TDS along the Suez Canal's beach. The GW's initial concentration was set at 500 mg/l. as illustrated in Fig. 14 [13].

Physical boundaries" are created by the presence of a large surface water body, by the presence of a hydrogeological unit with extremely low permeability (e.g., fault filled with clay gouge, rock unit, ice, or permafrost zone), in a small-scale model, by a manufactured barrier. By a system of natural drainage (for instance, a position on top of a mountain around by deep valleys).

10.4 Model Parameterization

The determining of the hydraulic parameters allocated to the model domain is an essential element of representing the hydrogeological conceptual model. Hydraulic conductivity and/or transmissivity, confined and unconfined aquifer storage qualities (specific storage and specific yield), and aquifer compressibility are among the properties included in the saturated flow model.

10.4.1 Model Hydraulic Parameters

The hydraulic parameter values are crucial for controlling the flow characteristics in any hydrological system. Tables 9 and 10 provided the initial hydraulic parameter values for the NDA area. These values, which were gathered from earlier investigations and some computations, were used as model input data [10, 54].

The transport model employed the same hydraulic and hydrological parameters as the preceding flow model for recharging and discharge (visual MODFLOW classic interface). The scale-dependent values were determined as, the longitudinal dispersivity (α_L) is 100 m, the lateral dispersivity (α_T) equal 10 m, the vertical dispersivity (α_V) is 1.00 m, and the diffusion coefficient value (D^*) equal 10^{-4} m^2/day [13].

Table 10 The hydrological characteristics of the Nile Delta's Quaternary aquifer based on previous studies [10]

Main hydraulic units	Hydraulic conductivity K (m/day)	Transmissivity T (m^2/day)	Storage coefficient S (m/day)	Specific yield S_s (1/m)	Porosity n (%)	Effective porosity n_{eff} (%)
[52]	75	15,000–75,000	10^{-4} to 10^{-3}	–	25–40	–
[53]	119	–	10^{-4} to 10^{-3}	0.15	30	
[14]	70–100	–	–	–	–	–
[36]	50	–	–	–	–	–

10.4.2 Model Recharge and Abstraction

Aquifer recharge and discharge are vital factors in the flow and the movement of SWI. The bed levels and water levels at the beginning and end of the canals and the calculating corresponding conductance are shown in Table 11 [17], plus extra water used for irrigation, with daily rates varying between 0.25 and 0.80 mm as illustrated in Fig. 2 [6], the second by average annual precipitation of 25 mm [17], and the third by inter-aquifer flow to GW. The Nile Delta aquifer's GW abstraction activities include governmental production wells for irrigation water (available) and production wells for drinking water (not available). The distribution of the research area's abstraction wells was determined by data that was obtained by (RIGW), and in 2008, the total abstraction rate was 2.78×10^9 m^3/year induced in the built NDA model as illustrated in Fig. 8.

Table 11 Water elevations in canals [17]

	Start		End		Conductance	
Name	Stage level (m)	Bottom level (m)	Stage level (m)	Bottom level (m)	Start (m^2/s)	End (m^2/s)
Ismalia canal	18.5	15.5	7.6	4.6	500	1,200
Rashid branch	18.6	15.6	1.35	−0.1	800	800
Damieta branch	19.91	16.91	1.86	0	800	800
Nubaria canal	9.75	6.75	2.92	0	3,200	3,200
Salehia canal	9.91	6.91	6.99	6.3	350	350
Sayedia canal	11.34	8.34	5.4	2.4	325	325
Bahr mois canal	15.375	12.375	7.9	4.4	475	475
Boaheia canal	12.18	9.18	5.3	2.3	365	365
Bahr sager	7.87	4.87	0.8	−2.2	338	338
Hagar canal	5.9	2.9	−1.87	−4.12	1,300	1,300
Fareash canal	7	4	0.37	−2.63	267	267
Mohamadia	5.59	2.59	0.58	−2.42	410	410
East kahnat	7.85	4.85	4.275	1.275	410	410
Canal project naser	1.94	−1.06	−0.09	−3.09	189	189
Riah tawfiki	16.09	9.06	6	4.2	800	800
Sharkawia canal	6	4.2	−0.8	−2.2	469	469
El Nile canal	3.8	−0.8	1.7	0.64	265	265
Bahr tiro	6.16	3.16	0.68	−2.2	504	504
Mit yazied canal	9	6	4.41	1.41	254	254
Bahr shebin	13.52	10.52	3.52	0.52	489	489
Riah nasri	18.3	14.81	1.6	0.079	800	800
Atf canal	15.25	12.25	10.8	7.8	87	87

10.4.3 Properties of the Model Hydraulic Package

Water levels in canals and drains are one of the main elements influencing the water level for a GW system. Irrigation and drainage networks existed in the ND, which affect the GW level in the aquifer of ND.

River Package

The two branches of the Nile are considered as the main surface water sources for the ND, which include major control structures such as barrages and regulators, Additionally, it distributes river water to Rayahs and major canals, this mechanism acts as a controlled boundary head. The two branches' hydraulic characteristics -Rayahs and the main canals- are used as conceptual input components through the river package interface, which includes the bed's conductance, thickness, and vertical hydraulic conductivity in addition to the river's stage and level. Some canals have been completely, and others have just partially been penetrated in the upper layer of semi-confining clay. River nodes can simulate either a source or a sink thanks to this feature, because of how the GW and surface water systems interact that is dependent on the vertical value of the hydraulic conductivity and the head difference between them. The conductance value for a river, stream, drain, and the overall head boundary conditions regulates the amount of seepage from the surface system. A mathematical expression can be used to calculate it.

$$C = \frac{L \times W \times K}{M} \tag{1}$$

where C is the conductance of the river's bed (L^2/T), L river reach length in each grid cell (L), W width of the riverbed in each grid cell (L), K vertical hydraulic conductivity of river's bed (L/T), and M each grid cell's riverbed thickness (L).

Drainage Package

Using the drain stage for drain elevation and the bed layer's hydraulic conductivity, the main drain in the study area was simulated like the river package.

10.5 Model Calibration

A calibration process compares the calculated head from the model to the ones measured in 2008 using real measure values. The field data for piezometric heads and the model results for the computed heads values are comparable, this provides a wealth of reality for future GW level predictions.

10.5.1 Head Calibration

In the research region, 60 observation wells have been distributed in the ND region. It was based on the piezometric head data for RIGW in 2008 [6] that was available as illustrated in Fig. 15.

The model was repeatedly run till the root mean square error (RMSE) among the simulated and observed heads by RIGW in 2008 reached 0.42 m. During the simulation, the hydraulic conductivity was also calibrated, and the ratio of hydraulic conductivity between the vertical and horizontal directions was 1:10. Figure 16 reveals that the first layer's calibrated hydraulic conductivity value ranged between 4.22×10^{-8} and 2.82×10^{-7} m/s. Figure 17 shows the calibrated values of layers from 2–11 horizontal hydraulic conductivity which ranged from 4.22×10^{-4} to 28.21×10^{-4} m/s [17]. Figure 18 shows a relationship between the observed levels of GW for 60 records from various observation wells in 2008 and the simulated levels obtained by RIGW. The RMSE and correlation coefficient (R2) corresponding to the GW levels modeled and observed were 0.42 and 0.9925.

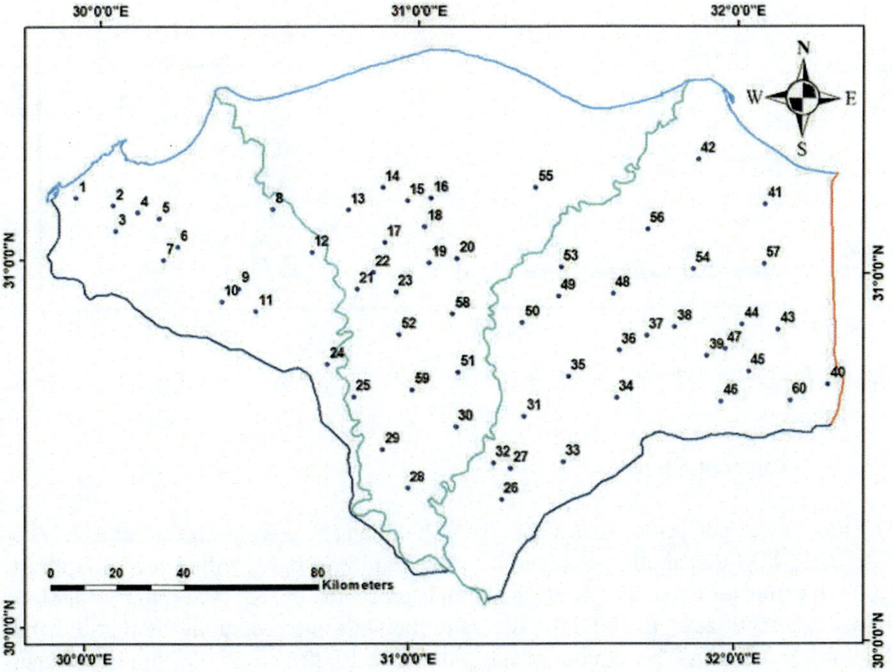

Fig. 15 Distribution of observation wells in the research area for model calibration

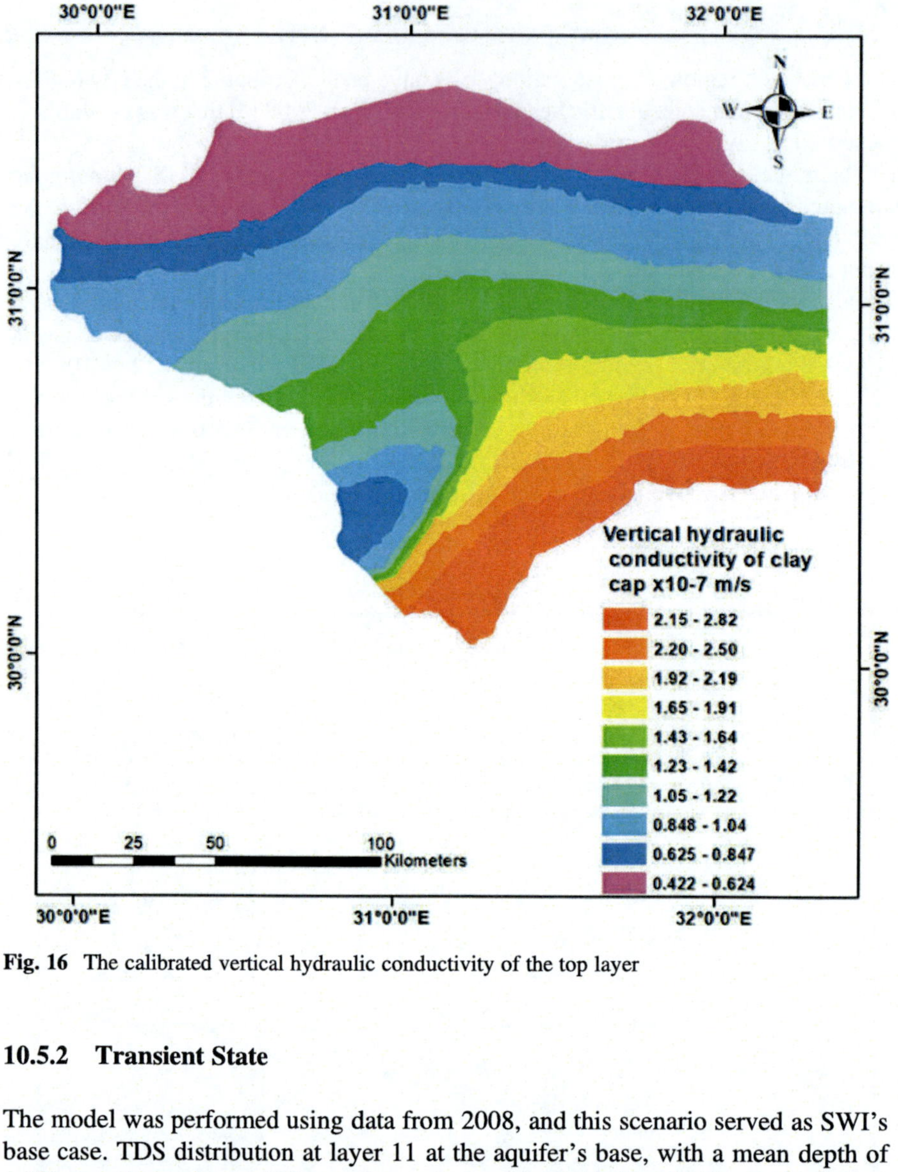

Fig. 16 The calibrated vertical hydraulic conductivity of the top layer

10.5.2 Transient State

The model was performed using data from 2008, and this scenario served as SWI's base case. TDS distribution at layer 11 at the aquifer's base, with a mean depth of 1,000 m to the north and 200 m to the south, is presented in Fig. 19. Figure 20 depicts three cross sections of the ND: the first is in the west, the second at the middle, and the third at the east. The Equi-concentration lines 1,000 and 35,000 ppm represent the freshwater (FW) and saltwater (SW) lines, respectively. As shown in Fig. 21a, the SW line 35,000 ppm at the first cross section shows saline water, and it moved further 40 km from the coastline. It was identified at the aquifer's base (at layer 11). The FW line 1,000 ppm, which separates the freshwater from the SW, additionally moved into 68.5 km. The dispersion zone is 50 km wide, as illustrated in Fig. 21b. At

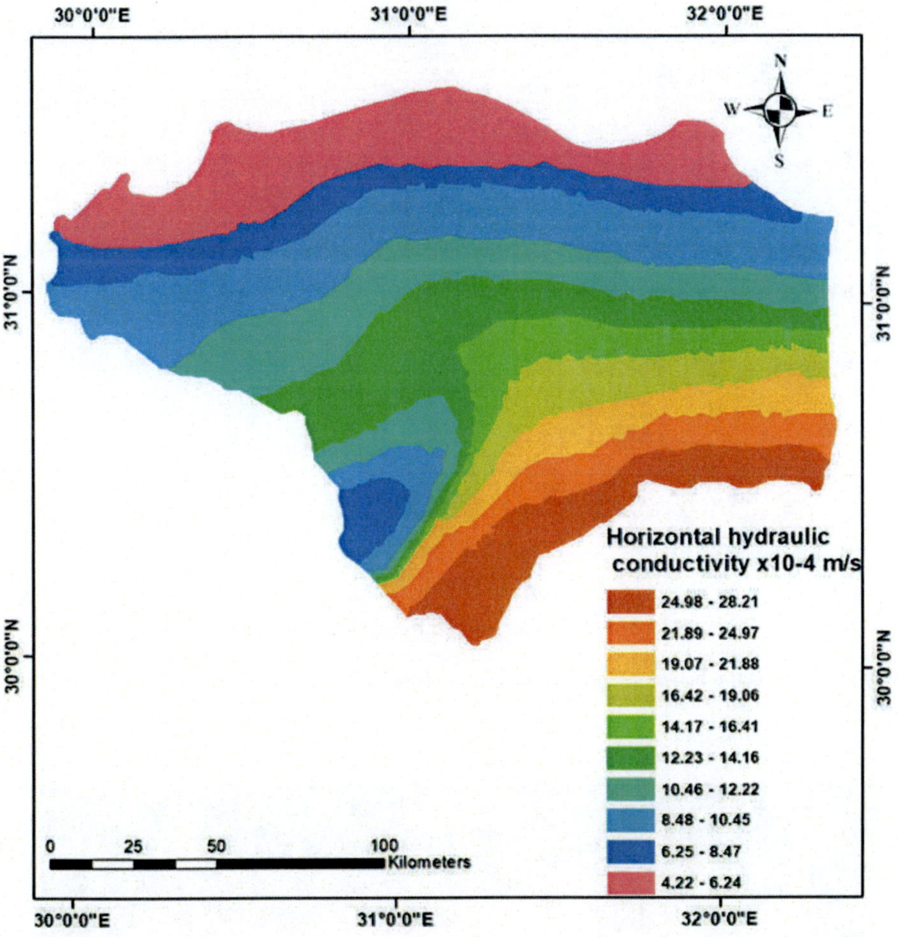

Fig. 17 The calibrated horizontal hydraulic conductivity of the quaternary aquifer

the second cross section, the SW line 35,000 ppm advanced inside to 43 km from the coastal line, identified at the aquifer's bottom; the FW line 1,000 ppm moved into 93 km, identified at the aquifer's bottom. As demonstrated in Fig. 21c, at the third cross section, the SW line 35,000 ppm moved inside to a point 50 km from the coastal line. At the ND aquifer bottom, the FW line 1,000 ppm moved inside to a point 81 km, and the dispersion zone is 31 km wide.

Between the simulated salinity of GW and the actual GW salinity of latest investigations, a calibration process was made at layer 11 which visualizes the base of the aquifer for both investigations. The Equi-concentration lines' progress at 35,000 and 1,000 ppm inland serves as the basis for the calibration. The calibration revealed good agreement between the results of recent research and the simulated study as illustrated in Table 12.

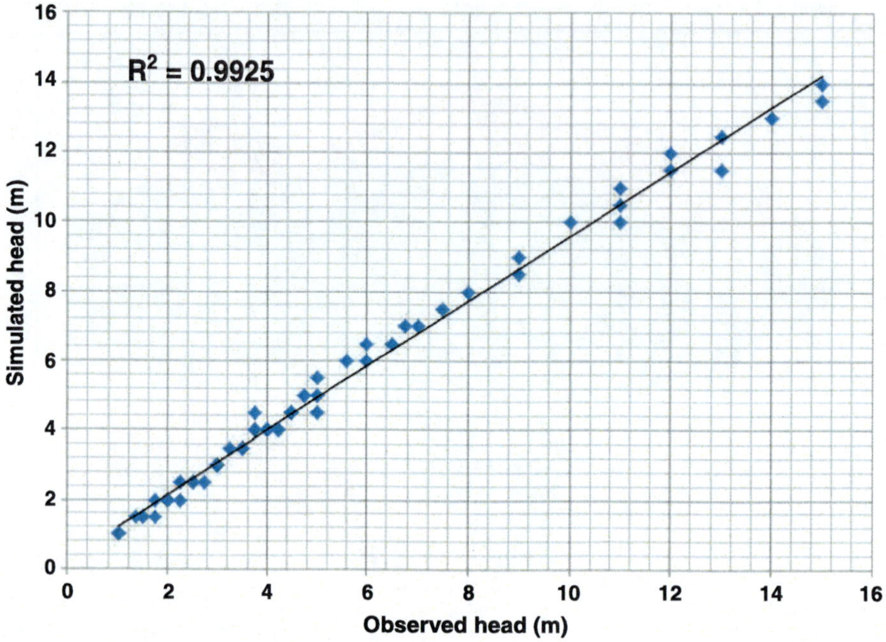

Fig. 18 The relation between the observed and the simulated groundwater level for the year 2008

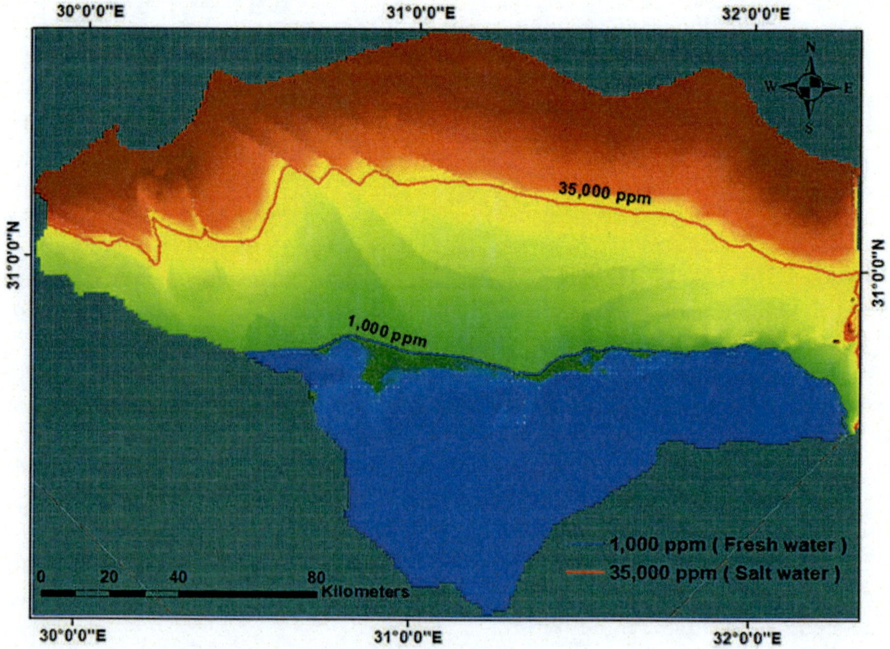

Fig. 19 Horizontal TDS distribution at layer (11) of the NDA

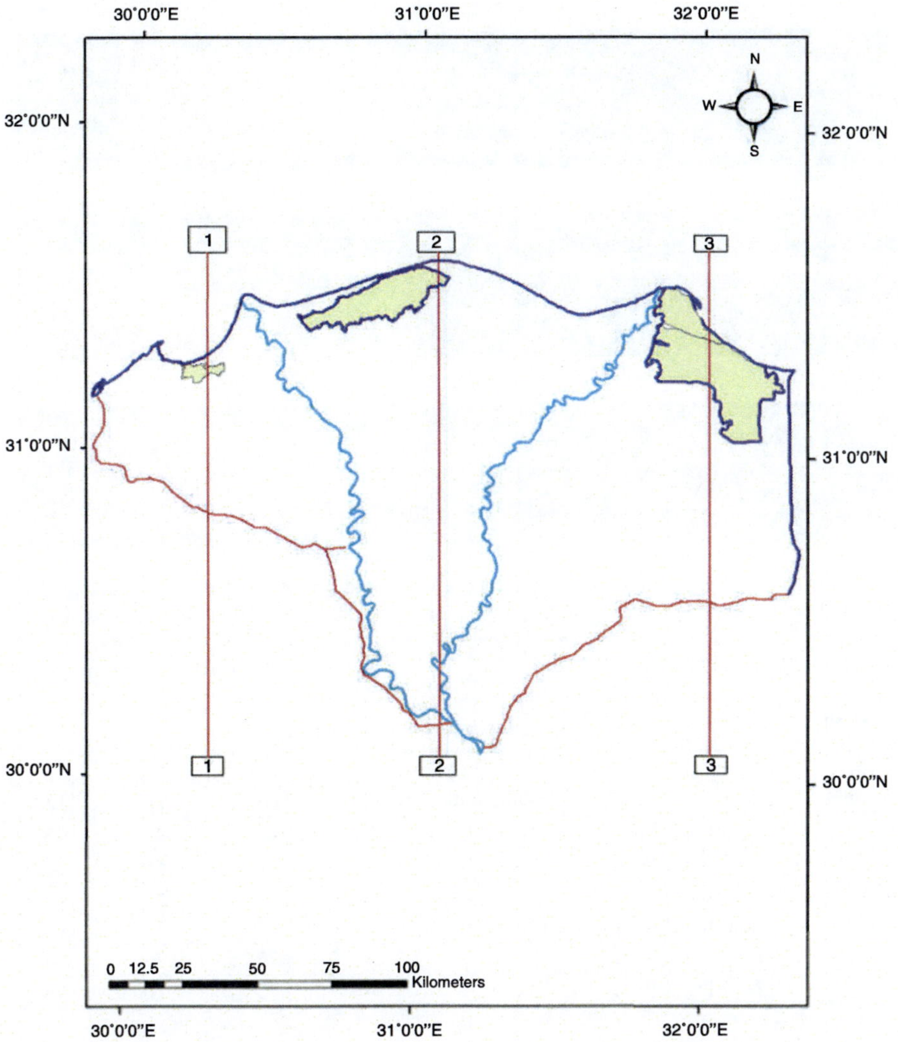

Fig. 20 Nile Delta aquifer's typical cross sections

10.5.3 Heads of Flow and Direction

The output results of the GW head calculation are illustrated in Fig. 22, from 17 m in the south to zero meters in the north; it shows how the GW level varies. The fundamental idea is that GW flows from high to low water heads. Figure 23 represents the GW flow direction and the velocity in the aquifer of ND from Cairo at head 16.00 m to the coastal (shoreline) at zero ones. At the clay cap layer, the minimum velocity is 0.0007 m/day. The NDA's average velocity equals 0.51 m/ day.

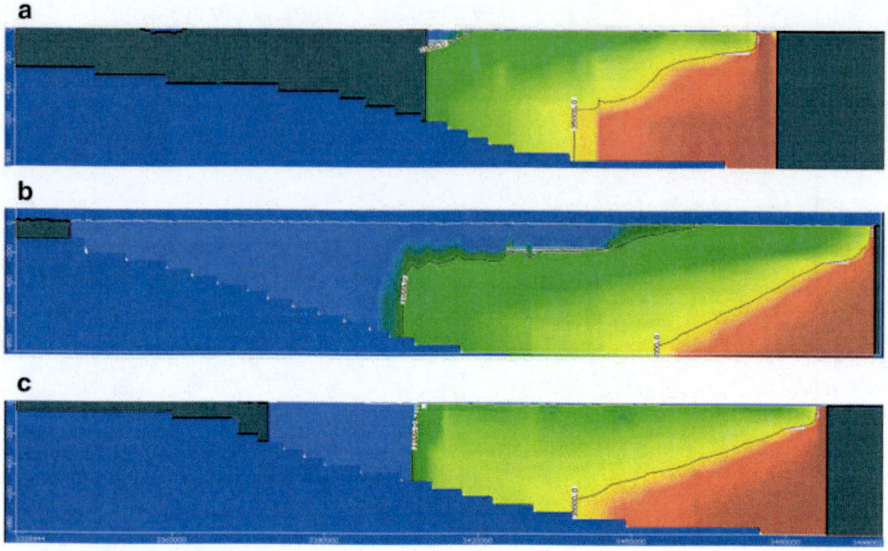

Fig. 21 (**a**) Nile Delta's aquifer's vertical salt concentration distribution for the basic case at cross sections 1 (west). (**b**) Nile Delta's aquifer's vertical salt concentration distribution for the basic case at cross section 2 (Middle). (**c**) Nile Delta's aquifer's vertical salt concentration distribution for the basic case at cross sections 3 (east)

Table 12 Comparison between recent studies and simulated studies

Study						
Position	Line	This study	[14]	[19]	[17]	[23]
West	Equi-line 35 (km)	40	37	48	40	23
	Equi-line 1 (km)	68.5	66	72.5	66	53
Middle	Equi-line 35 (km)	43	43	63.75	56	43.5
	Equi-line 1 (km)	93	75	93.75	105	68
East	Equi-line 35 (km)	50	50	76.25	75	33.5
	Equi-line 1 (km)	81	86	90.75	100	78

11 Advance of Shoreline Due to Climate Change

For simulating the behavior of the saltwater/freshwater interface, numerical modeling techniques have been examined. Consequently, GW potential laterally and upward movements and interface dynamics in the ND region under the impact of anticipated SLR by 0.25 m and 0.5 m rise. Due to the lack of data about the topography of the study area, it needs high accuracy of the Digital Elevation Model (DEM). So, the location of the new shoreline due to 0.25 m and 0.5 m of sea level rise is obtained based on approximation, dependent on the previous studies. For logic results the most recent studies are used. Armanuos et al. [69] used (MODFLOW + SEAWAT) and determined that Equi-concentration line

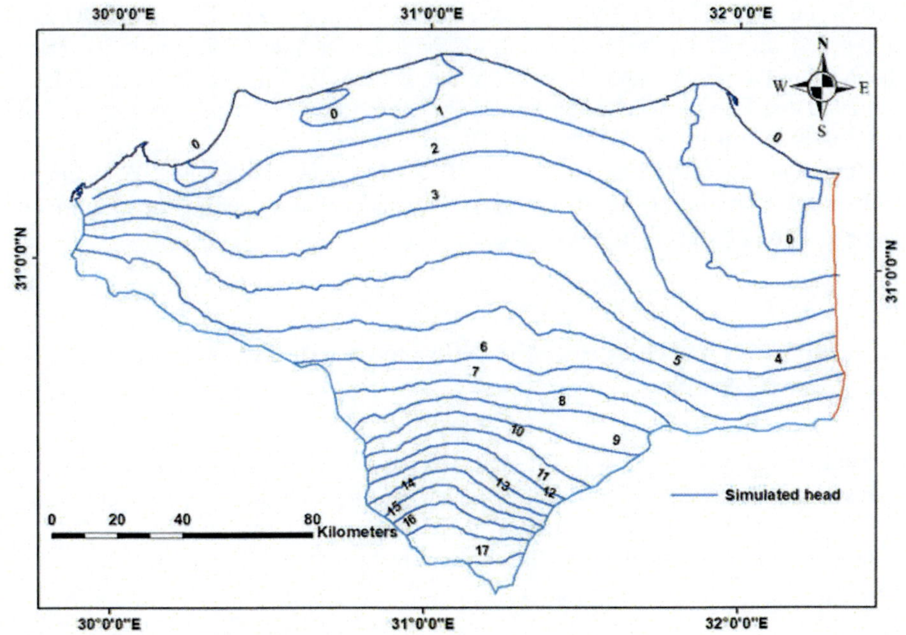

Fig. 22 Aerial view of the NDA's calculated groundwater level

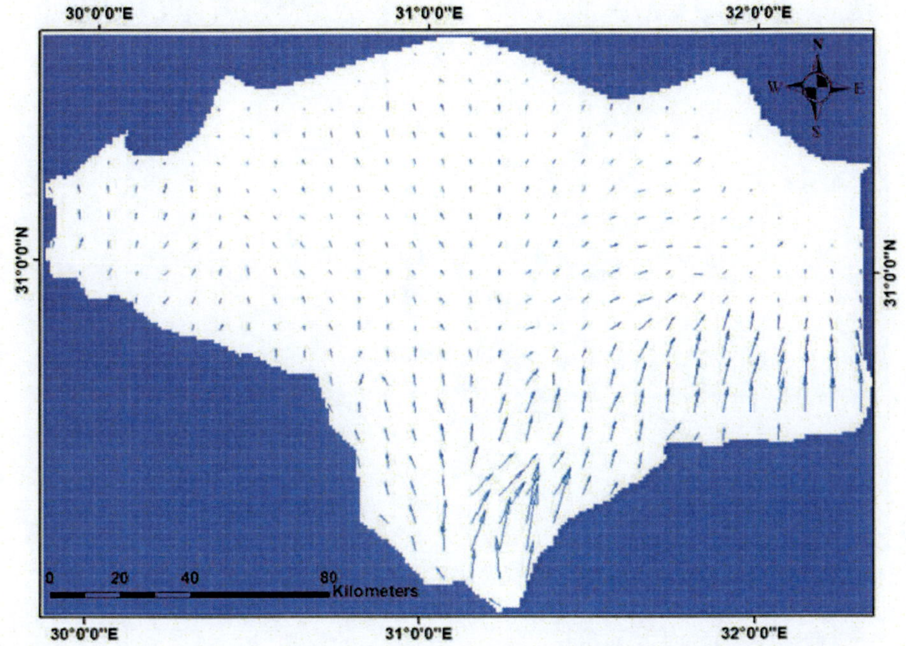

Fig. 23 Aerial view of contour velocity and the groundwater in the NDA's direction

35,000 ppm moved into 44.19 km in the east, 58.35 km in the middle, and 79.11 km in the west. Equi-concentration line 1,000 ppm moved into 67.90 km in the east, 106.9 km in the middle, and 105.5 km in the west for 0.25 m rise of sea level. It is estimated that Equi-concentration line 35,000 ppm intruded inland 44.51 km in the east, 59.09 km in the middle, and 81.10 km in the west. Equi-concentration line 1,000 ppm moved into 68 km in the east, 107.4 km in the middle, and 105.56 km in the west for 0.5 m rise of sea level. This data was analyzed for getting the location of the new shoreline as shown in Fig. 24a, b, c.

12 Saltwater Intrusion in the Nile Delta Aquifer (the Base Case Scenario)

Distributions of SWI from four different perspectives of the aquifer of ND are the primary outputs of the transport model. The first view of the TDS distribution at layer 3, as shown in Fig. 25a, with an average aquifer thickness varying from 225 m in the north to 50 m in the south. The second perspective at layer 6, which has an

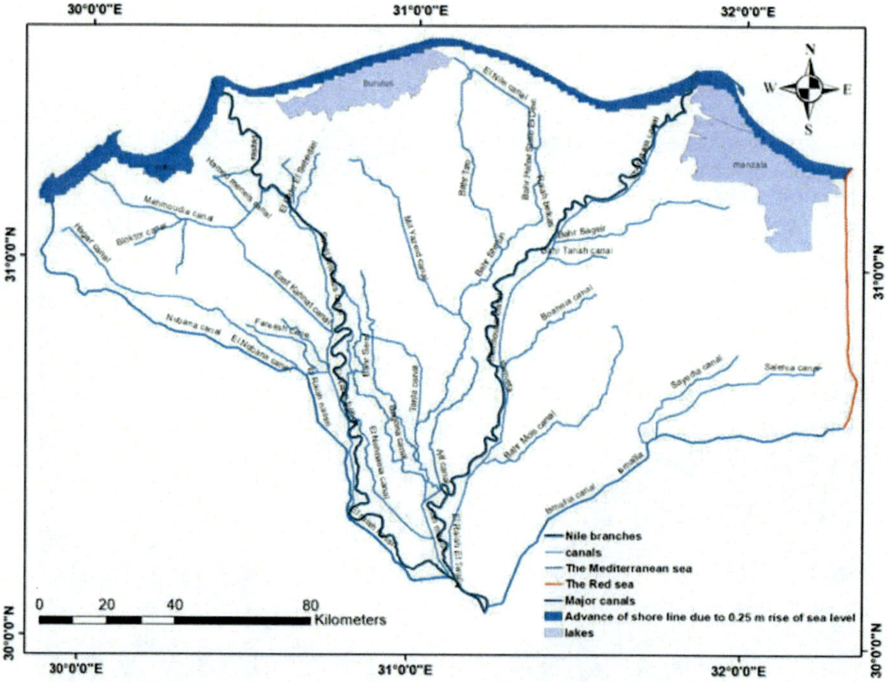

Fig. 24 (**a**) Affected areas due to 0.25 m SLR. (**b**) Affected areas due to 0.5 m SLR. (**c**) Location of shoreline due to 0.25 m and 0.5 m SLR

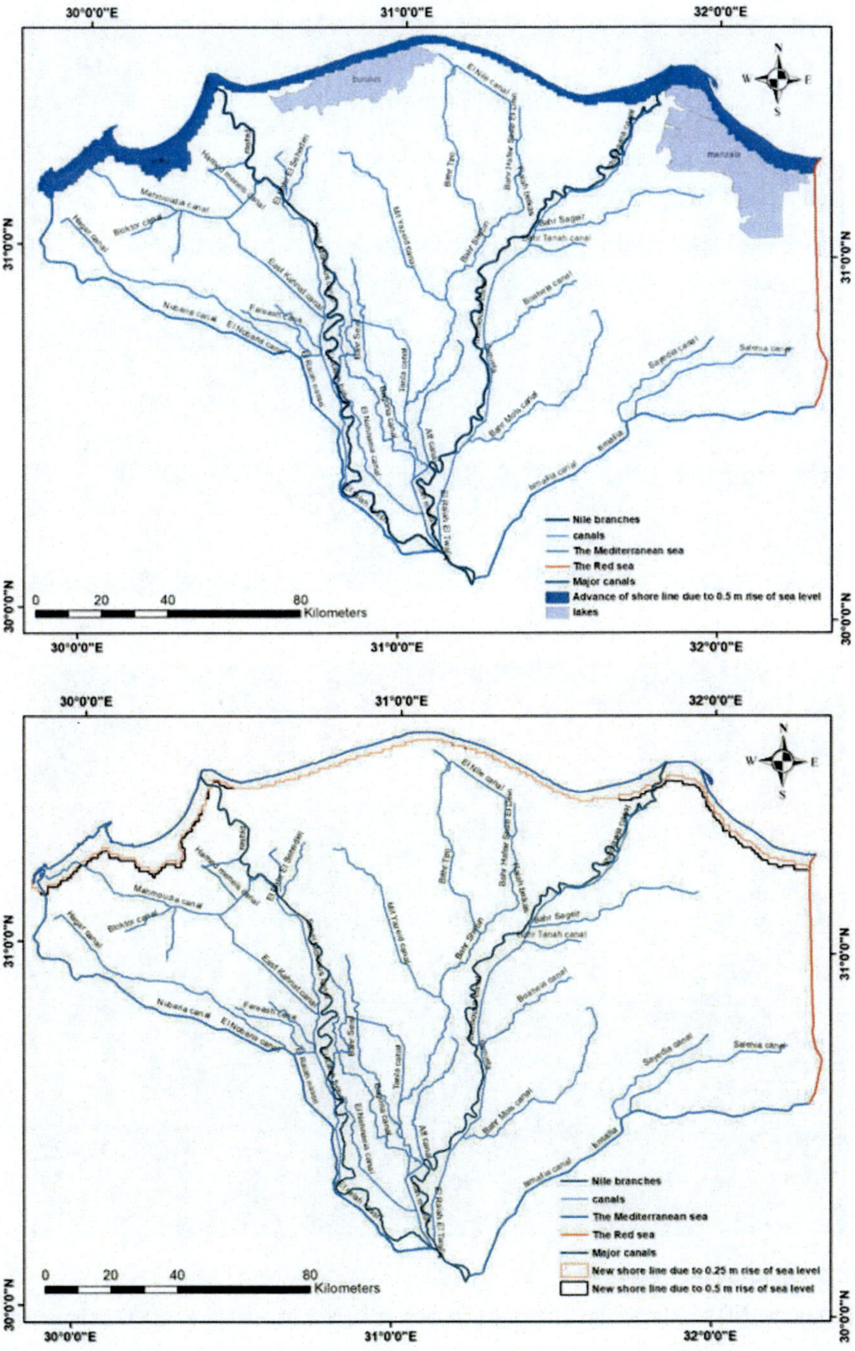

Fig. 24 (continued)

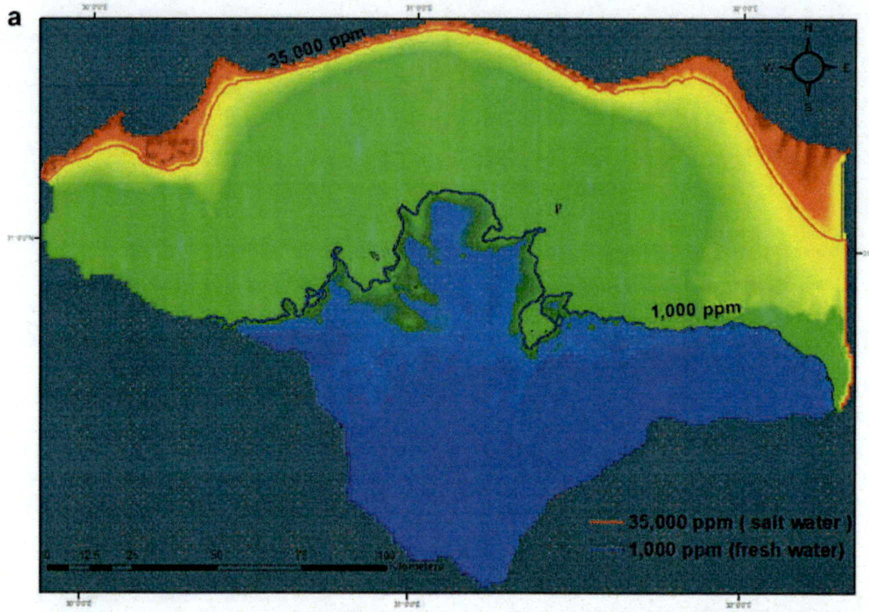

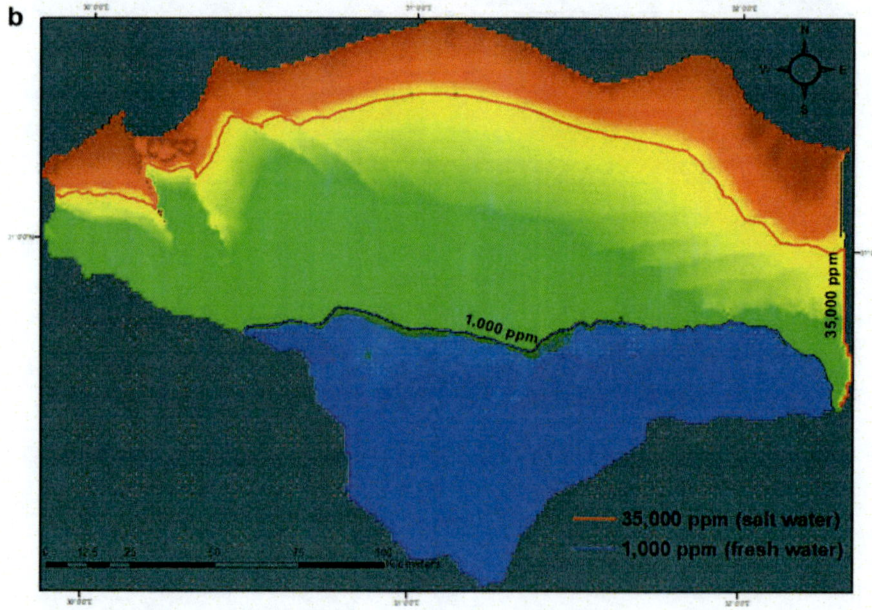

Fig. 25 (**a**) TDS horizontal distribution at layer 3 of the NDA. (**b**) TDS horizontal distribution at layer 6 of the NDA. (**c**) TDS horizontal distribution at layer 9 of the NDA. (**d**) TDS horizontal distribution at layer 11 of the NDA

c

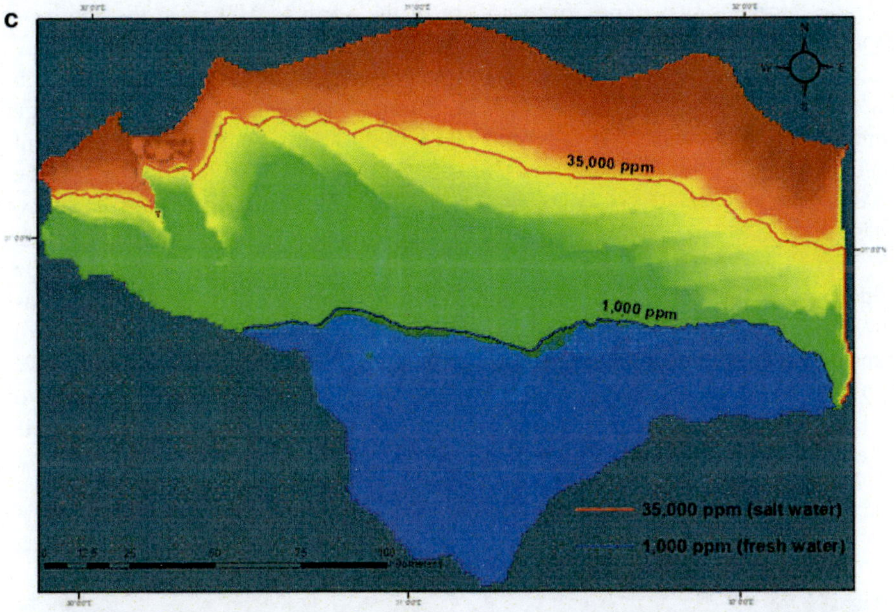

d

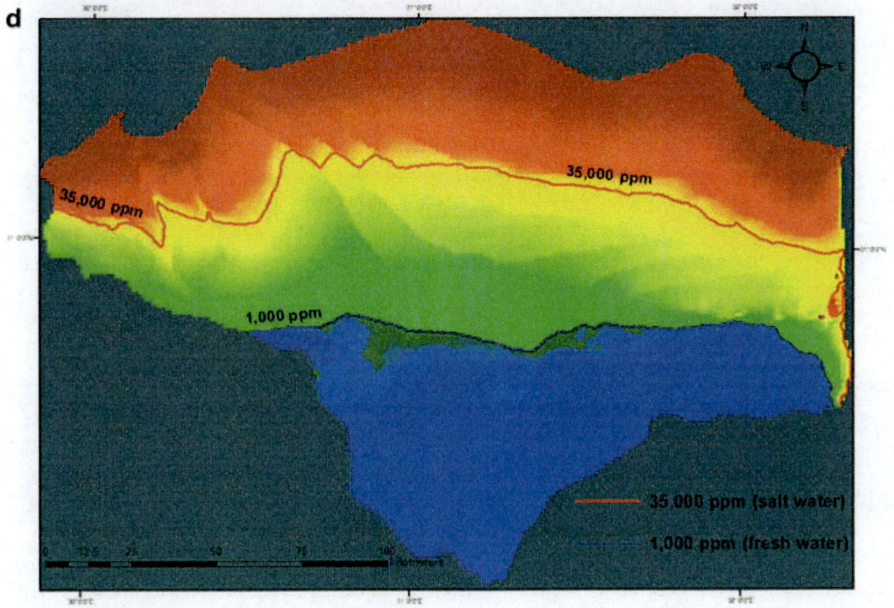

Fig. 25 (continued)

average thickness that varies from 450 m in the north to 100 m in the south, is depicted in Fig. 25b. The third perspective at layer 9 is also shown, with an average aquifer thickness of 150 m in the south and 675 m in the north, as depicted in Fig. 25c, d represents the final view of the aquifer's bottom, which has an average depth of 900 m in the north (at sea) and 200 m in the south. For the aquifer's current state (base case), three cross sections were taken as presented in Fig. 20. The SW line 35,000 ppm moved 40 km inland in accordance with section 1 toward the west, whereas the FW line 1,000 ppm moved 68.5 km inland from the coastline. Thus, as shown in Fig. 26a, the mixing zone between the FW and SW is 28.5 km. In the middle region, at section 2, the SW line 35,000 ppm moved 43 km inland, whereas the FW line 1,000 ppm moved 93 km inside from the seashore line, creating a 50 km transition zone, as shown in Fig. 26b. It is obvious that at section 3 in the east the SW line 35,000 ppm shifted into a distance of 50 km, the FW line 1,000 ppm shifted into 81 km from the coastline so the length of the transition zone between FW and SW is 31 km as depicted in Fig. 26c.

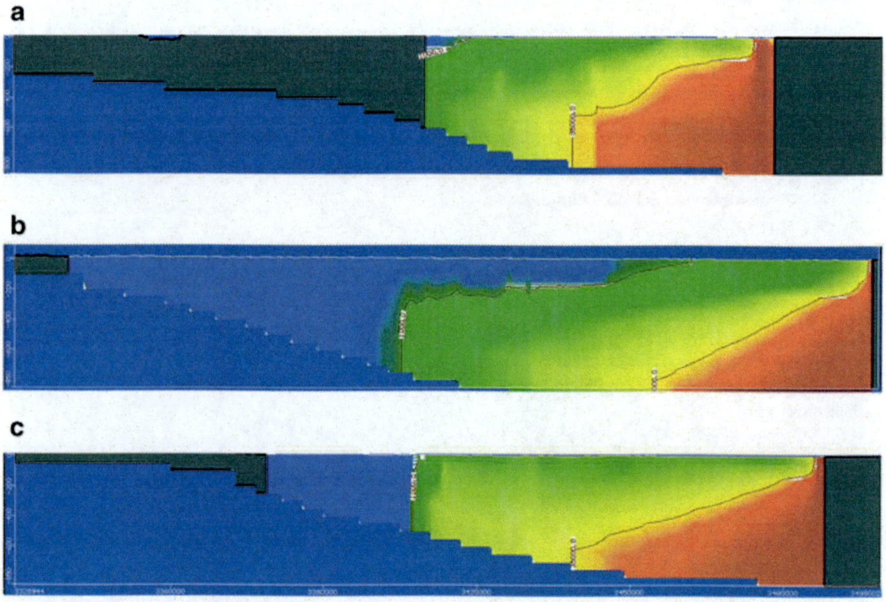

Fig. 26 (**a**) Vertical salt concentration distribution of the aquifer of ND for basic case at cross section 1(west). (**b**) Vertical salt concentration distribution of the aquifer of ND for basic case at cross section 2 (middle). (**c**) Vertical salt concentration distribution of the aquifer of ND for basic case at cross sections 3 (east)

13 Creep of Saltwater Interface in the Nile Delta Aquifer

An analysis is conducted on the 60-year history of SW's migration into the NDA. The current study's FW line of 1,000 ppm was compared to the Equi-concentration lines of 1,000 ppm for the years 1960, 1980, and 1992, derived from Morsy (2009), as illustrated in Fig. 27. In the research area, three parts were collected, as Fig. 20 shows. The first is in the west, when 1,000 ppm of the FW line moved inland at intervals of 55, 59, 44, and 68.5 km from the coastline. The second, central part demonstrated how the 1,000 ppm FW line moved inland over 72.5, 70, 65, and 93 km. The FW line 1,000 ppm moved inward with a distance of 83.5 km, 90 km, 93 km, and 81 km for the years 1960, 1980, 1992, and 2022 (current study), correspondingly, according to the third part, which is located in the east. While the creep of SW decreases in the east, it increases in the west and central of the ND.

14 Impact of Climate Change on Seawater Intrusion in the Nile Delta Aquifer

Several climate change models predicted that SLR would increase, last of them is adopted by the IPCC in 2014 is known as Representative Concentration Pathway (RCP) is a greenhouse gas concentration (not emissions) trajectory. RCP predicted

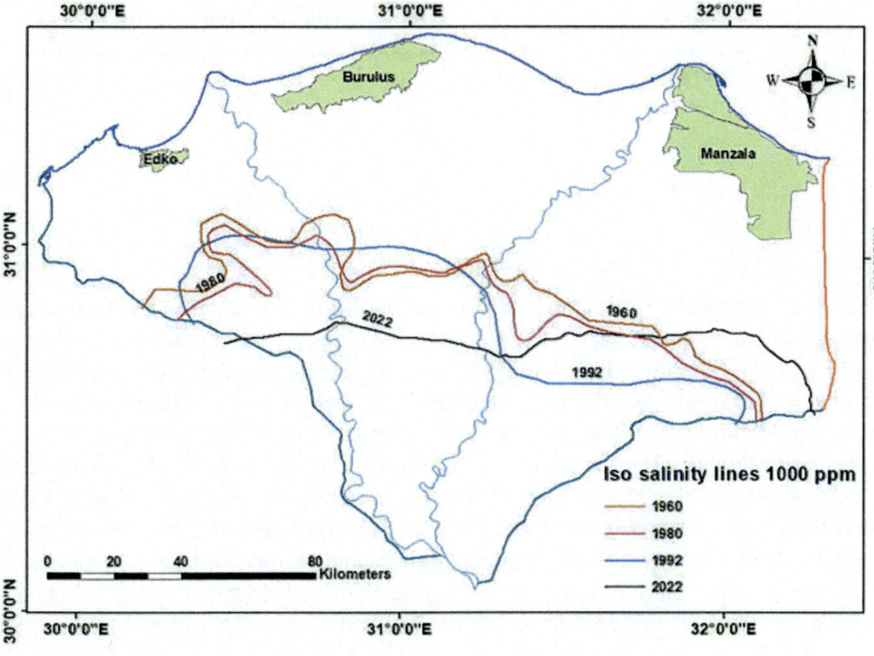

Fig. 27 Creep of saltwater in 1960, 1980, 1992, and 2022 (current study)

Table 13 A combination of SLR and changing the abstraction rate

Scenario	State
The first scenario	0.25 m rise of sea level and the abstraction rate maintained the same of base case
The second scenario	0.25 m rise of sea level and decreasing the abstraction rate to half
The third scenario	0.25 m rise of sea level and doubling the abstraction rate
The fourth scenario	0. 5 m rise of sea level and the abstraction rate maintained the same of base case
The fifth scenario	0. 5 m rise of sea level and decreasing the abstraction rate to half
The sixth scenario	0. 5 m rise of sea level and doubling the abstraction rate

the mean SLR of 63 cm by the year 2100. Also, some areas may get less rainfall than forecast, which will drop the amount of water in the irrigation and drainage network. The demand on GW as an alternate supply will increase as the population density rises. Climate change will have an impact on these parameters; hence this published report focuses on SLR and changed well extraction. Numerous studies are needed to determine the effects of certain variables, including SLR and well pumping rate, on the intrusion of SW in the NDA. These studies are essential for protecting the aquifer's GW from this risk, which could damage the water's quality. Earlier studies predicted that rising temperatures would result in higher SLR, whereas rising human density would result in higher GW extraction. The current study simulates a combination of SLR and abstraction rate from groundwater in the following six different scenarios as illustrated in Table 13. The Equi-concentration lines 1,000 ppm and 35,000 ppm of each scenario are compared to the Equi-concentration lines 1,000 ppm and 35,000 ppm of the base case.

14.1 Saltwater Intrusion Distribution for Scenario No. 1 (0.25 m SLR + the Same Abstraction)

In the first scenario, the discharge rate is assumed to be the same as the base case (Fig. 28), with a 0.25 m SLR. Compared to the SW line 35,000 ppm in the base case, the FW line 35,000 ppm moved into by 4.3 km. As seen in Fig. 29a, the freshwater's FW line at 1,000 ppm progressed on land at 3 km greater than the base case's FW line at 1,000 ppm at section 1, which is situated west of Delta. The SW line 35,000 ppm moved inland at section 2, in the central of the NDA, by 2.944 km more than the SW line 35,000 ppm of the base scenario. In comparison to the FW line 1,000 ppm of the basic scenario, Fig. 29b, the 1,000 ppm FW line moved into aquifer by 3 km. The SW line 35,000 ppm advanced at section 3, which is in the east, 4.3 km farther than the FW line 35,000 ppm of the base scenario. The FW line at 1,000 ppm moved 5.1 km further inland than the base case's FW line at 1,000 ppm,

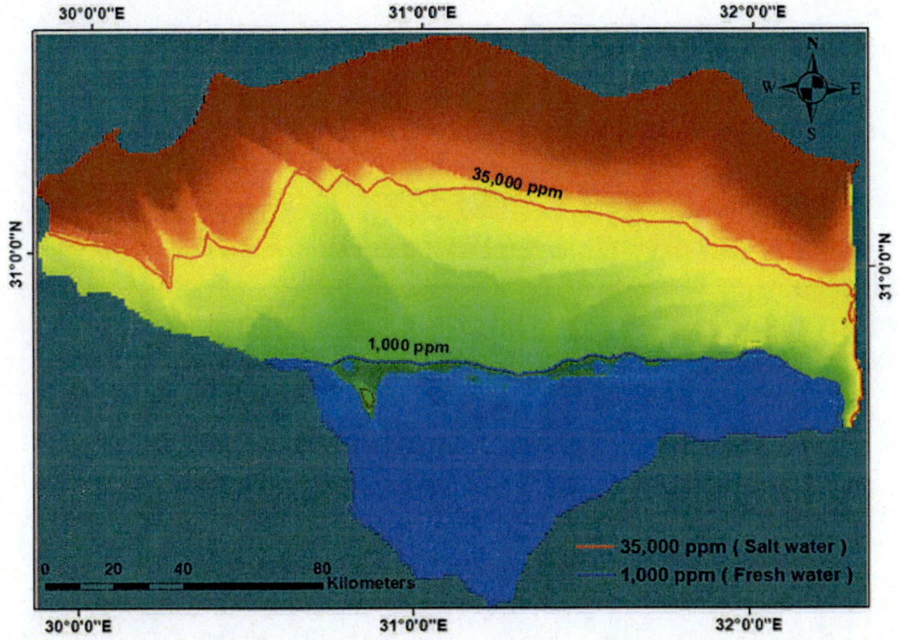

Fig. 28 TDS areal distribution at 25 cm SLR

as seen in Fig. 29c. The advancement of the SW line at 35,000 ppm and the FW line at 1,000 ppm from the base example is shown in Table 14.

14.2 Saltwater Intrusion Distribution for Scenario No. 2 (0.25 m SLR + Half Abstraction)

The second scenario considers 0.25 m SLR and half abstraction of the base case (Fig. 30). The SW line 35,000 ppm of the SW lagged toward the shoreline by 14.16 km less than the SW line 35,000 ppm of the base case. The FW line 1,000 ppm of the FW lagged toward the sea by 4.066 km less than the FW line 1,000 ppm of the base case at section 1, located west of Delta, Fig. 31a. At section 2 which is located in the middle of NDA, the SW line 35,000 ppm lagged toward the coast by 0.8 km less than the SW line 35,000 ppm of the base case. The FW line 1,000 ppm lagged toward the coastline by 15.842 km less than the FW line 1,000 ppm of the base case, Fig. 31b. At section 3 which is in the east, the SW line 35,000 ppm lagged to the shoreline by 8.832 km less than the SW line 35,000 ppm of the base case. The FW line 1,000 ppm lagged toward the sea by 1.122 km less than the FW line 1,000 ppm of the base case, Fig. 31c. Table 15 shows the retreat of the SW line 35,000 ppm and the FW line 1,000 ppm less than the base case.

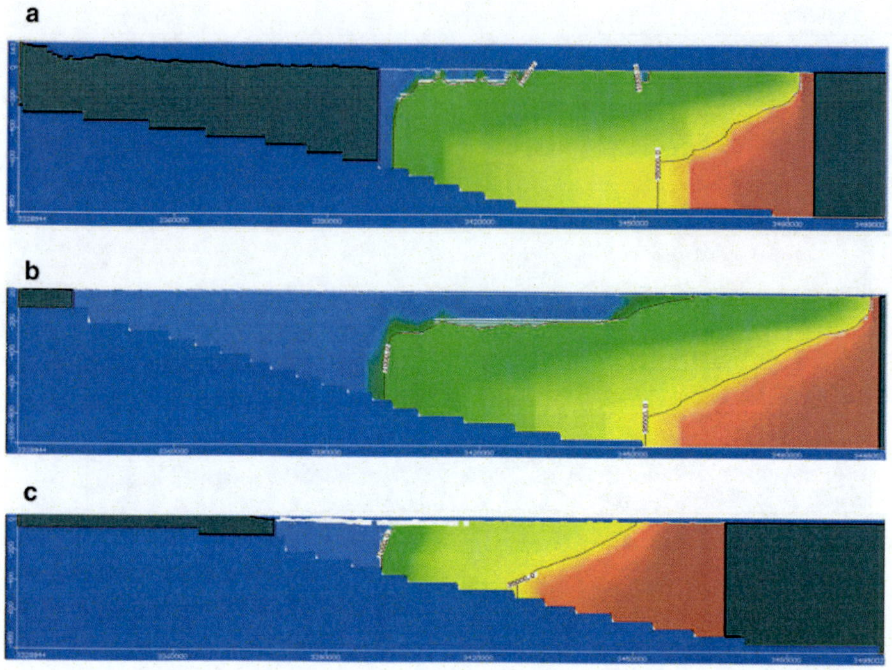

Fig. 29 (**a**) Vertical TDS distribution in the NDA at SLR by 25, in the west. (**b**) Vertical TDS distribution in the NDA at SLR by 25, in the middle. (**c**) Vertical TDS distribution in the NDA at SLR by 25, in the east

Table 14 The advancement SW line 35,000 ppm and FW line 1,000 ppm more than the base case for scenario 1

Location	Advance of line 35,000 ppm from the base case	Advance of line 1,000 ppm from the base case
West	+4.3	+3
Middle	+2.944	+3
East	+4.3	+5.1

14.3 Saltwater Intrusion Distribution for Scenario No. 3 (0.25 m SLR + Double Abstraction)

In the third scenario, the discharge rate is doubled from the base case value (32), and a 0.25 m SLR is considered (Fig. 32). As shown in Fig. 33a, the FW line 1,000 ppm of the freshwater developed in land 18.786 km greater than the FW line 1,000 ppm of the base case at section 1, situated west of ND. The SW line 35,000 ppm of the SW advanced inside by 46.685 km greater than the SW line 35,000 ppm of the base case. Section 2, which is in the central of the NDA, shows that the SW line 35,000 ppm moved inside by 57.761 km more than the base case's SW line 35,000 ppm, and the

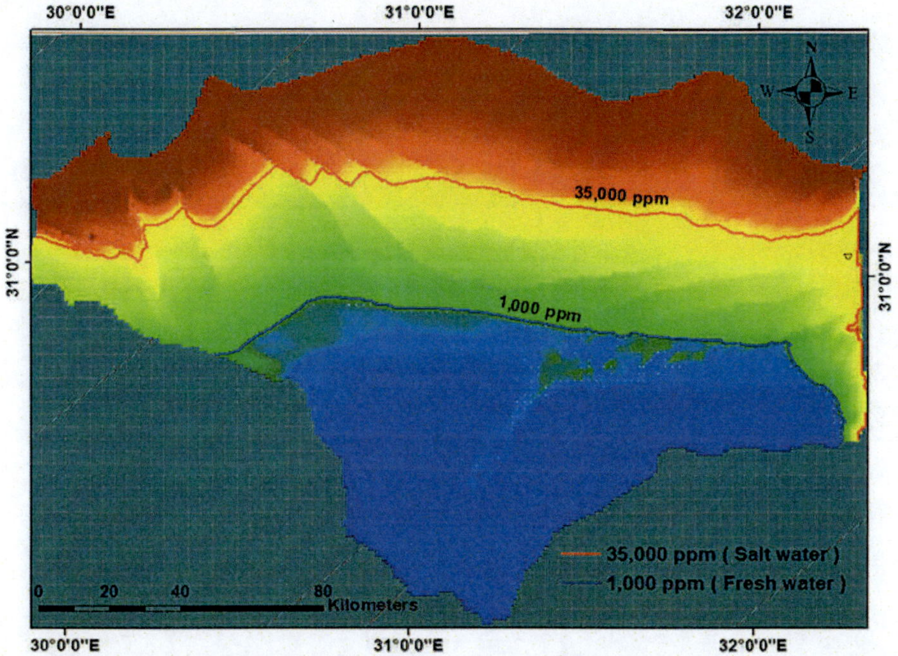

Fig. 30 TDS areal distribution at (25 cm SLR and half abstraction)

FW line 1,000 ppm migrated inland by 52.293 km greater than the base case's FW line 1,000 ppm (Fig. 33b). The SW line 35,000 ppm advanced 38.975 km further at section 3, positioned in the east, then the SW line 35,000 ppm of the base case. Similarly, the FW line 1,000 ppm moved 22.712 km further into the interior than the FW line 1,000 ppm of the base case, as depicted in Fig. 33c. Table 16 illustrates the progression of SW line 35,000 ppm and FW line 1,000 ppm for scenario no. 3 from the base case.

A comparison between base case and first three scenarios is illustrated in Fig. 34 to reveal the advancement and lag of FW line 1,000 ppm.

14.4 Saltwater Intrusion Distribution for Scenario No. 4 (0.5 m SLR + the Same Abstraction)

As illustrated in Fig. 35, the fourth scenario assumes a 0.5 m SLR and the same discharge rate as the base case. As illustrated in Fig. 36a, the FW line 1,000 ppm of the freshwater moved in land by 3 km more than the FW line 1,000 ppm of the base case at section 1, located in the west of the ND. The SW line 35,000 ppm developed inside by 5.3 km greater than the SW line 35,000 ppm of the base case. Section 2, in

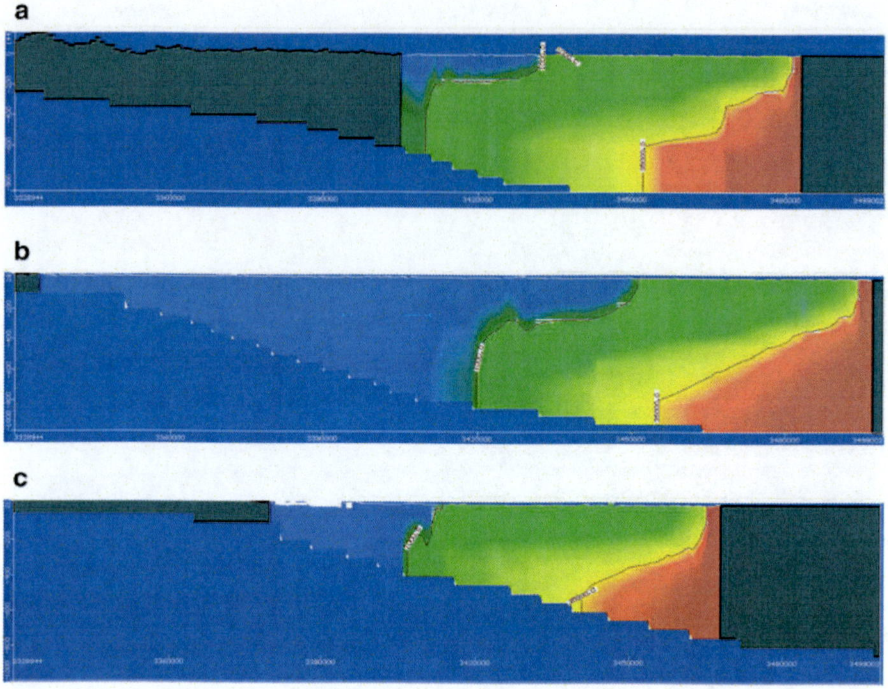

Fig. 31 (**a**) Vertical TDS distribution in the NDA at (SLR by 25 and half abstraction), in the west. (**b**) Vertical TDS distribution in the NDA at (SLR by 25 and half abstraction), in the middle. (**c**) Vertical TDS distribution in the NDA at (SLR by 25 and half abstraction), in the east

Table 15 Retreat of SW line 35,000 ppm and FW line 1,000 ppm from the base case for scenario 2

Location	Location of line 35,000 ppm from the base case	Location of line 1,000 ppm from the base case
West	−14.16 km	−4.066 km
Middle	−0.8 km	−15.842 km
East	−8.832 km	−1.112 km

the central of the NDA, had an SW line 35,000 ppm that moved inland 3 km further than the base case's SW line 35,000 ppm, and an FW line 1,000 ppm that moved inland 4 km further than the base case's FW line 1,000 ppm, as can be seen in Fig. 36b. The SW line 35,000 ppm in section 3, which is east of the NDA, advanced 6.2 km farther than the base case's SW line 35,000 ppm. Compared to the FW line 1,000 ppm of the basic case, the 1,000 ppm FW line advanced inland by 5.4 km, as presented in Fig. 36c. Table 17 illustrates the progression of SW line 35,000 ppm and FW line 1,000 ppm of scenario no.4 compared with the base case.

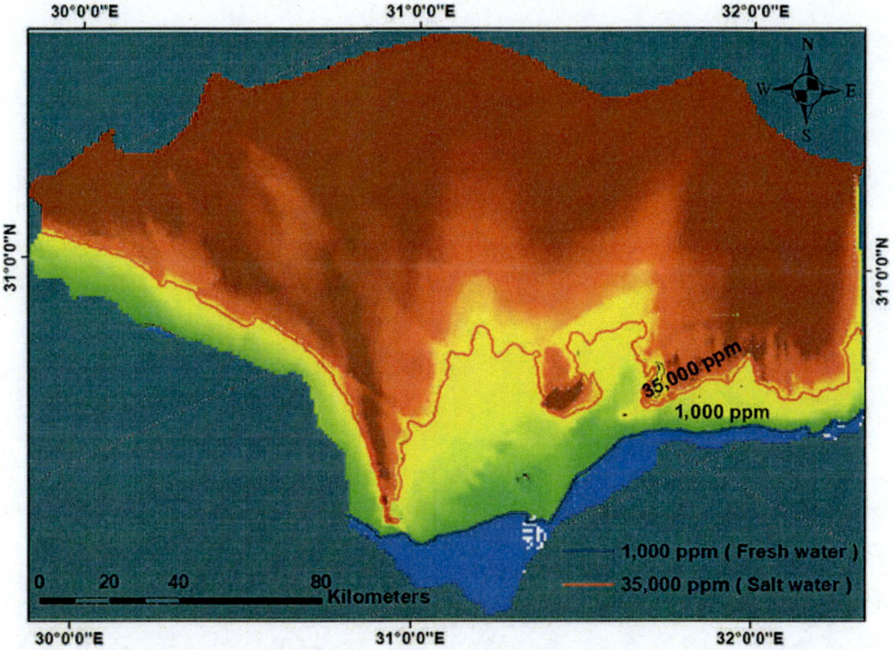

Fig. 32 TDS areal distribution at (25 cm SLR and double abstraction)

14.5 Saltwater Intrusion Distribution for Scenario No. 5 (0.5 m SLR + Half Abstraction)

The fifth scenario considers 0.5 m SLR and half abstraction of the base case, as shown in Fig. 37. The SW line 35,000 ppm of the SW lagged toward the shore line by a distance of 13.88 km less than the SW line 35,000 ppm of the base case, the FW line 1,000 ppm of the FW lagged toward the sea by a distance of 3.365 km less than the FW line 1,000 ppm of the base case at sect.1, located in the west of the NDA, as presented in Fig. 38a. At section 2 which is located in the middle of NDA, the SW line 35,000 ppm lagged toward the coast by 0.4 km less than the SW line 35,000 ppm of the base case, the FW line 1,000 ppm lagged toward the coastline by 3.365 km less than the FW line 1,000 ppm of the base case, as shown in Fig. 38b. At section 3 which is located in the east of the NDA, the SW line 35,000 ppm lagged to the shoreline by 7.851 km less than the SW line 35,000 ppm of the base case. The FW line 1,000 ppm lagged toward the sea by 1.121 km less than the FW line 1,000 ppm of the base case, as shown in Fig. 38c. Table 18 shows the retreat of SW line 35,000 ppm and FW line 1,000 ppm less than the base case.

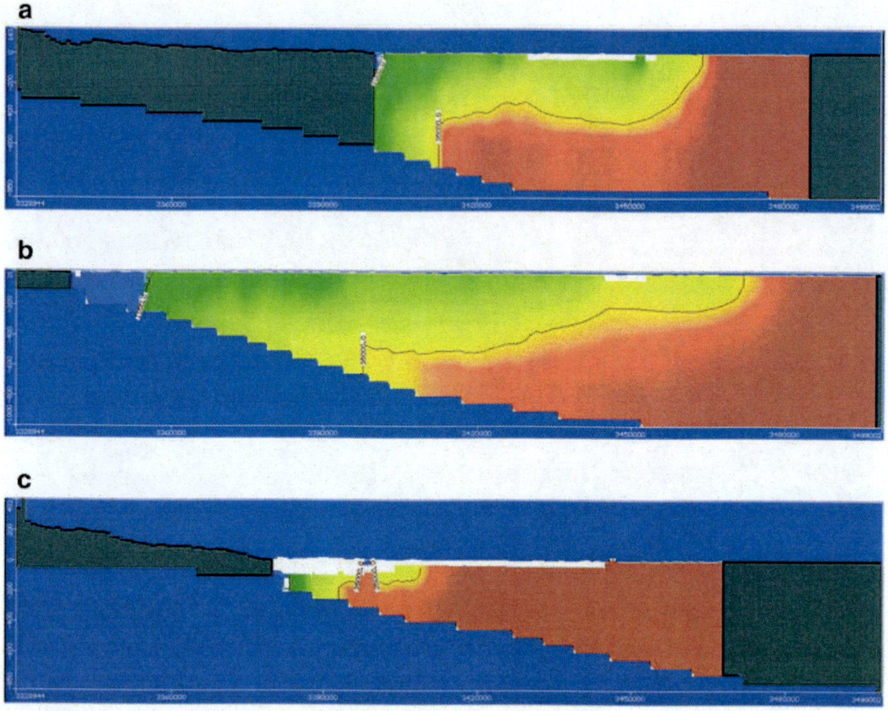

Fig. 33 (**a**) Vertical TDS distribution in the NDA at (SLR by 25 and double abstraction), in the west. (**b**) Vertical TDS distribution in the NDA at (SLR by 25 and double abstraction), in the middle. (**c**) Vertical TDS distribution in the NDA at (SLR by 25 and double abstraction), in the east

Table 16 The advancement SW line 35,000 ppm and FW line 1,000 ppm from the base case for scenario 3

Location	Advance of line 35,000 ppm from the base case	Advance of line 1,000 ppm from the base case
West	+46.685 km	+18.786 km
Middle	+57.761 km	+52.293 km
East	+38.975 km	+22.712 km

14.6 Saltwater Intrusion Distribution for Scenario No. 6 (0.5 m SLR + Double Abstraction)

For the sixth scenario, which considers 0.5 m SLR, the discharge rate is doubled from the base case, as shown in Fig. 39. The SW line 35,000 ppm of the SW moved into NDA by a distance of 48.968 km more than the SW line 35,000 ppm of the base case, the FW line 1,000 ppm of the FW advanced in land by a distance of 20.048 km more than the FW line 1,000 ppm of the base case at sect.1, located in the west of the NDA, as shown in Fig. 40a. At section 2 which is located in the middle of NDA, the

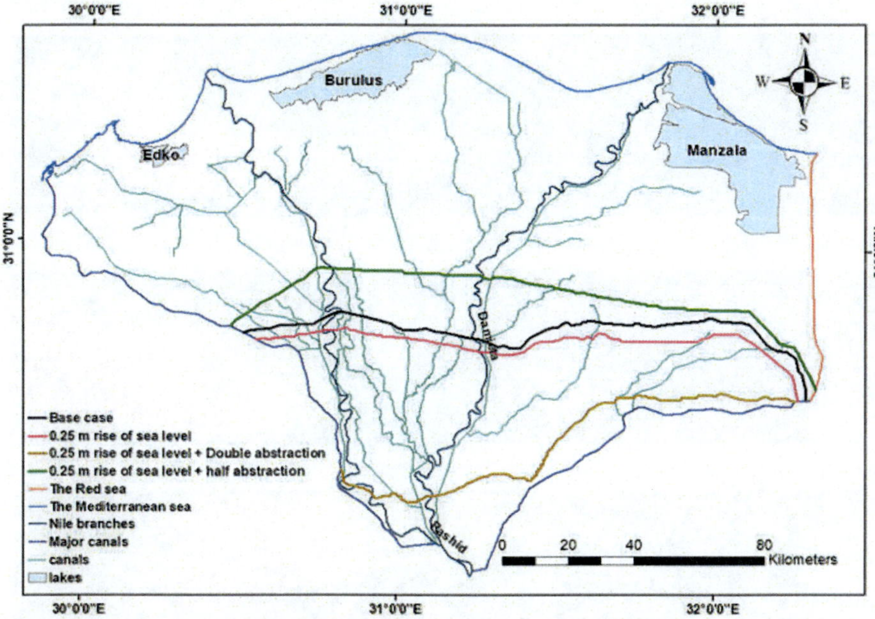

Fig. 34 Comparison between FW line 1,000 ppm of base case and first three scenarios

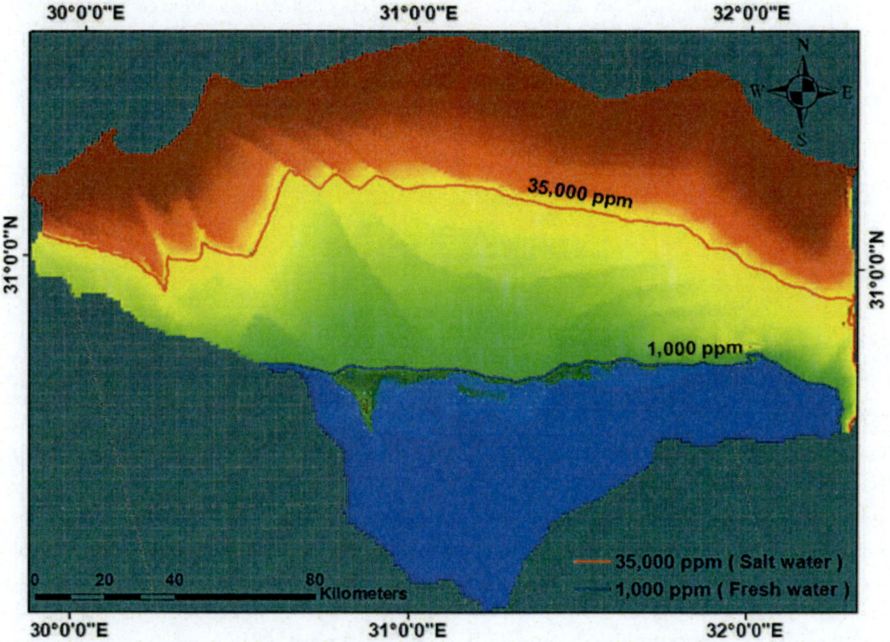

Fig. 35 TDS areal distribution at 50 cm SLR

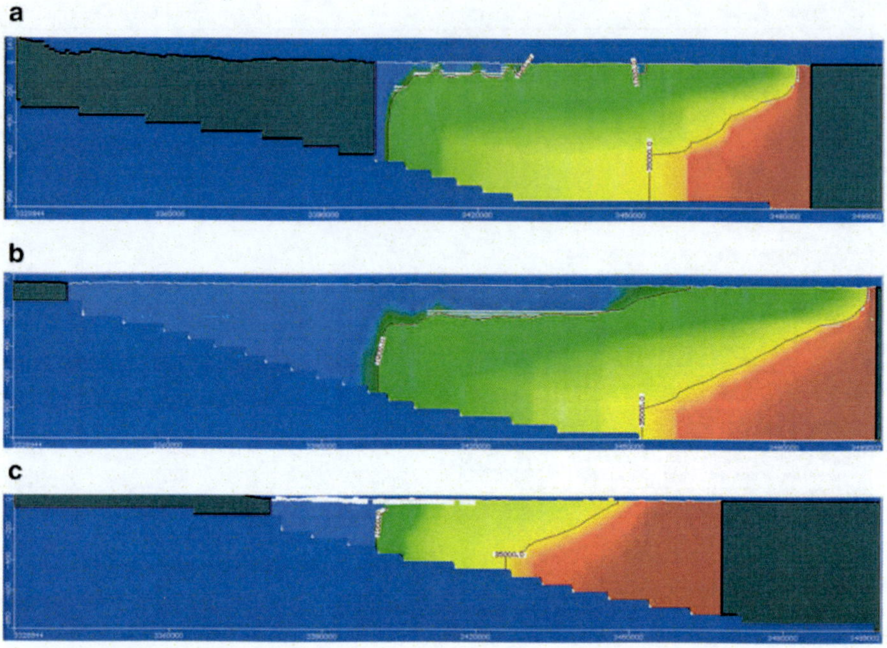

Fig. 36 (**a**) Vertical TDS distribution in the NDA at SLR by 50 cm more than the intrusion of the FW line 1,000, in the west. (**b**) Vertical TDS distribution in the NDA at SLR by 50 cm more than the intrusion of the FW line 1,000, in the middle. (**c**) Vertical TDS distribution in the NDA at SLR by 50 cm more than the intrusion of the FW line 1,000, in the east

Table 17 The advancement SW line 35,000 ppm and FW line 1,000 ppm from the base case for scenario 4

Location	Advance of line 35,000 ppm from the base case	Advance of line 1,000 ppm from the base case
West	+5.3 km	+3 km
Middle	+3 km	+4 km
East	+6.2 km	+5.4 km

SW line 35,000 ppm moved into the NDA by 57.77 km more than the SW line 35,000 ppm of the base case, the FW line 1,000 ppm moved into NDA by 53.013 km more than the FW line 1,000 ppm of the base case, as presented in Fig. 40b. At section 3 which is located in the east, the SW line 35,000 ppm progressed by 40.273 km more than the SW line 35,000 ppm of the base case. The FW line 1,000 ppm progressed inland by 22.8 km more than the FW line 1,000 ppm of the base case, as shown in Fig. 40c. Table 19 illustrates the progression of SW line 35,000 ppm and FW line 1,000 ppm of scenario no.6 compared with the base case.

A comparison between base case and last three scenarios is illustrated in Fig. 41 to reveal the advancement and lag of FW line 1,000 ppm.

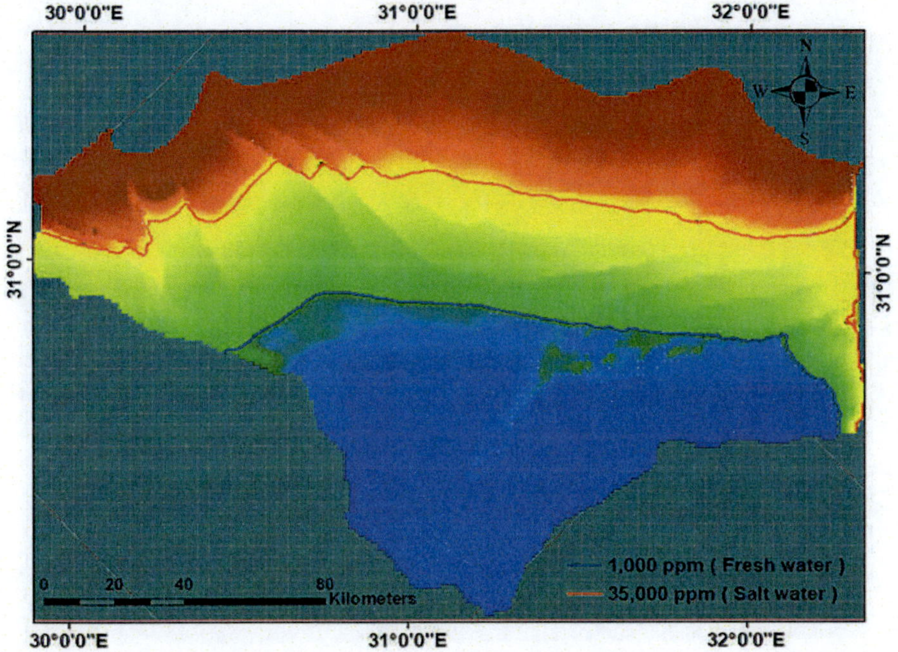

Fig. 37 TDS areal distribution at (50 cm SLR and half abstraction)

15 Conclusion

Egypt is one of the countries whose surface water resources from the Nile River will be affected by this scarcity, the construction of the Grand Ethiopian Renaissance Dam, and a drop in rainfall due to climate change. Additionally, the anticipated SLR and rising human density will impact GWL. So, the primary goals of this study are to employ MODFLOW to create a comprehensive three-dimensional GW model for the NDA to study the GW head, under various climate change scenarios, such as a combination of sea level rising and varying GW pumping. The SEAWAT code was applied to simulate the intrusion of SW in the NDA under different climate change scenarios. The primary conclusion and recommendation from the current study are presented in the following section.

The following conclusions can be presented for the studied scenarios based on the numerical investigation of the current study:

The base case: The SW line 35,000 ppm moved 40 km inland in the west of the NDA, whereas the FW line 1,000 ppm migrated 68.5 km inland from the coastline, according to results based on data from earlier research from 2008. In the middle, the 1,000 ppm FW line moved 93 km inland from the beach line, while the 35,000 ppm SW line introduced 43 km inland. The SW line 35 progressed 50 km inland in the east, whereas the FW line 1,000 ppm advanced 81 km interior from the coast.

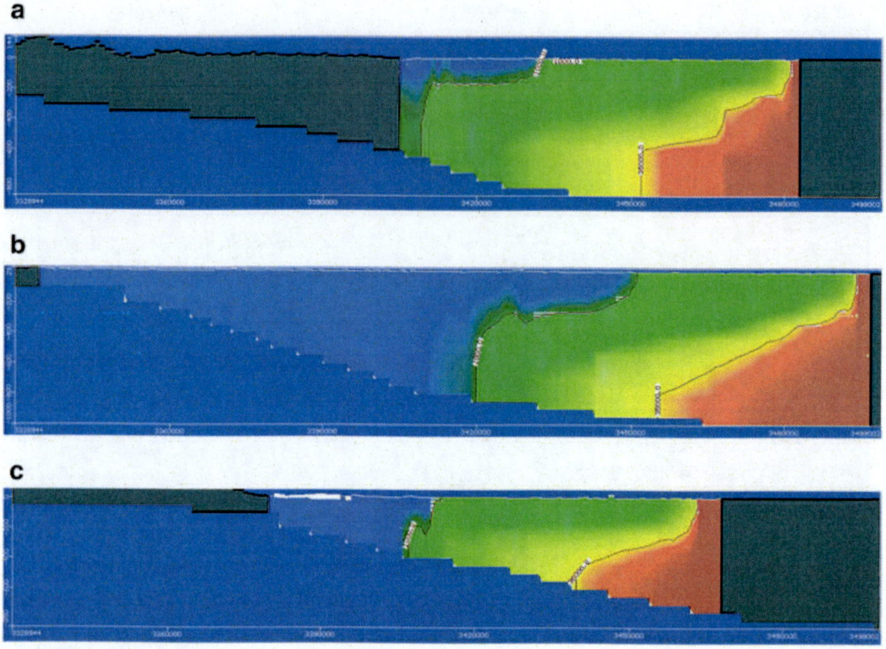

Fig. 38 (**a**) Vertical TDS distribution in the NDA at (SLR by 50 cm and half abstraction), in the west. (**b**) Vertical TDS distribution in the NDA at (SLR by 50 cm and half abstraction), in the middle. (**c**) Vertical TDS distribution in the NDA at (SLR by 50 cm and half abstraction), in the east

Table 18 Retreat of SW line 35,000 ppm and FW line 1,000 ppm from the base case for scenario 5

Location	Location of line 35,000 ppm from the base case	Location of line 1,000 ppm from the base case
West	−13.88 km	−3.365 km
Middle	−0.4 km	−15.842 km
East	−7.851 km	−1.121 km

Scenario 1: considers a 0.25 m SLR and uses the same discharge rate as the base case. The findings demonstrated that, in comparison to the base case, the SW line 35,000 ppm, which corresponds to the SW, progressed 4.3 km further inland, and the FW line 1,000 ppm, which represents the freshwater, advanced 3 km farther inland at the west of the ND. In comparison to the base scenario, the SW line 35,000 ppm migrated 2.944 km farther inland in the central of the NDA, while the FW line 1,000 ppm migrated 3 km further inland. In comparison to the base case, the SW line 35,000 ppm moved 4.3 km further inland in the east. In comparison to the default case, the 1,000 ppm FW line advanced 5.1 km farther inland.

Scenario 2: considers a 0.25 m SLR and half abstraction of the base case. The results showed that, in the west of NDA, the freshwater FW line, at 1,000 ppm, lagged regarding the sea by 4.066 km when compared to the base case, and the SW

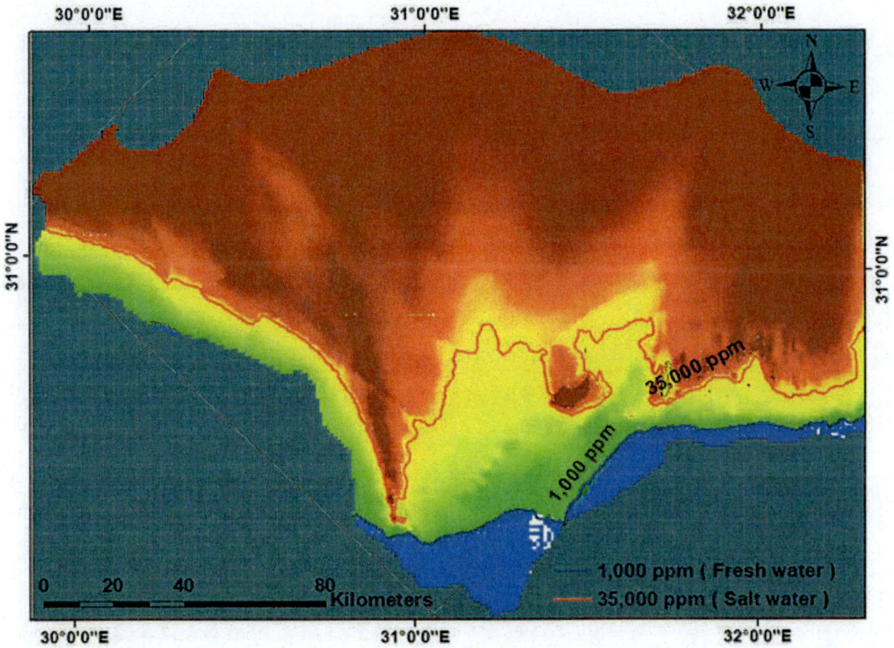

Fig. 39 TDS areal distribution at 50 cm SLR and double abstraction

line, at 35,000 ppm, represented the Southwest, which came in toward the shoreline by 14.16 km. Within the NDA, the SW line 35,000 ppm trailed 0.8 km in the direction of the coast when compared to the base scenario, while the FW line 1,000 ppm trailed 15.842 km in the same direction. Compared to the basic case, the SW line 35,000 ppm in the east was 8.832 km behind the coastline. In comparison to the base case, the FW line 1,000 ppm lagged 1.122 km toward the sea.

Scenario 3: considers a 0.25 m SLR and doubles the discharge rate compared to the base case. In comparison to the base case, the results showed that the SW line 35,000 ppm, which represents the SW, developed 46.685 km further inland, and the FW line 1,000 ppm, which reflects the freshwater, moved 18.786 km farther inland in the west of NDA. The SW line 35,000 ppm migrated 57.761 km further inland in the middle of the NDA than the base case, and the FW line 1,000 ppm migrated 52.293 km farther inland than the base scenario. Compared to the basic case, the SW line 35,000 ppm advanced 38.975 km further inland in the east. In comparison to the base case, the FW line 1,000 ppm advanced 22.712 km farther inland.

Scenario 4: considers a 0.5 m SLR and uses the same discharge rate as the base case. In contrast to the base case, the results showed that the SW line 35,000 ppm, which indicates the SW, advanced 5.3 km further inland in the west of NDA, while the FW line 1,000 ppm, which represents freshwater, developed 3 km farther inland. The SW line 35,000 ppm moved 3 km further inland in the middle portion of the NDA than the base case, while the FW line 1,000 ppm moved 4 km further inland

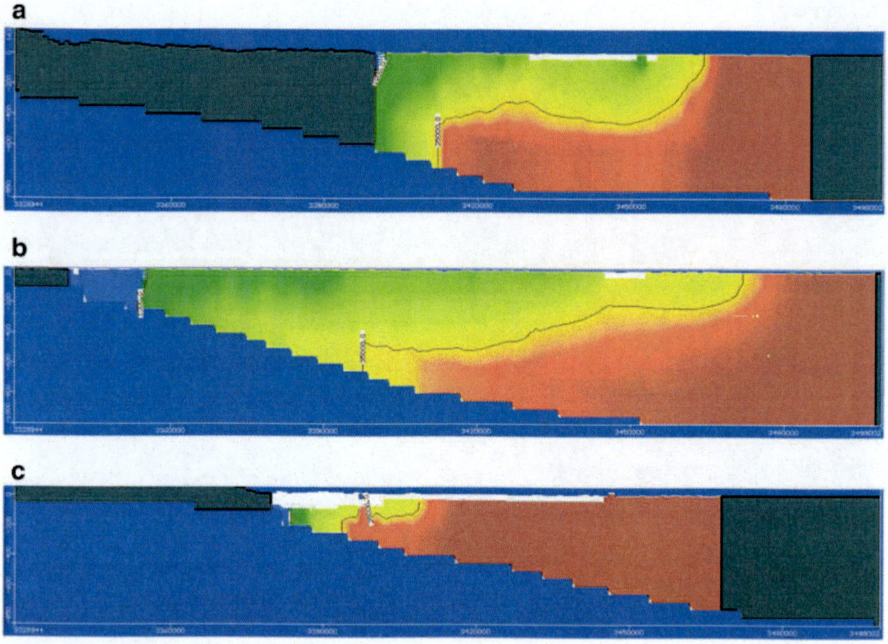

Fig. 40 (**a**) Vertical TDS distribution in the NDA at (SLR by 50 cm and double abstraction), in the west. (**b**) Vertical TDS distribution in the NDA at (SLR by 50 cm and double abstraction), in the middle. (**c**) Vertical TDS distribution in the NDA at (SLR by 50 cm and double abstraction), in the east

Table 19 The advancement of SW line 35,000 ppm and FW line 1,000 ppm from the base case for scenario 6

Location	Advance of line 35,000 ppm from the base case	Advance of line 1,000 ppm from the base case
West	+48.968 km	+20.048 km
Middle	+57.770 km	+53.013 km
East	+40.273 km	+22.8 km

than the base case. Compared to the basic case, the SW line 35,000 ppm moved 6.2 km further inland in the east. In contrast to the base case, the FW line 1,000 ppm advanced 5.4 km further inland.

Scenario 5: considers a half abstraction of the base scenario and a 0.5 m SLR. According to the outcomes, in the west of NDA, the FW line 1,000 ppm, which represents the freshwater, came into the sea by 3.365 km compared with the base case, while the SW line 35,000 ppm, which reflects the SW, came in toward the shoreline by 13.88 km. In comparison to the base scenario, the SW line 35,000 ppm trailed 0.4 km to the coast in the middle of the NDA, while the FW line 1,000 ppm lagged 3.365 km near the coast base. Compared to the basic case, the SW line

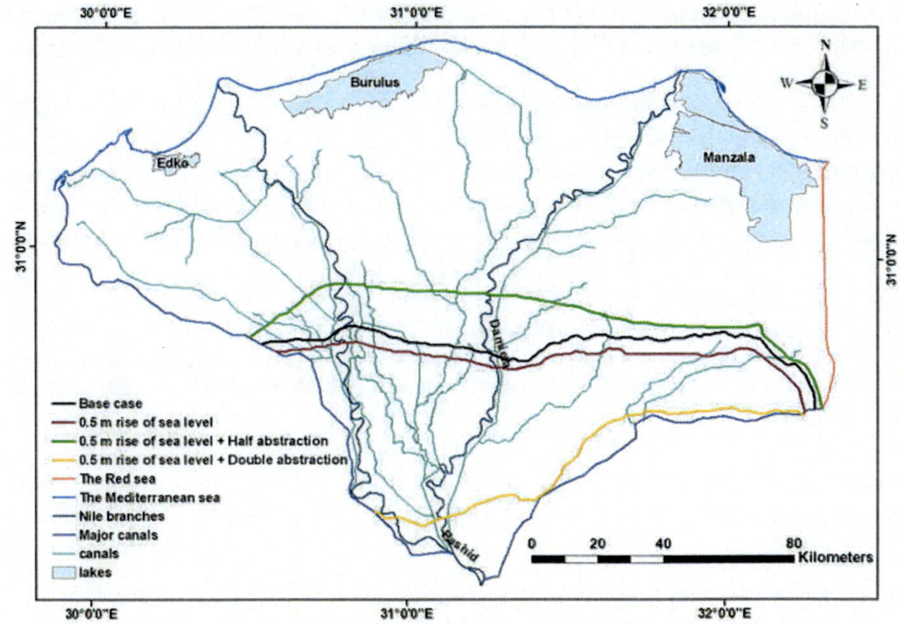

Fig. 41 Comparison between FW line 1,000 ppm of base case and last three scenarios

35,000 ppm in the east was 7.851 km behind the coastline. In comparison to the base case, the FW line 1,000 ppm lagged 1.121 km toward the sea.

Scenario 6: considers a 0.5 m SLR and doubles the discharge rate compared to the base case. The findings showed that, in the west of NDA, the SW lines 35,000 ppm, corresponding to the SW, and 1,000 ppm, which indicate freshwater, both developed more inland relative to the base case by 48.968 km and 20.048 km, respectively. In comparison to the base scenario, the SW line 35,000 ppm migrated 57.77 km further inland in the central of the NDA, while the FW line 1,000 ppm moved 53.013 km farther inland. In comparison with the base case, the SW line 35,000 ppm moved 40.273 km further inland in the east. In comparison to the default case, the FW line 1,000 ppm advanced 22.8 km farther inland.

16 Recommendation

The NDA's GW resources should be managed according to the following suggestions, and future study can consider the following recommendation into account:

1. Considering the real drainage network into the GW modeling for the Nile Delta, to better depict and simulate the interplay between groundwater and surface water.

2. Updating the numerical simulation of the intrusion of SW by using the most recent observations of GW level, canals water levels, GW salinity, and recent pumping rates.
3. Utilizing both the numerical and experimental models, to examine several SWI control scenarios, considering the climate scenario.
4. Restrict the abstraction rate from some wells at the north of ND, in places where the rate of withdrawal is high, and to provide another water source to reduce dependence on wells water, to reduce the creep of saltwater toward the agricultural lands in the middle and the south of the ND.

Acknowledgments The authors would like to thank the Faculty of Engineering, Tanta University, Tanta, Egypt for the support in completing this research. This book chapter is a part of the master's degree thesis of the second author. The authors would like to thank the editors for reviewing this book chapter.

References

1. Abd-Elhamid HF (2010) A simulation-optimization model to study the control of seawater intrusion in coastal aquifers. https://doi.org/10.1002/nag.1068
2. Hamza KI (2004) Numerical analysis of saltwater upconing below a pumping well. Proceedings of the 18th salt water intrusion meeting, Cartagena, Spain
3. Armanuos AM (2017) Experimental and numerical study of saltwater intrusion in the Nile Delta aquifer, Egypt, PhD thesis, Egypt-Japan University of Science and Technology
4. Said R, Beheri S (1961) Quantitative geomorphology of the area east of Cairo. Bull Soc Geograph Egypt 3. 34 p
5. MWRI (2013) Adaptation to climate change in the Nile Delta through integrated coastal zone management, Ministry of Water Resources and Irrigation
6. Morsy WS (2009) Environmental management to groundwater resources for Nile Delta Region PhD thesis, Faculty of Engineering, Cairo University
7. Armanuos AM, Elgaafary KA, Gado TA (2023) Landfill site selection using MCDM methods and GIS in the central part of the Nile Delta, Egypt. Environ Monit Assess 195:1407. https://doi.org/10.1007/s10661-023-11946-8
8. Metwally MI, Armanuos AM, Zeidan BA (2023) Comparative study for assessment of groundwater vulnerability to pollution using DRASTIC methods applied to Central Nile Delta, Egypt. Int J Energ Water Res 7:175–190. https://doi.org/10.1007/s42108-022-00198-w
9. Monem MM (2009) Study the environmental impacts on groundwater system in Nile Delta PhD thesis, Faculty of Engineering, Ain shams University
10. Mabrouk B, Farhat Abd-Elhamid H, Badr M, Ludwig R (2013) Adaptation to the impact of sea level rise in the northeastern Nile Delta, Egypt. EGU general assembly conference abstracts. https://doi.org/10.1023/A:1009684210570
11. Gaamea O (2000) Behavior of the transition zone in the Nile Delta aquifer under different pumping schemes. PhD thesis, Faculty of Engineering, Cairo University
12. El Didy S, Darwish M (2001) Studying the effect of desalination of Manzala and Burullus Lakes on salt water intrusion in the Nile Delta. Water Science, National Water Research Center
13. Shamrukh M, Corapcioglu MY, Hassona FA (2001) Modeling the effect of chemical fertilizers on ground water quality in the Nile Valley aquifer, Egypt. Groundwater 39(1):59–67. https://doi.org/10.1111/j.1745-6584.2001.tb00351.x

14. Sayed M, Nour El-Din M, Nasr F (2004) Impacts of global warming on precipitation patterns on the Nile basin. Second Regional Conference on Action Plans for Integrated Development
15. Armanuos AM, Ibrahim MG, Mahmod WE, Negm A, Yoshimura C, Takemura J, Zidan BA (2017) Evaluation of the potential impact of Grand Ethiopian renaissance dam and pumping scenarios on groundwater level in the Nile Delta aquifer. Water Supply 17(5):1356–1367. https://doi.org/10.2166/ws.2017.037
16. Armanuos A, Ahmed K, Sanusi Shiru M, Jamei M (2021) Impact of increasing pumping discharge on groundwater level in the Nile Delta aquifer, Egypt. Knowl Based Eng Sci 2(2): 13–23. https://doi.org/10.51526/kbes.2021.2.2.13-23
17. Armanuos AM, Negm A (2018) Integrated groundwater modeling for simulation of saltwater intrusion in the Nile Delta aquifer, Egypt. In: Groundwater in the Nile Delta. Springer, pp 489–544. https://doi.org/10.1007/698_2017_184
18. Armanuos AM, Negm A, Yoshimura C et al (2016) Application of WetSpass model to estimate groundwater recharge variability in the Nile Delta aquifer. Arab J Geosci 9:553. https://doi.org/10.1007/s12517-016-2580-x
19. Abdelaty IM, Abd-Elhamid HF, Fahmy MR, Abdelaal GM (2014) Investigation of some potential parameters and its impacts on saltwater intrusion in Nile Delta aquifer. J Eng Sci 42(4):931–955. https://doi.org/10.21608/JESAUN.2014.115039
20. RIGW (1992) Hydrogeological map of Nile Delta, scale 1:500000, 1st edn., ND
21. Abd-Elhamid H, Javadi A, Abdelaty I, Sherif M (2016) Simulation of seawater intrusion in the Nile Delta aquifer under the conditions of climate change. Hydrol Res 47(6):1198–1210. https://doi.org/10.2166/nh.2016.157
22. Nofal E, Amer M, El-Didy S, Fekry A (2015) Sea water intrusion in Nile Delta in perspective of new configuration of the aquifer heterogeneity using the recent stratigraphy data. J Am Sci 11(6):281–292
23. Elshinnawy H, Zeidan B, Ghoraba S (2015) Impact of hydraulic conductivity uniformity on seawater intrusion in the Nile Delta aquifer, Egypt. Google scholar link
24. Wassef R, Schüttrumpf H (2016) Impact of sea-level rise on groundwater salinity at the development area western delta, Egypt. Groundw Sustain Dev 2–3:85–103. https://doi.org/10.1016/j.gsd.2016.06.001
25. Van Engelen J, Verkaik J, King J, Nofal ER, Bierkens MF, Oude Essink GH (2019) A three-dimensional palaeohydrogeological reconstruction of the groundwater salinity distribution in the Nile Delta aquifer. Hydrol Earth Syst Sci 23(12):5175–5198. https://doi.org/10.5194/hess-23-5175-2019
26. Elshinnawy IA, Almaliki AH (2021) Vulnerability assessment for sea level rise impacts on coastal Systems of Gamasa Ras El Bar Area, Nile Delta, Egypt. Sustainability 13(7):3624. https://doi.org/10.3390/su13073624
27. Armanuos AM, Emara SR, Shalby A, Metwally MI, John AP, Negm A (2023) Groundwater in the Nile Delta aquifer, Egypt: assessment, modelling and management with climate change in the Core. In: Ali S, Armanuos AM (eds) Groundwater in arid and semi-arid areas. Earth and environmental sciences library. Springer, Cham. https://doi.org/10.1007/978-3-031-43348-1_11
28. Ali S, Armanuos AM (2023) Introduction to "Groundwater in Arid and Semi-Arid Areas". In: Ali S, Armanuos AM (eds) Groundwater in arid and semi-arid areas. Earth and environmental sciences library. Springer, Cham. https://doi.org/10.1007/978-3-031-43348-1_1
29. Mabrouk M, Jonoski A, Solomatine D, Uhlenbrook S (2013) A review of seawater intrusion in the Nile Delta groundwater system–the basis for assessing impacts due to climate changes and water resources development. Hydrol Earth Syst Sci Dis 10(8):10873–10911. https://doi.org/10.5194/hessd-10-10873-2013
30. Wahaab RA, Badawy MI (2004) Water quality assessment of the River Nile system: an overview. Biomed Environ Sci 17(1):87–100
31. Sefelnasr A, Sherif M (2014) Impacts of seawater rise on seawater intrusion in the Nile Delta aquifer, Egypt. Groundwater 52(2):264–276. https://doi.org/10.1111/gwat.12058

32. Amer M, Ridder N (1988) Land drainage in Egypt. Drainage Research Institute, Water Research Center, Cairo, p 376
33. Warner JW, Gates TK, Attia FA, Mankarious WF (1991) Vertical leakage in Egypt's Nile Valleys estimation and implications. J Irrig Drain Eng 117(4):515–533. https://doi.org/10.1061/(ASCE)0733-9437(1991)117:4(515)
34. El Ramly IM (1997) Hydrogeological and water quality characteristics of the saturated zone beneath the various land uses in the Nile Delta region, Egypt. Freshwater Contamin 243:255–261
35. SWS (2010) User's manual. 460 Phillip street – suite 101 Waterloo, ON, Canada, N2L 5J2
36. RIGW (2002) Nile Delta groundwater modeling report, Research Institute for Groundwater, Kanater El-Khairia, Egypt
37. Kashef AAI (1983) Salt-water intrusion in the Nile Delta. Groundwater 21(2):160–167. https://doi.org/10.1111/j.1745-6584.1983.tb00713.x
38. Agrawala S, Moehner A, El Raey M, Conway D, Van Aalst M, Hagenstad M, Smith J (2004) Development and climate change in Egypt: focus on coastal resources and the Nile. Org Econ Co-operation Dev 1:1–68
39. El-Raey M (1994) Vulnerability assessment of the coastal zone of Egypt to the impacts of sea level rise. Phase II, US Country Study Program
40. Shoemaker WB (2004) Important observations and parameters for a salt water intrusion model. Ground Water 42(6):829–840. https://doi.org/10.1111/j.1745-6584.2004.t01-2-.x
41. CORI (2008) Coastal vulnerability to climate changes and adaptation assessment for coastal zones of Egypt, Final Report. Wikipedia, http://en.wikipedia.org
42. Sakr S, Attia F, Millette J (2004) Vulnerability of the Nile Delta aquifer of Egypt to seawater intrusion. International conference on water resources of arid and semi-arid regions of Africa, issues and challenges, Gaborone, Botswana
43. IPCC (2007) "An Assessment of the Intergovernmental Panel on Climate Change", adopted section by section at IPCC plenary XXVII (Valencia, Spain, 12–17 November 2007), represents the formally agreed statement of the IPCC concerning key findings and uncertainties contained in the working group contributions to the fourth assessment report
44. RIGW (1992b) Groundwater resources and projection of groundwater development, water security project, (WSP), Research Inst. for Groundwater, Kanater El-Khairia
45. RIGW (2003) Monitoring of groundwater microbiological activities in the Nile Delta aquifer. A study completed for the National Water Quality and Availability Management project (NAWQAM), Kanater El-Khairia, Egypt
46. Ismail IA-EM (2014) Numerical and experimental study for simulating climatic changes effects on Nile Delta aquifer, PhD thesis, Faculty of Engineering, Zagazig University
47. Eid H (2001) Climate change studies on Egyptian agriculture, soils, water and environment research institute SWERI ARC. Ministry of Agriculture, Giza, Egypt
48. Eid HM, El-Marsafawy SM, Ouda SA (2007) Assessing the economic impacts of climate change on agriculture in Egypt: a Ricardian approach. World Bank Policy Research Working Paper(4293)
49. Elshinnawy I, Abo Zed A, Ali M, Deabes E, Abdel-Gawad S (2008) Coastal vulnerability to climate changes and adaptation assessment for coastal zones of Egypt. Ministry of Water Resources and Irrigation (MWRI). National Water Research Center (NWRC). Coastal Research Institute (CoRI). Cairo, Egypt
50. SNC (2010) Egypt's Second National Communication, Egyptian Environmental Affairs Agency (EEAA-May 2010), under the United Nations Framework Convention on Climate Change on Climate Change
51. EGSA (1997) Topographical maps cover Nile Delta scale 1:1500000 and 1:10000
52. RIGW (1992) Hydrogeological map of Egypt, Nile Delta scale 1:50000000. NWRC, Cairo, Egypt
53. Zaghloul Z, Taha A, Hegab O, El Fawal F (1977) The Neogene-quaternary sedimentary basins of the Nile delta. Egypt J Geol 21(1):1–19

54. El-Arabi M (2007) Environmental impact of new settlements in groundwater in a region in the Nile Delta, PhD thesis, Faculty of Engineering, Zagazig University Zagazig, Egypt
55. Diab M, Dahab K, El Fakharany M (1997) Impacts of the paleohydrological conditions on the groundwater quality in the northern part of Nile Delta, the geological society of Egypt. Geol J B 4112:779–795
56. Nossair A (2011) Climate changes and their impacts on groundwater occurrence in the northern part of East Nile Delta MSc thesis, Faculty of Science, Zagazig University, Egypt
57. Taha MS, Armanuos AM, Zeidan BA (2023) Impact of climate change on seawater intrusion, and shore line advance in Nile Delta, Egypt. Delta Univ Sci J 6(2):182–196. https://doi.org/10.21608/dusj.2023.318648
58. Mansour NI (2006) Recent hydrogeological studies on the eastern Nile Delta, Egypt, PhD thesis, Faculty of Science Helwan University, Egypt
59. RIGW (1990) Hydrogeological map for the Nile Delta area, scale 1:500000. Research Institute for Ground Water
60. Sherif M, Sefelnasr A, Javadi A (2012) Incorporating the concept of equivalent freshwater head in successive horizontal simulations of seawater intrusion in the Nile Delta aquifer, Egypt. J Hydrol 464:186–198. https://doi.org/10.1016/j.jhydrol.2012.07.007
61. Sherif MM, Al-Rashed MF (2001) Vertical and horizontal simulation of seawater intrusion in the Nile Delta aquifer. First international conference on saltwater intrusion and coastal aquifers, monitoring, modeling, and management, Essaouira, Morocco
62. Sherif MM, Hamza KI (2001) Mitigation of seawater intrusion by pumping brackish water. Transp Porous Media 43(1):29–44. https://doi.org/10.1023/A:1010601208708
63. Sherif MM, Singh VP (1999) Effect of climate change on sea water intrusion in coastal aquifers. Hydrol Process 13(8):1277–1287. https://doi.org/10.1002/(SICI)1099-1085(19990615)13:8%3C1277::AID-HYP765%3E3.0.CO;2-W
64. Wels C, Mackie D, Scibek J (2012) Guidelines for groundwater modelling to assess impacts of proposed natural resource development activities. Ministry of Environment, Water Protection & Sustainability Branch
65. Sherif MM, Singh VP, Amer AM (1988) A two-dimensional finite element model for dispersion (2D-FED) in coastal aquifers. J Hydrol 103(1–2):11–36. https://doi.org/10.1016/0022-1694(88)90003-0
66. Kumar C (2002) Groundwater flow models. Scientist 'E1' National Institute of hydrology Roorkee–247667 (Uttaranchal) publication
67. Lee S (2004) Investigation of seawater intrusion in coastal plain and karst hinterland aquifers of cape range, northwestern Australia. Proceeding of the 18th Salt Water Intrusion Meeting, Cartagena (Spain)
68. Stocker TF, Qin D, Plattner GK, Tignor MM, Allen SK, Boschung J, Nauels A, Xia Y, Bex V, Midgley PM (2014) Climate change 2013: the physical science basis. Contribution of working group I to the fifth assessment report of IPCC the intergovernmental panel on climate change
69. Armanuos AM, Eldin Ibrahim MG, Negm A, Takemura J, Yoshimura C, Mahmod WE (2022) Investigation of seawater intrusion in the Nile Delta aquifer, Egypt. J Eng Res 6(1):10–28. https://doi.org/10.21608/ERJENG.2022.113766.1046

Groundwater Quality Prediction in Upper and Middle Cheliff Plain, Algeria Using Artificial Intelligence

Yamina Elmeddahi, Salah Difi, Hemza Allali, and Ragab Ragab

Contents

1 Introduction 166
2 Materials and Methods 167
 2.1 Description of Case Study Area and Data 167
 2.2 Methodology 168
3 Results and Discussion 174
 3.1 Physico-Chemical Characteristics of the Water 174
 3.2 Spatial Distribution of Water Quality Index (WQI) 175
 3.3 Groundwater Quality Prediction 176
 3.4 Model Comparisons 178
4 Conclusion 180
5 Recommendations 182
References 182

Abstract Groundwater studies are very important for the establishment of a quantitative and qualitative management system that ensures the supply and protection of water in water-scarce semi-arid areas like Algeria. However, monitoring and conducting measurements of groundwater quality and quantity are difficult, costly, and time-consuming; therefore, the use of predictive models such as artificial intelligence (AI) is becoming an attractive alternative.

Y. Elmeddahi (✉), S. Difi, and H. Allali
Department of Hydraulic, Civil Engineering and Architecture Faculty, University of Hassiba Benbouali, Chlef, Algeria

Vegetal Chemistry-Water-Energy Laboratory (LCV2E), Chlef, Algeria
e-mail: y.elmeddahi@univ-chlef.dz; s.difi@univ-chlef.dz; h.allali@univ-chlef.dz

R. Ragab
UK Centre for Ecology and Hydrology, Maclean Building, Crowmarsh, Gifford, Wallingford, Oxon, UK
e-mail: Ragab@icid.org; Rag@ceh.ac.uk

Shakir Ali and Abdelazim Negm (eds.), *Groundwater Quality and Geochemistry in Arid and Semi-Arid Regions*, Hdb Env Chem (2024) 126: 165–184, DOI 10.1007/698_2023_1048,

Published online: 8 December 2023

For this purpose, three approaches based on machine learning techniques were used to improve the prediction of water quality in the Upper and Middle Cheliff plain in Algeria. In this study, the most dominant parameters of the water quality index (WQI) that were extracted by principal component analysis (PCA) were used in multilayer perceptron neural network (MLPNN), support vector regression (SVR), and decision tree regression (DTR) models. Various combinations of input data were investigated and models were evaluated in terms of prediction performance, using several statistical criteria. Various potential physicochemical water quality variables were considered for the calculation of the water quality index (WQI) in the study area. This work will be helpful to decision-makers and water authorities for sustainable groundwater resource management and planning.

Keywords Artificial intelligence, Cheliff Plain, Algeria, Groundwater management, Groundwater quality, Water quality index

1 Introduction

To properly manage water resources in a region characterized by water scarcity aggravated by drought and population growth, water resources studies are very important to provide a quantitative and qualitative management system [1].

In the study area, groundwater is one of the most important natural resources for various water uses due to the limited surface water availability and reduced precipitation, exacerbated by climate change [1–3]. Groundwater quality surveillance is a key component of water resources planning. Predicting water quality will help in better-protecting water resources from pollution.

Several researchers have widely investigated groundwater and surface water quality assessments, suggesting a quality index (WQI) with various definitions and formats [4–6]. The WQI applies a statistical method to represent water quality variables in a single value that is in accordance with WHO norms [7]. Various models were proposed in the literature to predict the WQI. Recently, various tools based on machine learning have been implemented and used in water quality prediction.

Novel machine learning algorithms coupled with the principal component analysis (PCA) method, comprising linear regression, gradient boosting regression, random forest regression, and support vector regression proposed by Islam Khan et al. [8]. While other research applied multiple models like gradient boosting and Lasso regression for predicting water quality using four important parameters [9].

The objective of the current study is: (1) to evaluate the ability of machine learning models and (2) to determine the most appropriate model for assessing groundwater quality for the Middle and Upper Cheliff alluvial plain in Algeria, based on the water quality index (WQI).

2 Materials and Methods

2.1 Description of Case Study Area and Data

The Upper and Middle Cheliff plains are part of the Cheliff-Zahrez watershed which covers more than 22% of the area of northern Algeria. The study area lies between 36° 01′and 36° 20′ north latitude and 0° 58′and 02° 30′ east longitude (Fig. 1).

The watershed of the Upper and Middle Cheliff includes a total of 11 sub-basins. Its area is approximately 10,701 km^2. This basin is drained by the Cheliff wadi, which crosses over 349 km.

Many geological formations have petro-physical characteristics favorable to groundwater storage in the study area. The oldest are attributed to the Jurassic age and the most recent correspond to Quaternary alluvium. The groundwater potential in the Upper and Middle Cheliff plains is estimated at 43 hm^3/year according to the ANRH [10, 11]. This water potential is the primary source of water supply and irrigation.

A semi-arid continental climate characterizes the plains of the Upper and Middle Cheliff with very hot dry summers and cold rainy winters. Temperatures can exceed 40°C in July and drops as low as 5°C in February.

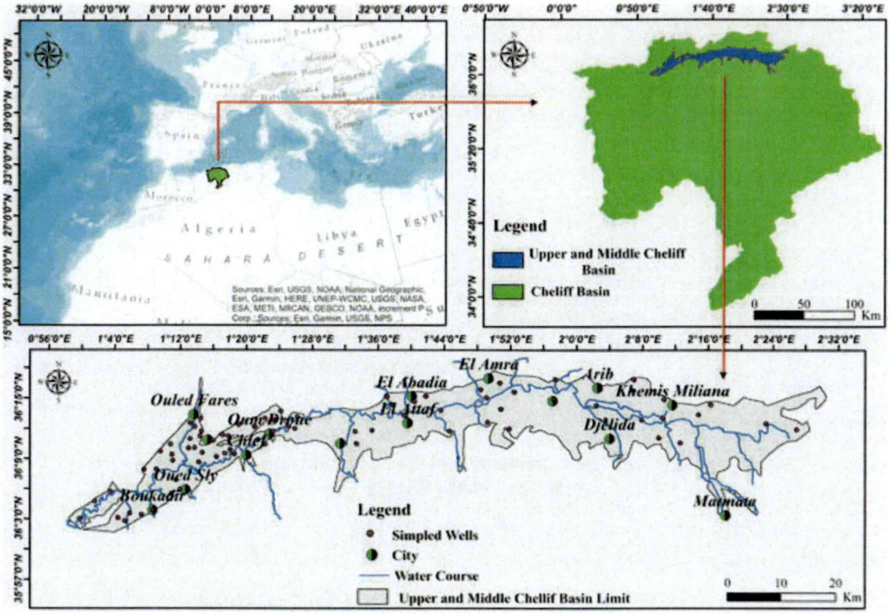

Fig. 1 Location map of the study area

2.2 Methodology

Our proposed approach is based first on the preparation of the dataset using feature extraction techniques (PCA). For WQI prediction, three machine learning algorithms were used, namely MLPNN, SVR, and DTR models. The flowchart is illustrated in Fig. 2, which shows the different steps used. The procedure and methodology reported in Elmeddahi and Ragab [5] were followed in this case study, however, with completely new data sets obtained from the National Agency for Water Resources (ANRH) of Algeria.

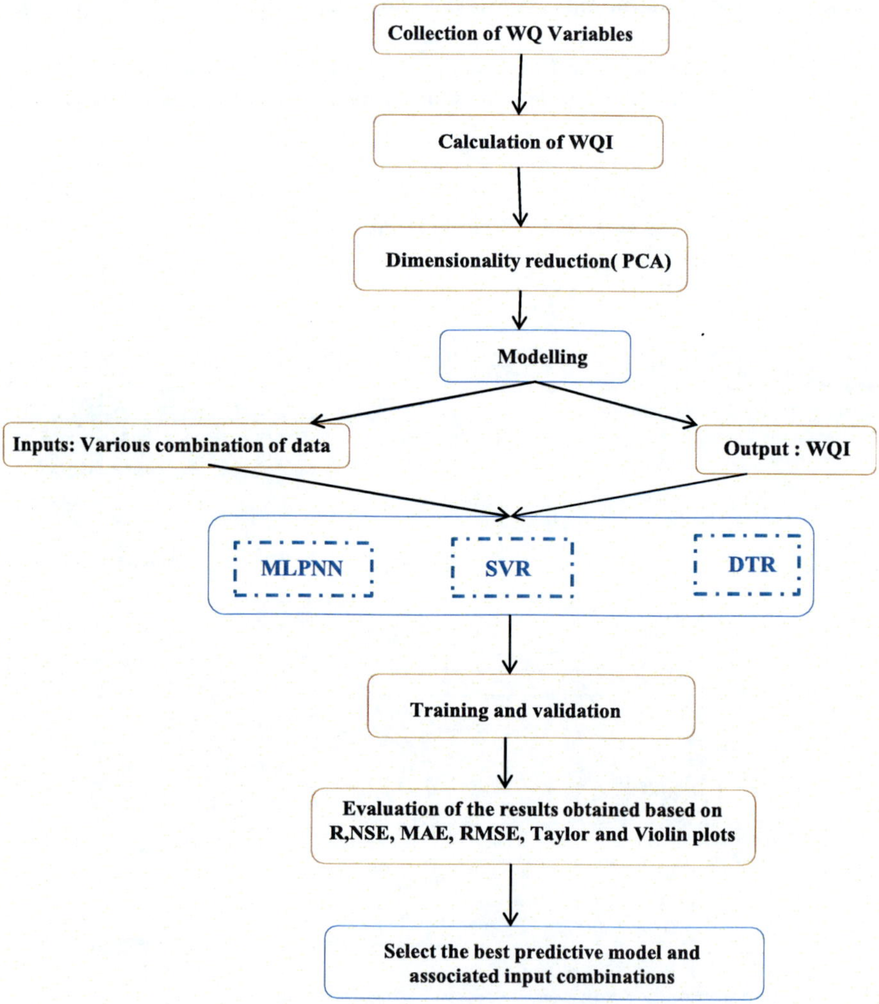

Fig. 2 Flowchart of the proposed methodology for WQI prediction

2.2.1 Calculation of the Groundwater Quality Index (WQI)

The estimation of the water quality index aims to convert the selected water parameters into a single scaled value, based on the weighted arithmetic indicator to identify the quality of water and its suitability for drinking purposes. The data were collected over a 5-year period, from 2016 to 2020, from 87 observation wells in the area [5]. The computation of the WQI consists firstly in determining the weighted value (*wi*) of each parameter according to their relative influence on the overall quality of drinking water. A weighting value, between 1 and 5, has been assigned to each factor according to its importance [12]. The highest weight of 5 was assigned to electrical conductivity (EC) and nitrate, because of their direct effect on drinking water quality. Bicarbonate was given a minimum weight of 1 because it has less importance in the assessment of water quality. The remaining parameters were weighted between 2 and 4 according to their influence on the drinking water quality assessment (Table 1).

The relative weight (*wi*) of each parameter was calculated, after determining the quality rating scale (*qi*) for each parameter.

Then, the WQI was calculated by summing the sub-index of all parameters by the following equations:

$$Wi = \frac{wi}{\sum_{i=1}^{n} wi} \tag{1}$$

$$qi = \frac{Ci}{Si} \times 100 \tag{2}$$

$$SI_i = W_i \times q_i \tag{3}$$

$$\text{WQI} = \sum_{i=1}^{n} SI_i \tag{4}$$

where W_i is the relative weight, w_i is the parameter weight, n is the number of parameters, (q_i) is the quality ranking, (C_i) is the concentration of each chemical parameter in the analyzed sample in milligrams per liter, and (S_i) is the admissible limit of each parameter in drinking water according to the WHO [7] standard. The calculated WQI values are generally classified into five groups, which are represented in Table 2.

Principal component analysis (PCA) techniques for feature extraction and selection were applied. For this purpose, a comparative analysis based on the performance of three models with different parameter combinations: multilayer perceptron neural network (MLPNN), support vector regression (SVR), and decision tree regression (DTR) has been carried out. The physicochemical water quality variables such as electrical conductivity (EC), pH, Ca^{2+}, Mg^{2+}, Na^{+}, K^{+}, Cl^{-}, $SO_4{}^{2-}$, $NO_3{}^{-}$, and $HCO_3{}^{-}$ were considered for computing the water quality index (WQI).

Table 1 Statistics description of the water quality parameters

Parameter	Unit	WHO water quality standard [7]	Max	Min	Mean	Coefficient of variation	Standard deviation	Weight
pH		8.5	9.40	6.00	7.86	0.08	0.61	3
EC	µS/cm	1,500	9,270.00	360.50	2,461.02	0.83	2,042.65	5
Na^{+}	mg/l	150	1,645.00	50.50	320.44	0.88	283.87	3
Ca^{2+}	mg/l	100	582.00	12.00	200.85	0.61	123.37	2
Mg^{2+}	mg/l	75	238.00	1.20	78.44	0.63	49.42	3
K^{+}	mg/l	12	32.00	0.00	5.33	0.90	4.85	2
Cl^{-}	mg/l	200	3,110.00	62.00	615.26	0.43	542.21	4
SO_4^{2-}	mg/l	250	1,215.00	0.50	376.68	0.80	304.57	4
NO_3^{-}	mg/l	50	126.00	0. 0	42.17	0.83	34.92	5
HCO^{-}_3	mg/l	300	558.00	30.50	308.39	0.33	102.49	1
WQI			444.00	25.54	116.48	0.68	79.54	

Table 2 Groundwater classification based on the water quality index

Class	WQI value	Type of water quality	Samples per category (%)
1	<50	Excellent	15
2	50–100	Good	39
3	100.1–200	Poor	37
4	200.1–300	Very poor	06
5	>300	Unsuitable for drinking	03

2.2.2 Principal Component Analysis (PCA) for Dimensionality Reduction

Dimensionality reduction is the procedure of decreasing the number of variables across the most influential features and preserving as much detail of the original data. Feature extraction consists of converting the data in the high-dimensional space into a lower-dimensional space that comprises most of the valuable information [13].

There are many feature extraction methods, the most widely used of which is PCA.

The principal component analysis is a robust unsupervised linear transformation technique that is being used widely to convert the original set of correlated variables into a smaller number of uncorrelated variables that describe all the variances. The uncorrelated variables resulting from the PCA are labeled principal components (PCs).

The numbers of PCs were selected based on the Kaiser criterion using the first eigenvectors sorted in descending order of eigenvalues. Correlation coefficients between variables and factors are often used to select the most dominant parameters in water quality (Table 3).

2.2.3 Prediction Models

In this work, eight combinations of 10 parameters were selected as inputs (EC, pH, Ca^{2+}, Mg^{2+}, Na^+, K^+, Cl^-, $SO_4{}^{2-}$, $NO_3{}^-$, and $HCO_3{}^-$) and one as output (WQI) as shown in Table 4. For the testing and validation process, 70% and 30% of collected data were utilized, respectively.

Multilayer Perceptron Neural Networks (MLPNN)

The multilayer perceptron neural network (MLPNN) is one of the most applied approaches in various prediction domains [14–17]. The multilayer perceptron is structured in several layers through which information flows from the input layer to a single output layer. The three layers (an input layer, a hidden layer, and an output layer) are interconnected by the weight of their neurons (Fig. 3). The trial and error method was chosen to determine the number of hidden nodes.

Table 3 Descriptive statistics of the created principal components

	F1	F2	F3	F4	F5
Variables eigenvectors obtained through the PCA application					
pH	−0.289	−0.615	0.155	0.395	0.588
EC µ/cm	**0.974**	−0.129	−0.040	0.112	0.016
Calcium	**0.827**	0.325	0.081	0.264	−0.006
Magnésium	0.629	0.479	−0.269	−0.245	0.323
Sodium	**0.896**	−0.373	0.052	0.028	−0.124
Potassium	0.627	−0.559	0.349	−0.062	−0.001
Chlorures	**0.916**	−0.274	−0.133	0.103	−0.154
Sulfates	0.580	0.595	0.079	−0.191	0.353
Nitrates	−0.044	**0.703**	0.117	0.650	−0.113
Bicarbonates	0.003	0.266	**0.908**	−0.189	−0.035
pH	−0.289	−0.615	0.155	0.395	0.588
Eigenvalues					
	4.48	2.18	1.09	0.81	0.63
Percent of total variance proportion					
	44.83	21.77	10.93	8.08	6.27
Cumulative percent of total variance proportion					
	44.83	66.59	77.52	85.60	91.87

Table 4 The input combinations of different models

MLPNN	DTR	SVR		Input combination	Output
MLPNN1	DTR1	SVR1		EC, Ca, Na, Cl	WQI
MLPNN2	DTR2	SVR2		EC, Ca, Na, Cl, NO3	WQI
MLPNN3	DTR3	SVR3		EC, Ca, Na, Cl, NO_3 HCO_3	WQI
MLPNN4	DTR4	SVR4		EC, Mg, SO_4, Ca, pH, HCO_3, NO_3	WQI
MLPNN5	DTR5	SVR5		EC, Ca, Na, Cl, NO_3, Mg	WQI
MLPNN6	DTR6	SVR6		EC, Ca, Na, Cl, NO_3, Mg, K	WQI
MLPNN7	DTR7	SVR7		EC, Ca, Na, Cl, NO_3, Mg, K, pH	WQI
MLPNN8	DTR8	SVR8		EC, Ca, Na, Cl, NO_3, Mg, K, pH , SO_4 HCO_3	WQI

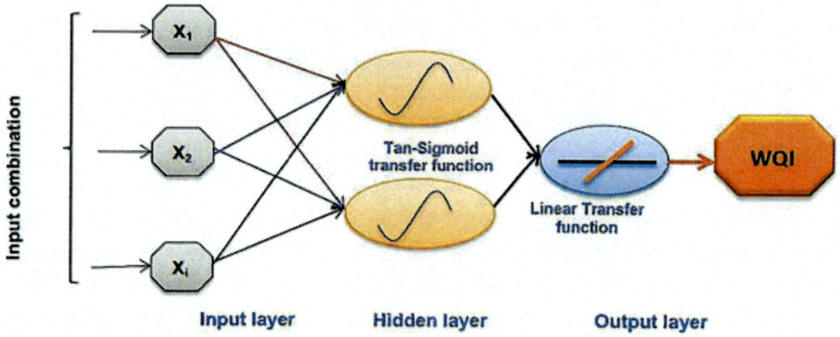

Fig. 3 Architecture of multilayer perceptron neural networks

The back-propagation algorithm was applied in the training phase to adjust the connection weights to minimize the error by calculating the difference between the calculated and predicted values. These weights were combined and processed further via an activation function and then transferred to the back layers. The tangent sigmoid and the linear transfer functions are the most commonly used activation functions.

Decision Tree Regression (DTR)

Decision tree regression (DTR) is an efficient classical approach to supervised learning that is composed of several branches and nodes [18]. A node with outer edges is called an inner node while the remaining nodes are called leaves.

The decision tree structure is based on a hierarchically organized set of data clusters, known as nodes that are interconnected through the tree branches. At the top of the organizational hierarchy is the root of the tree (i.e., the root node) which holds all cases/observations that are contained in a dataset being used for training [19].

The DTR algorithm subdivides the data into clusters and organizes them in a hierarchical tree-like structure with inner and terminal nodes representing the divisions and leaves, respectively [18].

Support Vector Regression (SVR)

Support vector regression (SVR) is a supervised machine learning technique that was suggested based on the concept of support vector machine (SVM) by Vapnik [20]. SVR is a robust algorithm for solving linear and nonlinear problems through kernel functions that are applied to convert low-dimensional data into high-dimensional data. SVR offers both error minimization and hyperplane individualization flexibility to maximize the margin. It has been used successfully in various fields of water resources engineering with high accuracy of prediction for different purposes. The architecture of SVR is illustrated in Fig. 4. The mathematical equation of the SVR model can be written as:

$$f(X_i) = W^T \varnothing(X_i) + b \tag{5}$$

where $f(X)$ is the general form of nonlinear regression function, and $\phi(X)$ defines a nonlinear mapping function in terms of the input variable matrix X. W and b are the weighing vector and the bias parameter, respectively.

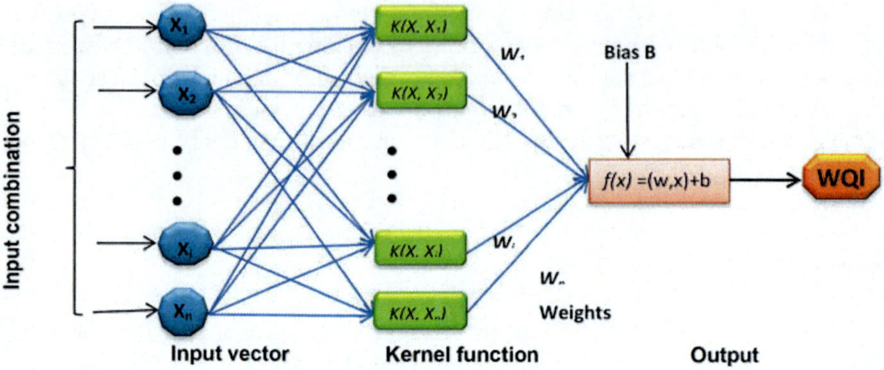

Fig. 4 Support vector regression architecture

2.2.4 Model Performance Evaluation

The performance evaluation criteria used in this study are the correlation coefficient (R), mean absolute error (MAE), root mean square error (RMSE), and Nash-Sutcliffe efficiency (NSE). In addition, the different models are also compared using scatter plots, Taylor plots, and Violin plots. Scatter plots graphically represent the consistency between observed and predicted values, whereas Taylor plots compare RMSE, and standard deviation to visually determine the most accurate predictive model [21]. Finally, Violin plot diagrams present the prediction results of probability density function.

3 Results and Discussion

3.1 Physico-Chemical Characteristics of the Water

Table 1 presents a descriptive statistical summary of the data set and the parameters that affect water quality. The pH varied from 6.00 to 9.40 with an average value of 7.86 that falls within the allowable limits of WHO standards [7].

There is a wide range of variation in the EC values, between 360.50 and 1,2130.00 µs/cm, with an average value of 2,461.02 µs/cm. The distribution of EC value revealed that about ~70% of the samples had more than the permissible limit of EC level in the study area with a very high coefficient of variation (83%) which reflects the wide dispersion of the data. The high EC levels were probably the result of anthropogenic groundwater pollution or water-rock interaction (i.e., geological conditions of the aquifer). It has also been reported that the EC can be highly influenced by the chemical composition of the water and its temperature [22].

The sodium concentration indicated that over 37% of the samples were below the WHO safe limit, with an average of approximately 320.44 mg/l. The calcium and magnesium enrichment of the water is possibly due to leaching from the study area's calcic and magnesia rock formations. Nine sampled water points (about 10%) provided water with potassium concentrations above the drinking water standard of 12 mg/l, however, 90% of the observations were below the WHO standard. It can be suggested that this was due to the alteration of potassium clays and the dissolution of chemical fertilizers Nitrogen-Phosphorus-Potassium known as (NPK) used by farmers in agriculture. The levels may also be associated with the discharge of wastewater effluents.

Most of the sampled water points (76%) have chloride concentrations exceeding the WHO standard, varying from 62.00 to 3,110.00 mg/l. The higher values of Cl^- can be explained by Triassic breakup in the form of NaCl-rich white gypsum efflorescence on the left bank of the Wadi Cheliff, and continued groundwater interaction with the Miocene marl substrate [23]. Also, halite dissolution and gypsiferous formation are responsible for some runoff's salinity and consequently the aquifers' salinity.

Also the sources of SO_4^{2-} in groundwater can be attributed to several factors such as dissolution and oxidation of sulfate minerals, industrial and domestic effluent discharges, and leaching from waste deposits. The concentration value of SO_4^{2-} varies from 0.50 to 1,215 mg/l with an average value of 376.68 mg/l, 45% of the samples were found to be below the permitted limit. As for the nitrate, 34% of samples had concentrations above the WHO standard threshold of 50 mg/l. The sources of nitrate in groundwater were probably industrial wastewater; nitrogen fertilizers, surface water infiltration, standing water, and agricultural wastewater use. The presence of bicarbonate (HCO_3^-) in natural water is affected by the level of dissolved carbon dioxide, temperature, pH, cations, and some soluble salts [24].

3.2 *Spatial Distribution of Water Quality Index (WQI)*

The water quality index (WQI) is an appropriate indicator of water use, especially as drinking water, and is a useful tool for managing groundwater resources. The WQI values for the year 2020 ranged from 25.54 to 444.00 with a mean value of 116.48 and a high coefficient of variation (68%), indicating a large variation in the quality of the groundwater samples (Table 1). Based on the WQI map for 2020 (Fig. 5), five classes have been identified (Table 2).

The excellent and good water quality was found in the northeast, central El Amra, the locality of Oum Drou and Ouled Fares, and in the southwest of the case study area where 15% and 39% of the samples are classified as excellent and good, respectively. However, a poor category with 37% of samples was observed in the densely populated area, while very poor groundwater (6%) and unsuitable for consumption (3%) is observed in the Oued Fodda region.

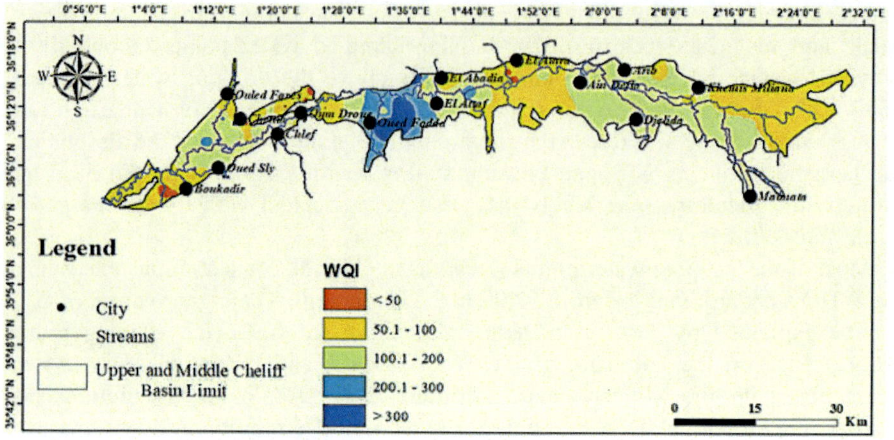

Fig. 5 Spatial distribution map of WQI

3.3 *Groundwater Quality Prediction*

Different combinations of possible input parameters (from MLPNN1 to MLPNN8 as shown in Table 4) were selected from the dimension reduction applied by PCA. Three models, i.e., MLPNN, DTR, and SVR, were applied and compared to demonstrate the proposed model's feasibility and allow the determination of the suitable input variables to construct a good model for predicting the WQI.

The models were evaluated using the R, NSE, RMSE, and MAE performance metrics, the Scatter plots, the Taylor's diagram, and Violin plots. The numerical results are summarized in Table 5, in which the best models' results have been given in bold.

3.3.1 Multilayer Perceptron Neural Network Model

The numbers of neurons have been selected in hidden layer to predict the output data based on trial and error. According to the results presented in Table 5, and in the training and validation set, the model MLPNN1 (multilayer perceptron neural network) performs appropriately and produced the lowest value of RMSE and MAE, and the highest value of R with RMSE = 9.937, MAE = 7.625 and R = 0.993, and RMSE = 9.962, MAE = 8.388 and R = 0.986 in the training and validation phase, respectively. The best Nash-Sutcliffe for both phases was 0.986 and 0.971. Generally, the results obtained by the MLPNN8, MLPNN2, and MLPNN6 models are satisfactory when compared with the other models, but with somewhat high errors in the validation phase. The MLPNN7 yielded the highest RMSE and MAE values of 23.44 and 12.201, respectively.

Table 5 Performances of different prediction models

	Training				Validation			
Models	R	NSE	RMSE	MAE	R	NSE	RMSE	MAE
MLPNN models								
MLPNN1	**0.993**	**0.986**	**9.937**	**7.625**	**0.986**	**0.971**	**9.962**	**8.388**
MLPNN2	0.997	0.993	7.099	5.384	0.985	0.962	11.450	7.655
MLPNN3	0.997	0.994	6.424	4.590	0.968	0.920	16.553	10.788
MLPNN4	0.998	0.995	3.442	2.480	0.982	0.937	14.664	7.072
MLPNN5	0.998	0.998	2.523	1.814	0.970	0.916	16.988	10.499
MLPNN6	0.998	0.998	1.985	1.556	0.991	0.969	10.354	7.992
MLPNN7	0.999	0.999	0.457	0.302	0.955	0.840	23.444	12.201
MLPNN8	0.999	0.999	0.977	0.602	0.993	0.971	10.038	6.831
DTR models								
DTR1	0.965	0.931	22.329	17.519	0.185	0.055	60.220	43.263
DTR2	0.964	0.929	22.582	17.666	0.120	0.201	64.249	48.233
DTR3	0.964	0.929	22.582	17.666	0.120	0.201	64.249	48.233
DTR4	0.969	0.940	20.864	13.031	0.730	0.447	43.598	23.072
DTR5	0.969	0.940	20.864	13.031	0.730	0.447	43.598	23.072
DTR6	**0.969**	**0.940**	**20.864**	**13.031**	**0.753**	**0.485**	**42.063**	**22.533**
DTR7	0.907	0.822	35.812	15.673	0.815	0.402	45.327	29.675
DTR8	0.907	0.822	35.812	15.673	0.815	0.402	45.327	29.675
SVR models								
SVR1	0.955	0.701	46.482	40.383	0.796	0.276	49.880	44.292
SVR2	0.964	0.738	43.475	38.325	0.852	0.475	42.466	37.902
SVR3	**0.963**	**0.766**	**41.060**	**33.336**	**0.882**	**0.553**	**39.224**	**33.945**
SVR4	0.952	0.660	49.530	44.742	0.851	0.418	44.731	41.806
SVR5	0.939	0.619	52.421	45.660	0.866	0.150	54.063	50.468
SVR6	0.925	0.572	55.548	45.876	0.754	0.089	55.977	50.669
SVR7	0.916	0.555	56.652	47.830	0.712	0.253	50.692	46.054
SVR8	0.946	0.653	50.056	42.026	0.857	0.299	49.105	45.939

3.3.2 DTR Model

The results presented in Table 5 show that the prediction performance of the DTR model in the training phases is similar for all the different DTR combinations considered, with a correlation coefficient ranging from 0.907 to 0.969. During the validation phase, the results indicated that the DTR6 combination is the best model with the highest NSE and R values and the lowest error (R = 0.753, NSE = 0.485, RMSE = 42.063, and MAE = 22.533). Although the performance of the DTR4 to DTR8 models is close to that of the DTR6 combination, the models DTR1 to DTR3 give weak results in the validation phase with RMSE error (60.220 to 64.249), low Nash-Sutcliffe values (0.055), and a correlation coefficient of only 0.120. It can also be noticed that generally, the prediction performances are higher in the learning phase than in the validation phase (Table 5): this is obvious because during the training phase the prediction error was minimized.

3.3.3 SVR Model

The lowest values of RMSE and MAE range from 41.060 to 33.336 in the training stage and 39.224 to 33.945 in the validation phase. The highest values of Nash-Sutcliffe (NSE) are 0.766 and 0.553 in the training and in the validation phase, respectively. Similar to the SVR3, the SVR1 and SVR2 models showed less satisfactory performance for both training stages. SVR1 gave R = 0.955/0.796, NSE = 0.701/0.276, RMSE = 46.482/49.880, and MAE = 40.383/44.292, for training/validation stage. Weak results are observed in the validation phase and for the remainder of the models, which is reflected by low Nash-Sutcliffe values (NSE) and high errors ranging from 50.669 to 55.063.

3.4 Model Comparisons

For all input combinations, the best prediction accuracies were obtained using MLPNN1 with four input parameters, i.e., EC, Ca, Na, and Cl, for which the highest R and NSE values and lowest RMSE and MAE values were obtained. In the training and validation phases, the MLPNN1 yielded the highest R and NSE values of 0.98/0.95 and 0.96/0.88, respectively, and the lowest RMSE and MAE values of 11.2/15.03 and7.89/10.22, respectively.

The DTR1, DTR2, and DTR3 produced the worst accuracy (R = 0.185; 0.120, NSE = 0.055; 0.201, RMSE = 60.220; 64.249 and MAE = 43.263; 48.338) in the validation stage.

As already discussed, the prediction performance of the different models is also displayed here using scatter plots, Taylor plots, and violin plots. The three models considered (MLPNN1, DTR6, and SVR3) generally show outstanding prediction performance (points close to the 1:1 line) in both training phases. However, in the validation phases, MLPNN1 performed better (R^2 closer to 1.0) compared to the other two models (Fig. 6). Moreover, in Fig. 7, the Taylor diagrams also show that, in both phases, the MPLNN1 model performed better in terms of prediction as its values were closer to the reference.

Finally, a comparison of the violin diagrams for the different models was found to be very similar to the results shown by the scatter plots and Taylor diagram which confirms what is already seen with other criteria of prediction performances (Fig. 8).

A comparative analysis of our results with those from previous studies has shown the robustness of the suggested model. Kouadri et al. [25, 26] applied the artificial neural network (ANN) combined with the principal component analysis (PCA) method to predict the groundwater quality index using 13 parameters in the ELMERK region (southeastern Algeria). Four inputs [mineralization, total hardness

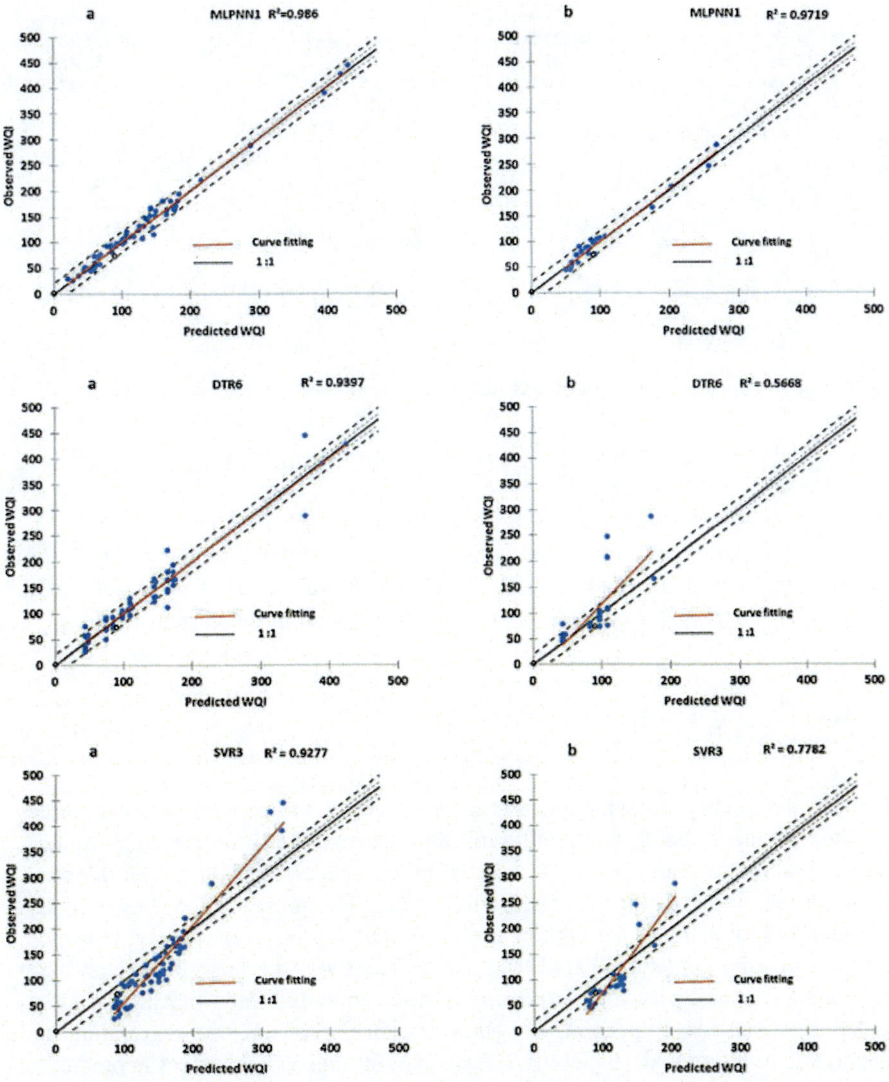

Fig. 6 Scatterplots of the predicted versus the observed WQI values of three models (MLPNN, DTR, and SVR) during training phase (**a**) and validation phase (**b**)

(TH), NO_3^-, and NO_2^-] were chosen according to PCA. The model exhibited an R-based performance of 0.997, which is close to the value obtained in our research (0.986). In other studies, several artificial intelligence algorithms are applied, including multilinear regression (MLR), random forest (RF), M5P tree (M5P), random subspace (RSS), additive regression (AR), artificial neural network (ANN), support vector regression (SVR), and locally weighted linear regression (LWLR) to predict the WQI in the Illizi region, southeast Algeria [25, 26]. Their results proved that the

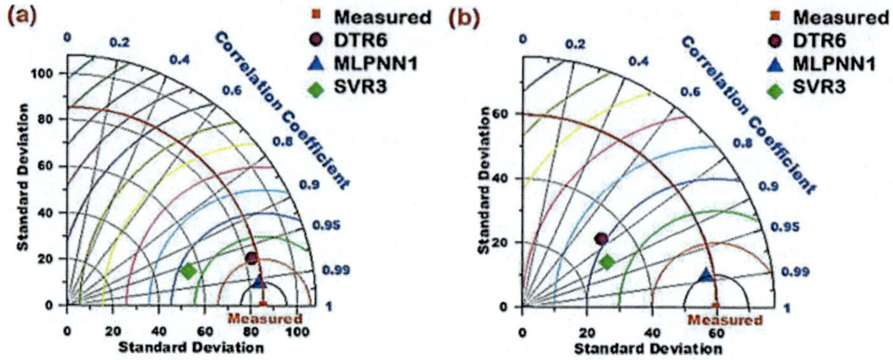

Fig. 7 Taylor diagram of three models (MLPNN, DTR, and SVR) (**a**: Training, **b**: Validation)

ANN model performed better than the other models, confirming the results obtained in our study.

In a recent study conducted by Elmeddahi and Ragab [5], the MLPNN model integrated with NSA approaches is significantly improved and outperforms models coupled with the PCA method with R and NSE values of 0.949, and 0.876, respectively. These values are less than the values found in the current work.

4 Conclusion

Groundwater quality assessment has an important role in improving water resources management, particularly in arid and semi-arid areas where surface water is limited, and water shortages have been exacerbated by climate change. Based on the water quality index, five groundwater classes have been determined in the present study: 15% and 39% of samples are ranked as excellent and good, respectively. However, the poor category contains 37% of the samples and the rest belong to the very poor and unfit for drinking category (9%). This study included three machine learning models (i.e., multilayer perceptron neural network, decision tree regression, and support vector regression), which were evaluated for their effectiveness in predicting the groundwater quality index in the Middle and Upper Cheliff plain of Algeria.

For this purpose, the feature selection approach (PCA) was implemented to identify the most significant variables for the WQI target.

The results revealed that the MPLNN1 model with four input parameters has the highest performance in the prediction of WQI at the training/validation phases (R = 0.993/0.986, NSE = 0.986/0.971, RMSE = 9.937/9.962 and MAE = 7.625/8.388). The rest of the MLPNN models have similar results, although the prediction performance was slightly less. This was followed by the SVR3, DTR6, and SVR2 models. The DTR model performed the worst among the three models considered.

PCA has proven to be a useful feature selection technique for its ability to prioritize inputs according to their importance. The proposed MPLNN combined

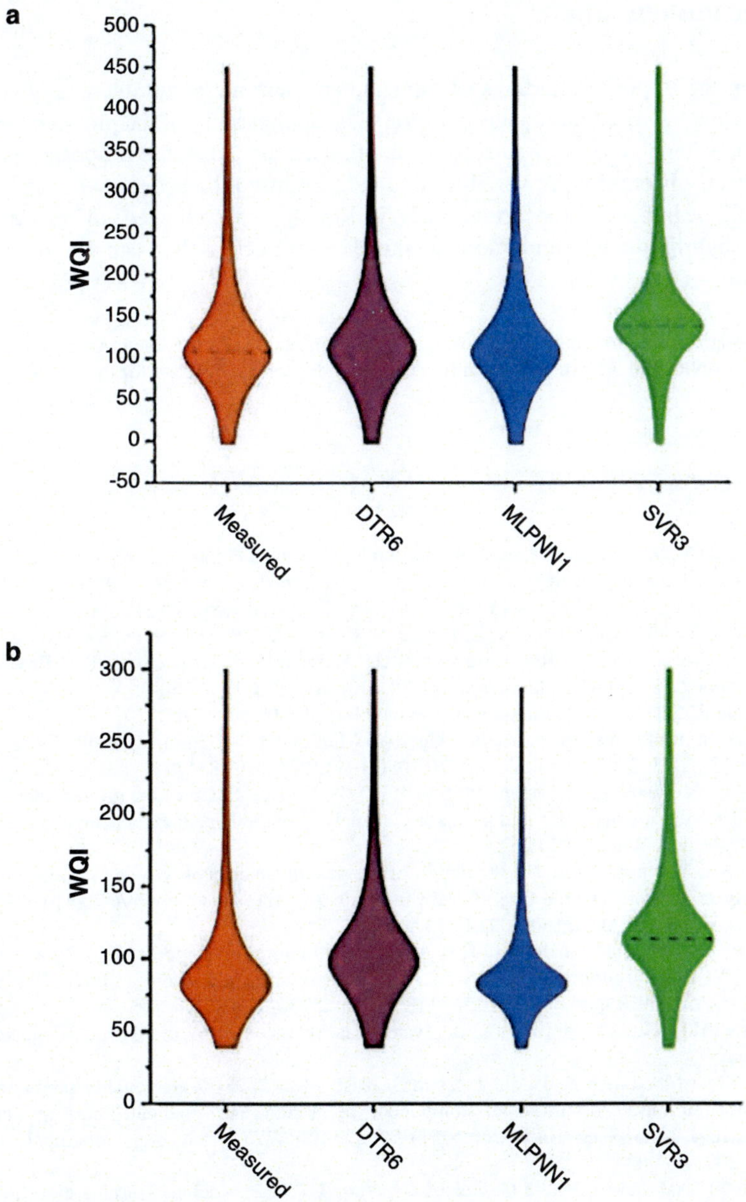

Fig. 8 Violin Plot of three models (MLPNN, DTR, and SVR) (**a**: Training, **b**: Validation)

model has improved the accuracy of WQI prediction with fewer input variables and thus can be utilized as a beneficial tool for prediction even with little available water quality data.

5 Recommendations

Based on the important findings of the chapter, there are some interesting suggestions that may warrant more consideration. The applicability of the proposed models should be extended to other sites for in-depth analysis. Also, the proposed methods with several different input variables can help improve the accuracy of the models and would be beneficial for further analysis. Finally, research needs to be conducted on new hybrid models and more advanced approaches that can improve WQI prediction.

Acknowledgments The authors acknowledge the support of the National Agency for Water Resources (ANRH) of Algeria, for providing the groundwater quality data for this study.

References

1. Elmeddahi Y, Mahmoudi H, Issaadi A, Goosen MFA, Ragab R (2016) Evaluating the effects of climate change and variability on water resources: a case study of the Cheliff basin in Algeria. Am J Eng Appl Sci 9(4):835–845. https://doi.org/10.3844/ajeassp.2016.835.845
2. Chidiac S, El Najjar P, Ouaini N, El Rayess Y, El Azzi D (2023) A comprehensive review of water quality indices (WQIs): history, models, attempts and perspectives. Rev Environ Sci Biotechnol 22(2):349–395. https://doi.org/10.1007/s11157-023-09650-7
3. Elmeddahi Y, Issaadi A, Mahmoudi H, Tahar Abbes M, Mattheus FAG (2014) Effect of climate change on water resources of the Algerian Middle Cheliff basin. Desalin Water Treat 52(10–12):2073–2081. https://doi.org/10.1080/19443994.2013.831777
4. Baba ME, Kayastha P, Huysmans M, Smedt FD (2020) Evaluation of the groundwater quality using the water quality index and geostatistical analysis in the Dier al-Balah Governorate, Gaza Strip, Palestine. Water 12(1):262
5. Elmeddahi Y, Ragab R (2022) Prediction of the groundwater quality index through machine learning in Western Middle Cheliff plain in North Algeria. Acta Geophys 70(4):1797–1814. https://doi.org/10.1007/s11600-022-00827-2
6. Patel PS, Pandya DM, Shah M (2023) A systematic and comparative study of water quality index (WQI) for groundwater quality analysis and assessment. Environ Sci Pollut Res 30(19): 54303–54323. https://doi.org/10.1007/s11356-023-25936-3
7. WHO (2011) Guidelines for drinking-water quality.4th edn. World Health Organization, Geneva
8. Islam Khan MS, Islam N, Uddin J, Islam S, Nasir MK (2022) Water quality prediction and classification based on principal component regression and gradient boosting classifier approach. J King Saud Univ Comput Inf Sci 34(8):4773–4781. https://doi.org/10.1016/j.jksuci.2021.06.003
9. Azrour M, Mabrouki J, Fattah G, Guezzaz A, Aziz F (2022) Machine learning algorithms for efficient water quality prediction. Model Earth Syst Environ 8(2):2793–2801. https://doi.org/10.1007/s40808-021-01266-6
10. ABH-CZ (2004) Hydraulic cadastre of the Cheliff-Downstream watershed of the Boughzoul dam – First part: Upper and middle Cheliff (Cadastre Hydraulique du bassin hydrographie du Cheliff-Aval du barrage de Boughzoul – Première partie: Haut et moyen Cheliff)
11. ANRH (2006) Assessment of the water resources of the middle Cheliff alluvial aquifer. Agence Nationale des Ressources Hydrauliques, Algeria. Internal Report

12. Elmeddahi Y, Moudjeber DE, Mahmoudi H, Goosen MFA (2022) Groundwater quality modeling using geostatistical methods and artificial neural networks: a case study of the Western Middle Cheliff alluvial plain in Algeria. Desalin Water Treat 255:145–156. https://doi.org/10.5004/dwt.2022.28335
13. Ali A, Gravino C (2022) Evaluating the impact of feature selection consistency in software prediction. Sci Comput Program 213:102715. https://doi.org/10.1016/j.scico.2021.102715
14. Difi S, Elmeddahi Y, Hebal A, Singh VP, Heddam S, Kim S, Kisi O (2022) Monthly streamflow prediction using hybrid extreme learning machine optimized by bat algorithm: a case study of Cheliff watershed, Algeria. Hydrol Sci J 68(2):189–208. https://doi.org/10.1080/02626667.2022.2149334
15. Ebadati N, Hooshmandzadeh M (2019) Water quality assessment of river using RBF and MLP methods of artificial network analysis (case study: Karoon River Southwest of Iran). Environ Earth Sci 78(17):1–12. https://doi.org/10.1007/s12665-019-8472-0
16. Mohanty S, Jha MK, Raul SK, Panda RK, Sudheer KP (2015) Using artificial neural network approach for simultaneous forecasting of weekly groundwater levels at multiple sites. Water Resour Manag 29(15):5521–5532. https://doi.org/10.1007/s11269-015-1132-6
17. Ucun Ozel H, Gemici BT, Gemici E, Ozel HB, Cetin M, Sevik H (2020) Application of artificial neural networks to predict the heavy metal contamination in the Bartin River. Environ Sci Pollut Res 27(34):42495–42512. https://doi.org/10.1007/s11356-020-10156-w
18. Chakraborty T, Chakraborty AK, Mansoor Z (2019) A hybrid regression model for water quality prediction. Opsearch 56(4):1167–1178. https://doi.org/10.1007/s12597-019-00386-z
19. Tien Bui D, Pradhan B, Lofman O, Revhaug I (2012) Landslide susceptibility assessment in Vietnam using support vector machines, decision tree, and nave bayes models. Math Probl Eng. https://doi.org/10.1155/2012/974638
20. Vapnik VN (1998) Statistical learning theory. Wiley, New York
21. Taylor KE (2001) In a single diagram, vol 106, pp 7183–7192
22. Bouderbala A (2019) The impact of climate change on groundwater resources in coastal aquifers: case of the alluvial aquifer of Mitidja in Algeria. Environ Earth Sci 78(24):1–13. https://doi.org/10.1007/s12665-019-8702-5
23. Elaid M, Hind M, Abdelmadjid B, Mohamed M (2020) Contribution of hydrogeochemical and isotopic tools to the management of Upper and Middle Cheliff Aquifers. J Earth Sci 31(5): 993–1006. https://doi.org/10.1007/s12583-020-1293-y
24. Khosravi R, Eslami H, Almodaresi SA, Heidari M, Fallahzadeh RA, Taghavi M, Khodadadi M, Peirovi R (2017) Use of geographic information system and water quality index to assess groundwater quality for drinking purpose in Birjand city, Iran. Desalin Water Treat 67:74–83. https://doi.org/10.5004/dwt.2017.20458
25. Kouadri S, Elbeltagi A, Islam ARMT, Kateb S (2021) Performance of machine learning methods in predicting water quality index based on irregular data set: application on Illizi region (Algerian southeast). Appl Water Sci 11(12):1–20. https://doi.org/10.1007/s13201-021-01528-9
26. Kouadri S, Kateb S, Zegait R (2021) Spatial and temporal model for WQI prediction based on back-propagation neural network, application on EL MERK region (Algerian southeast). J Saudi Soc Agric Sci 20(5):324–336. https://doi.org/10.1016/j.jssas.2021.03.004

Evolution of Groundwater in the Cheliff and Mitidja Aquifers (North Algeria) in Qualitative and Quantitative Terms

Faiza Hallouz and Mohamed Meddi

Contents

1 Introduction ... 186
2 Methodology ... 188
3 Study Area ... 189
3.1 Mitidja Plain ... 189
3.2 Cheliff Plain ... 194
4 Groundwater Quality and Quantity in the Cheliff and Mitidja Plains ... 199
4.1 Cheliff Plain ... 200
4.2 Mitidja Plain ... 203
5 Conclusion ... 207
6 Recommendations ... 208
References ... 209

Abstract Due to rapid population growth over time, the importance of freshwater is increasing, and the availability of this vital resource is becoming progressively more challenging. Consequently, the exploitation of untapped water sources becomes a necessity. With industrial development, groundwater is increasingly exposed to various contaminants by infiltrating polluted discharges. Water resources in Algeria, especially in the Cheliff and Mitidja plains, are limited, vulnerable, and unevenly

---

F. Hallouz (✉)
Department of Ecology and Environment, Faculty of Sciences of Nature and Life and Earth Sciences, Djilali BOUNAAMA University of Khemis Miliana, Ain Defla, Algeria

National Higher School for Hydraulics (ENSH), Water and Environmental Engineering Laboratory, Blida, Algeria
e-mail: F.hallouz@univ-dbkm.dz

M. Meddi
National Higher School for Hydraulics (ENSH), Water and Environmental Engineering Laboratory, Blida, Algeria
e-mail: m.meddi@ensh.dz

Shakir Ali and Abdelazim Negm (eds.), *Groundwater Quality and Geochemistry in Arid and Semi-Arid Regions*, Hdb Env Chem (2024) 126: 185–214, DOI 10.1007/698_2023_1052,

Published online: 12 December 2023

distributed spatially. This precarious situation inevitably calls for new measures to be taken to utilize these resources sustainably. Natural factors, such as drought or geological constraints, affect the supply and distribution of drinking water. Therefore, it is imperative to quantify and analyze the quantity and quality of water supplies, and to devise methods for managing this resource to ensure its long-term viability. The water quality in these regions has significantly deteriorated in recent years due to unregulated urban discharges, intensive use of chemical fertilizers, and poorly managed farms. These factors alter the chemical composition of water, rendering it unsuitable for intended purposes. This chapter offers a scientific overview of the current groundwater status in the Cheliff and Mitidja regions, focusing on both qualitative and quantitative aspects.

Keywords Algeria, Cheliff and Mitidja, Chemism of water, Groundwater pollution, Groundwater quality, North Africa

1 Introduction

Water is an essential element for the survival of humans, animals, and plants, which stands as a cornerstone of life itself. Ensuring water security is paramount for the continuation of human existence. Worldwide, the importance of water scarcity and the increasing demand for this essential resource have become more prominent. Groundwater, well known for its reliable availability and acceptable quality, plays a crucial role in numerous industrial and agricultural processes, especially in areas where rivers and lakes are in short supply. Across the globe, nations face the challenge of ensuring groundwater resources' quality and long-term sustainability. Recent studies by Zhang et al. [1] and Bose et al. [2] underline the pressing need for comprehensive strategies to prevent contamination, thereby averting potential ecological disasters and safeguarding public health. The ramifications extend beyond environmental concerns, intertwining with economic growth and societal well-being. Stretching vast distances, this ancient underground water source has enabled human settlement and agricultural pursuits in an unforgiving landscape. Here, the careful management of groundwater is emblematic of sustainability and a testament to human ingenuity [3].

In arid and semi-arid regions, the looming specter of drought further underscores the criticality of effective groundwater management [4]. The global scenario underscores the importance of safeguarding groundwater quality and availability. The exploitation of groundwater beyond its natural replenishment rate has been observed in places such as California in the United States, where agricultural demands have outpaced the available supply [5]. The Punjab region in northern India, for instance, relies extensively on groundwater for its agricultural output, a practice that demands vigilant monitoring to counterbalance the risk of over-extraction and subsequent land subsidence [6]. In Australia, the Great Artesian Basin stands as a testament to the enduring importance of groundwater in arid regions. In recent years, several regions of North Africa have grappled with alarming declines in groundwater levels, leading to concerns about long-term sustainability especially in Algeria [7].

Algeria is a country located in North Africa that faces several significant challenges. Among these challenges, we can cite drought, particularly during the 1940s and 1970s [8–16]. Drivers such as rapid population growth, urbanization, and climate change are significant factors that have profoundly impacted the region's economic and environmental stability. These problems have serious consequences on the economy and the quality of life of the Algerian population.

The Cheliff and Mitidja regions are two important regions in Algeria, known for their agricultural production and water resources. The Cheliff river stretches over 700 km and flows through several regions, including the Atlas Mountains. The Mitidja is a coastal plain that extends 80 km along the Mediterranean coast. These two regions are important for cities' agriculture, fishing, industry, and water supply. Both regions are important in Algeria which is rich in various natural resources, including water, fertile land, and minerals. However, these regions also have environmental problems, including freshwater pollution, land degradation, and biodiversity loss. These problems can adversely affect both the economy and the inhabitants' quality of life.

The groundwater resources in these two regions are estimated to be more than 2 billion m^3, with over 90% of these resources already exploited [17]. This over-exploitation has raised concerns about the sustainability of the water resources. One of the major concerns in the Cheliff and Mitidja regions is groundwater pollution due to the unscientific and excessive use of fertilizers in agriculture and certain industrial activities. Urbanization and demographic growth also contribute to pollution. The Cheliff and Mitidja plains are widely regarded as one of Algeria's most productive agricultural areas. They possess a large reservoir of groundwater, which serves as a vital source of irrigation and drinking water. Unfortunately, this valuable resource has experienced a significant reduction in natural recharge, leading to a substantial drop in water levels in certain regions. The primary causes of this reduction are the decrease in rainfall, amounting to about 20% since 1975 [8, 11, 16, 18–20], and continuous over-exploitation of groundwater [21].

The study addressed a research gap in understanding the long-term effects of pollution on the aquifers of the Cheliff and Mitidja plains in Algeria. Despite recognizing the effects of agricultural and industrial pollution, the specifics of pollutants and their long-term impacts are unexplored. To bridge this gap partially, this study's overall aim is to preserve water quality for drinking and irrigation in both plains. Therefore, it is essential to (1) monitor the groundwater resources and develop strategies for sustainable groundwater resource management, ensuring their availability for future generations, (2) understand the dynamics of the Cheliff and Mitidja aquifers in terms of both quality and quantity, and (3) have insights into the factors influencing groundwater evolution, including human activities and natural processes. Thus, the objective of this study was to investigate the evolution of groundwater in the Mitidja and Cheliff plains.

2 Methodology

The research focuses on the Cheliff and Mitidja aquifers situated in Northern Algeria, forming the geographical scope of the study. These aquifers hold immense significance as they are vital in supplying essential groundwater resources for agricultural and residential purposes. The selection of these aquifers as the study's focal point was influenced by their pivotal role in the local water supply and the existence of previous research efforts.

The methodology outlines a detailed approach in studying the evolution of groundwater in the Cheliff and Mitidja aquifers and is presented in Fig. 1. It involves reviewing the existing literature, collecting and analyzing data (such as climate, hydrology, and hydrography), comparing previous studies, conducting both quantitative and qualitative analyses, drawing conclusions, and providing

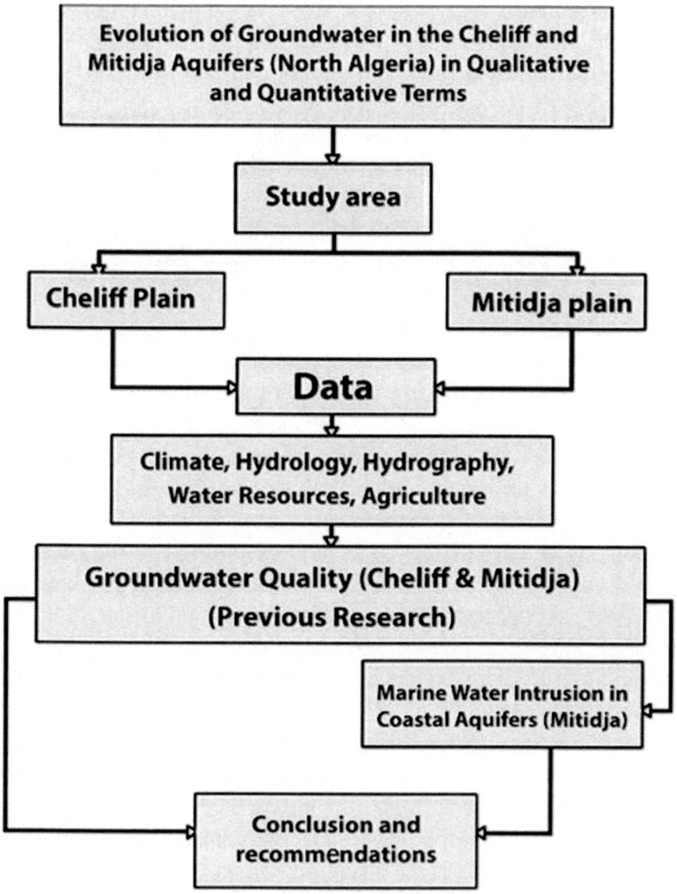

Fig. 1 Flowchart of the procedural stages devised in the research

recommendations for future research and groundwater management practices. Specific outcomes include:

- Quantitative understanding of groundwater flow patterns and trends over time
- Qualitative assessment of groundwater quality changes
- Identification of potential sources of contamination or deterioration in groundwater quality

3 Study Area

This study was conducted in two regions in Algeria: the Cheliff plain in the northwest and the Mitidja plain in the north (Fig. 2). These regions have distinct climatic and geomorphological contexts and are polluted due to excessive use of fertilizers in their agricultural plains. The Mitidja plain, particularly its eastern part (including the Rouiba and Réghaia regions), is affected by polluting industries, impacting the main aquifers studied in this research. The Cheliff River is distinguished from other wadis in Algeria due to its unique characteristics, originating from the highlands and draining a portion of these elevated areas. Its notable length and significant flow rates further differentiate it. The flow rates are notably influenced by the underlying geological structure of the regions it traverses [22]. The Mitidja plain, located in the northern region of Algeria, forms potential aquifers serving as the primary source of potable water for the central part of the country [23]. The study is important for safeguarding the quality of these waters, which are intended for drinking water supply and irrigation purposes.

3.1 Mitidja Plain

The Mitidja plain is the most significant sub-littoral plain in Algeria. It is situated in the hinterland of Algiers, the capital of Algeria, extending across the territories of Algiers, Boumerdès, Tipaza, and Blida. Encompassing an area of approximately 1,450 km^2 from Boumerdès in the east to Tipaza in the west (Fig. 2), it stretches over a length of 100 km with a width that varies from 8 to 18 km. This expansive plain is a part of the Algiers coastal basins, designated with code 02 by the National Agency for Hydraulic Resources (ANRH). It gently slopes toward the Sahel and the sea, with an average elevation of 100 m [24].

Geomorphologically, the Mitidja plain is divided into four sub-basins. Surface water outlets are located along the Mediterranean coast (Table 1 and Fig. 2), draining the main river valleys of the area, which belong to the Algerian coastal basin (code 02). Table 1 describes the main characteristics of the four sub-basins, with the largest ones being the Nador and Mazafran sub-basins [25].

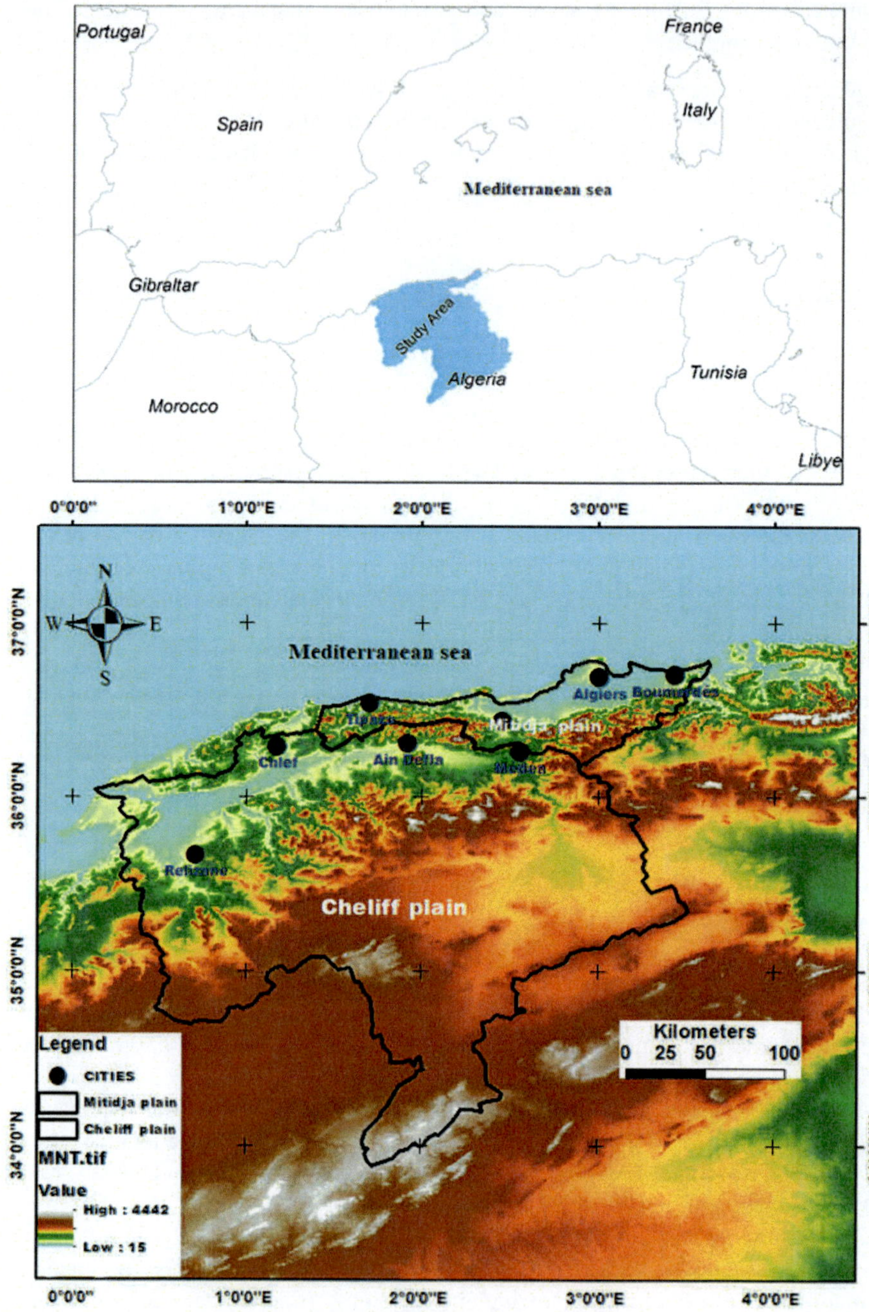

Fig. 2 Geographical location of the study plains (Cheliff and Mitidja)

Table 1 Sub-basins of the Mitidja plain [25]

Basin name	Area (km^2)	Description
Wadi Nador	230	Wadi Nador is the result of the confluence of several secondary wadis
Wadi Mazafran	1,860 1,270	About 60% of the basin area is mountainous. Wadi Mazafran is the result of the confluence of wadis Djer, Bouroumi, and Chiffa
Wadi El Harrach	380	Wadi El Harrach originates in the mountains of Tablat before emptying into the sea
Wadi Hamiz		Wadi Hamiz drains the basin with one of the Réghaia wadi

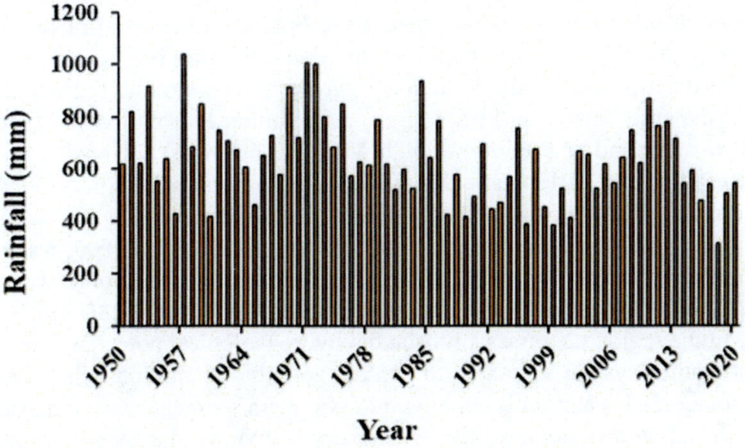

Fig. 3 Evolution of rainfall in Mitidja (1950–2020)

3.1.1 Climate

Between 1950 and 2020, precipitation in the Mitidja plain in northern Algeria exhibited interannual variability. It had a wet season before 1971 lasting nearly 4 years, followed by a prolonged drought period in the region (Fig. 3). Precipitation in the region has significantly declined (see Fig. 3). The Boufarik station recorded 1,006.97 mm of rainfall during the wet season. Typically, resource recovery takes place in November when rainfall exceeds evapotranspiration. Evapotranspiration accounts for approximately 80.13% of the total precipitation for the entire watershed. The Mitidja plain is influenced by the sub-humid Mediterranean climatic regime with continental influences, characterized by rainy and mild winters and hot summers. The study area receives about 640 mm rainfall per year on an average [26].

3.1.2 Hydrography

Hydrologically, the Mitidja plain is traversed by six major wadis that ensure the drainage of the mountainous watersheds of the Atlas Mountains [24, 27]. From east to west, these wadis are Hamiz, Djemaa, El Harrach, Chiffa, Bouroumi, and Djer. Downstream of the Mitidja plain, Wadi Bouroumi and Wadi Chiffa converge to form Wadi Mazafran. Wadi Harrach intersects with Wadi Djemaa near Baba Ali. To the east of Wadi Hamiz flows Wadi Réghaia, which is considered secondary. The wadis mentioned above are largely in hydraulic connection with the Mitidja aquifer, as their water can either infiltrate the aquifer or drain from it. The plain is divided into four river basins: Wadi Nador, Wadi Mazafran, Wadi Harrach, and Wadi Hamiz (Table 1). These wadis originate in the foothills of the Blidean Atlas. The rivers, particularly in their downstream reaches, contribute to the renewal of the alluvial aquifer of the Mitidja before flowing into the Mediterranean Sea.

The piezometric level in the Mitidja aquifer shows an alarming decline between 1990 and the early 2000s. In Mitidja East, a maximum fall of 18.6 m is recorded during low water and 15.9 m during high water, according to National Agency for Water Resources (ANRH) data. Moreover, during this period of groundwater regression, the Mitidja experienced rapid demographic change.

The number of inhabitants nearly doubled between 1987 and 2007, resulting in considerable demand for water [8]. In 2010, the volume of water mobilized exceeded 329 Mm^3, with surface water representing 53% and groundwater 28% [8].

The Mitidja region possesses a substantial hydrogeological potential, with access to a vast aquifer system estimated to contain around 500 million cubic meters of water resources. This aquifer is composed of two main segments: the Astian and the Quaternary alluvium. Notably, the Quaternary alluvium, the larger of the two, contributes a significant portion, supplying 295 million cubic meters annually, which accounts for 60% of the total available volume [24, 28].

3.1.3 Water Resources

The groundwater of the Mitidja aquifer supplies drinking water to many localities in the capital as well as several cities in the four wilayas – Algiers, Blida, Boumerdès, and Tipaza – located across large parts of the plain (Fig. 4). They also irrigate tens of thousands of hectares of agricultural land and supply almost to all industrial units.

According to the Hydrographic Basin Agency of Algiers [29], the groundwater resource of the Mitidja is estimated at 328 million m^3. This includes the main fields from which the aquifer is tapped: Mazafran I and II, Chebli, Barraki, Haouch Felit, and Hamiz [24].

Alongside the over-exploitation of the aquifer, exacerbated by prolonged drought, the piezometric level of the Mitidja aquifer has shown a continuous decline since 1980. Between 1980 and 2000, there was a significant and continuous drop in groundwater levels, with some areas experiencing declines of up to 30 m in their

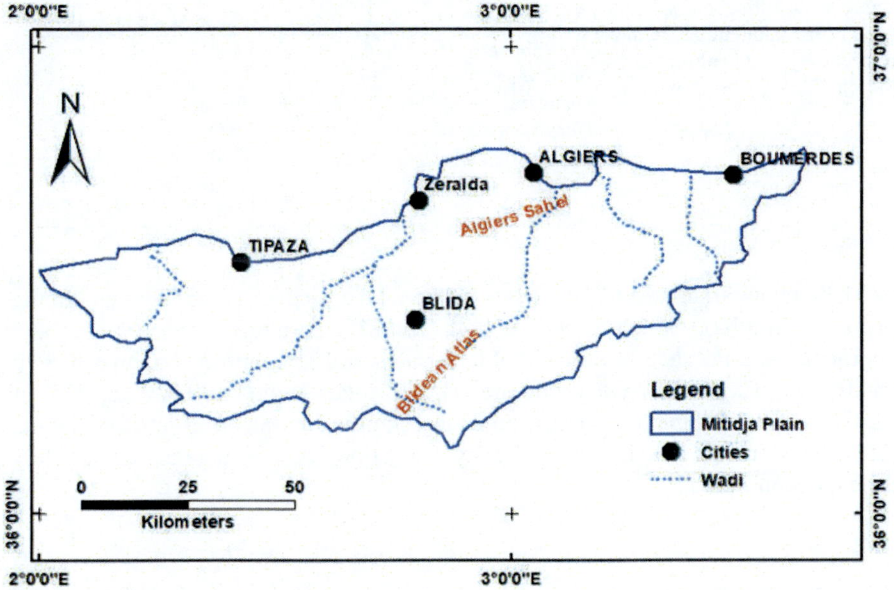

Fig. 4 Mitidja hydrographic region

aquifers. However, from 2000 to 2012, a stabilization of groundwater levels was also observed and reported by Meddi et al. [24]. Several factors may have contributed to this stabilization, including improved water management practices, initiatives to enhance groundwater recharge, potential changes in precipitation and drought conditions, increased awareness of groundwater conservation, technological advances, desalination adoption, research and monitoring efforts, and community engagement. Local geological and hydrological factors also play a crucial role in groundwater dynamics. Understanding these factors is crucial for sustainable groundwater management.

Moreover, the decline in water levels has facilitated the intrusion of saltwater into the eastern part of the Mitidja aquifer [30]. The aquifer of the plain of Wadi Nador in Tipaza serves as a stark example for several reasons: all boreholes tapping this aquifer (sixteen in total) have now been abandoned due to saltwater intrusion affecting the aquifer over a distance of more than 1.5 km from the coast [31].

3.1.4 Mitidja Agriculture

Soils in Mitidja are characterized by low organic matter content, with averages ranging between 0 and 3.5%. In Mitidja, fertilization rates vary widely depending on the type of crop, as detailed in Table 2 [25].

Table 2 Distribution quantities of ammonium nitrate with respect to types of culture in Mitidja ([25]; Open access)

Crop types	Applied amounts (kg/ha)
Vegetable crops	400
Dried vegetables	200
Citrus fruit	1,000
Trees	500
Vineyard	300
Cereals	200
Fillings	100

The total agrarian area (TAA: refers to the total agricultural land area within the plain of Mitidja), which includes both useful agricultural area (UAA) and unproductive land such as uncultivable terrains and structures, is approximately 164,000 ha. The useful agricultural area (UAA) totals around 100,000 ha; 60% of this is irrigated for intensive crops requiring significant water and fertilizer inputs [25]. The remaining land is covered by dry vegetation, where water delivery is solely dependent on rainfall.

3.2 *Cheliff Plain*

The Cheliff plain is located in west-central northern Algeria. The Cheliff-Zahrez hydrographic region (Fig. 2) consists of three river basins: the Cheliff, the Zahrez, and the coastal Dahra. Extending over 43,750 km^2, the Cheliff basin is the largest hydrographic basin in northern Algeria. This basin is subdivided into three watersheds: the upstream Cheliff basin ending at the Boughzoul dam, the Upper and Middle Cheliff basins, and the Lower Cheliff and Mina basins. The Cheliff valley, which is traversed by the Wadi Cheliff, is situated in the northern part of the Cheliff watershed and occupies 22% of the area in northern Algeria.

3.2.1 Climate

The climate of the Cheliff is characterized by hot, dry summers with occasional storms and mild, humid winters. This climate can be categorized as a semi-arid Mediterranean type [32] (Fig. 5). Existing climatic conditions play a crucial role in the hydrological and hydrogeological behavior of the basin. As shown in the data from 1950 to 2020 (Fig. 5), interannual rainfall in the Cheliff basin decreased from the early 1970s to the mid-2000s. Drought conditions reemerged starting in 2019.

The interannual rainfall map for the Cheliff basin illustrates the distribution of rainfall. Specifically, rainfall decreases as one moves inland from the sea, due to the gradual depletion of water vapor in the atmosphere. This happens as air currents lose their moisture while crossing over the mountain ranges [33]. Unusually, the Cheliff plains exhibit a semi-arid continental climate despite their proximity to the sea, being only about 50 km away on average [34]. The climate is notably harsh and has even

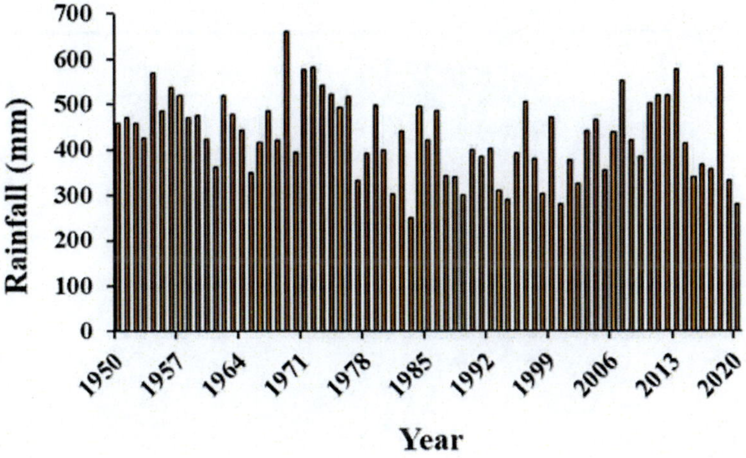

Fig. 5 Rainfall pattern in Cheliff (1950–2020)

been described as a "meteorological curiosity" for its unusual characteristics [35]. The area is called "Four du tell" or "portion of the Sahara lost in the tell." The temperature in the Cheliff region is significantly higher than in neighboring regions due to several factors, which can include its geographic location, local topography, and prevailing wind patterns. Often situated away from coastal or mountainous areas that typically have more moderate temperatures, the Cheliff may experience a more continental climate characterized by more significant temperature fluctuations.

In general, the average annual rainfall decreases from north to south and east to west. The annual rainfall is marked by strong temporal variability [36]. In the Upper and Middle Cheliff basins (Fig. 6) and the Dahra Mountains, the average annual rainfall exceeds 500 mm. The average rainfall varies between 300 and 500 mm/year in the Upper and Middle Cheliff plains, and between 350 and 400 mm/year in the Lower Cheliff and the Mina basins. The southern basin (Boughzoul) is extremely dry, with an average annual rainfall ranging between 170 and 260 mm/year [36].

3.2.2 Hydrographic Network

The hydrographic network of the Cheliff basin consists of a dense and intricate stream network, primarily composed of temporary watercourses (Fig. 7). The Wadi Cheliff is considered one of the major North African wadis. It is formed by the confluence of two large rivers: Wadi Touil and Nahr Ouassel. It has both the longest course and the highest flow in the region [37].

The Cheliff Wadi flows through a valley that runs parallel to the sea, featuring several meanders. Waters from Ebda and Ras Ouahrane flow into its right bank,

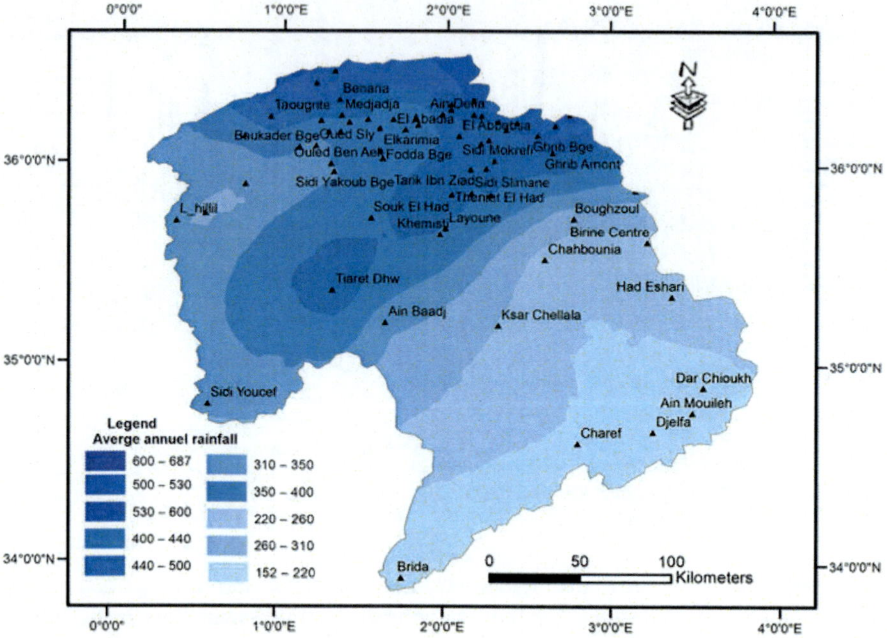

Fig. 6 Spatial distribution of average annual rainfall ([36]; Open access)

while Deurdeur, Harreza, Rouina Zeddine, Fodda, Sly, Rhiou, and Djedioula contribute to its left bank. The important tributary Wadi Mina also joins it further to the west [38]. The layout of the hydrographic network has been shaped by the structural elements that have influenced the region over geological time, particularly during the Quaternary period. The intricate hydrographic network was established due to major geological events that impacted the area and has been modified and reoriented over time due to tectonic evolution.

3.2.3 Groundwater Resources in the Catchment Area (Cheliff)

The Cheliff-Zahrez watershed comprises two distinct regions. In the northern part, it is characterized by the Cheliff valley, which is flanked by the imposing Tellian mountain ranges the Mount of Dahra to the north and the Ouarsenis massif to the south. These majestic mountain chains define the northern boundary of the watershed, showcasing the natural beauty of the region. In contrast, the southern section of the watershed encompasses the Zahrez basin, adding diversity to the landscape. This geographical diversity within the Cheliff-Zahrez watershed creates a unique and captivating environment, showcasing both the rugged charm of the mountains and the allure of the basin.

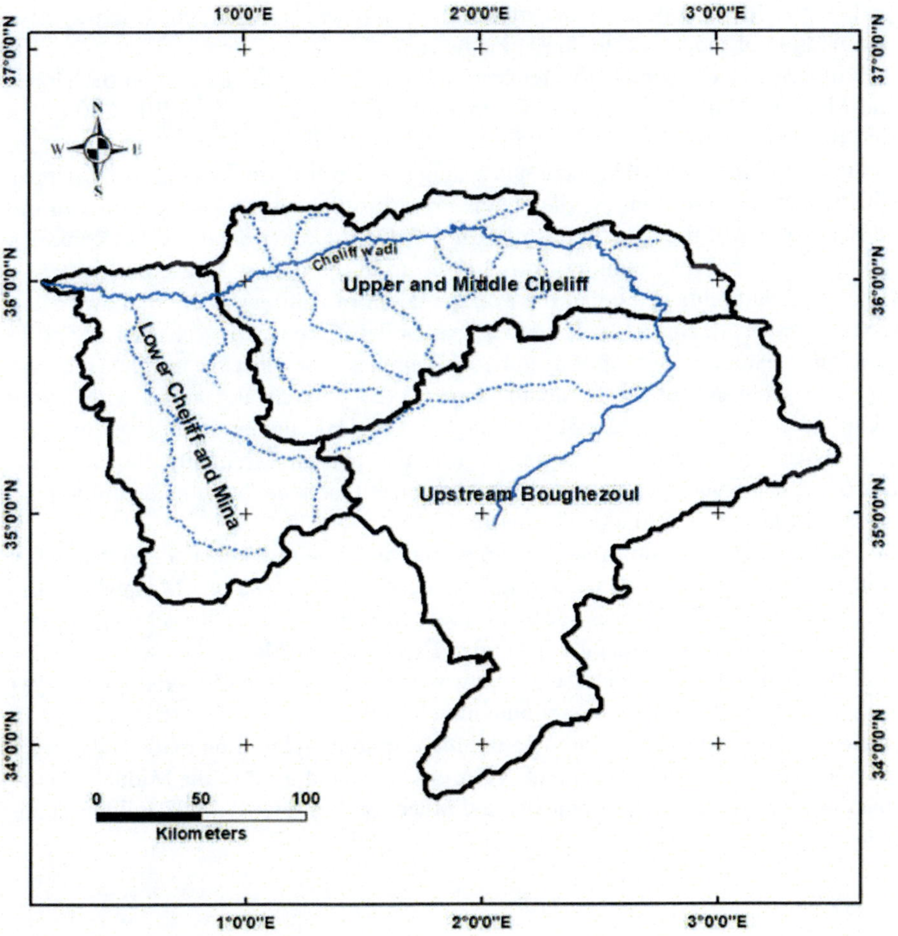

Fig. 7 Cheliff hydrographic region

Many geological formations in the area contain groundwater; the oldest of these are attributed to the Jurassic period, while the most recent correspond to Quaternary alluvium. In the northern zone, the two Tellian chains offer limited and unfeasible water resources found in limestone and sandstone formations. These formations are generally underdeveloped and are encased in larger, less permeable geological structures. The Cheliff-Zahrez region is home to 42 aquifers, boasting a water supply potential of 225 million cubic meters per year [17].

On the left bank, the groundwater levels are shallow. They vary between 3 and 60 m below ground in the plains and can reach depths of up to 80 m, particularly in the areas of Ain Soltane, Wadi Sly, and Boukadir. The aquifers are also shallow throughout the valley, with an average depth ranging from 10 to 30 m. This confirms that the groundwater accessed by these points originates from Quaternary alluvium

[39]. Water level alone does not indicate groundwater's origin, but it is part of a combination of tools used to determine its source.

Furthermore, the depth-to-water-level map of the alluvial aquifers in the Upper and Middle Cheliff displays a wide range of depths, ranging from 3 to 280 m, as illustrated in Fig. 9. The right bank features areas with deeper aquifers, such as Arib, located north of Khemis-Miliana and extending toward Ain Soltane in the Upper Cheliff aquifer. Additionally, the depths extend to the east toward El Amra in the alluvial aquifer of the Middle Eastern Cheliff. On the other hand, the left bank has shallow aquifer levels, ranging from 3 to 60 m in the plains. Measurements taken from multiple points show that the average depth of aquifers in the shallow valley ranges between 10 and 30 m. The water level or depth range, when studied alongside geological examinations of soil and rock samples, can provide insights into the origin of groundwater. If the depth range of the groundwater aligns with layers identified as Quaternary alluvium through geological studies, it suggests that the groundwater in those depths originates from that period. Essentially, the combination of the depth and the geological composition of that depth helps in identifying the age and origin of the groundwater source.

Madene et al. [39] confirmed these observations. The researcher added that in the area to the west of El Abadia (Fig. 8), depths exceed 80 m. Meanwhile, the maximum depth of 280 m is found in the northeastern part of the alluvial aquifer of Middle Cheliff, near the cities of Ouled Fares and Chattia.

It should also be noted that the groundwater resources in the Chlef region (wilaya) are primarily found in Quaternary alluvium, specifically Pliocene Astian sands, and in sedimentary rocks such as lithothamnium limestone. According to the Directorate of Water Resources, the volume of water extracted and used in the Middle Cheliff plain in 2014 from wells, boreholes, and other sources totaled 25.08 million cubic meters [41].

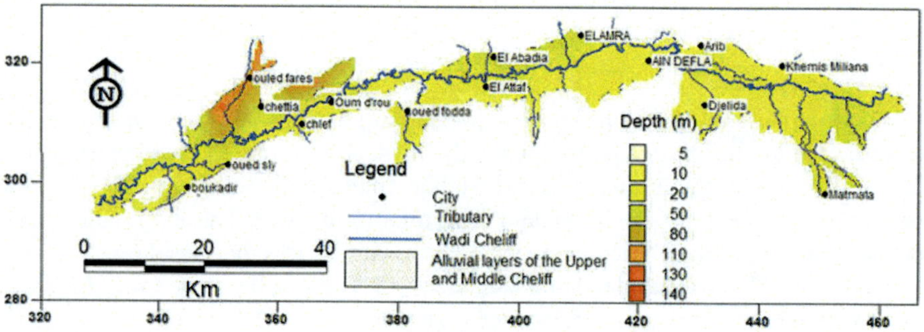

Fig. 8 Depth-to-water-level map of alluvial aquifers in Upper and Middle Cheliff ([39]; Open access)

4 Groundwater Quality and Quantity in the Cheliff and Mitidja Plains

Groundwater is the most heavily used resource in terms of volume. Between the years 1975 and 2000, Algeria experienced an intense and sustained drought, characterized by a 30% deficit in rainfall. This drought negatively impacted the flow patterns of rivers, the water levels in dam reservoirs, and groundwater recharge rates. These effects had serious repercussions on all socio-economic activities in the country. The main aquifers in the northern part of the country are currently being over-exploited, resulting in a decrease in the static water level, a reduction in available resources, and, most concerning, a deterioration in water quality that renders them unusable. Significant drawdowns in piezometric levels, exceeding 1 m per year in some cases, have been observed in numerous wells. Additionally, water quality has deteriorated in certain aquifers, particularly those located along the coast, such as the Mitidja and Annaba aquifers, due to marine intrusion [16, 42, 43]. Groundwater drawdowns have also been observed in the neighboring country of Morocco, where Del Vecchio and Kuper [44] have reported similar findings.

The renewal of this resource depends on aquifer recharge, which in turn is influenced by rainfall and evapotranspiration [44, 45]. An increase in evaporation and evapotranspiration, attributable to rising temperatures, could lead to decreased aquifer recharge and directly impact both the level and quality of the aquifers [46, 47].

Several recent studies have shown that climate trends correlate well with variations in groundwater levels [48, 49]. For example, in several regions of Algeria, groundwater levels have fallen by more than 30 m since 1975 due to drought and over-exploitation [24, 33, 42, 50, 51].

In Algeria, urban and industrial effluents are frequently discharged into the environment without appropriate treatment. These effluents are primary sources of water quality degradation, affecting both surface and groundwater. Agriculture also contributes to this degradation through the excessive use of pesticides, herbicides, and fertilizers, leading to groundwater pollution by nitrates [21, 25, 27, 39, 40, 42, 52–54].

Indeed, the Hydrographic Basin Agency, along with the Water Resources Directorate, Agricultural Services Directorate, and Directorate of Mines and Industry of the Departments of Ain Defla and Chlef, has identified potential sources of water pollution. They found that the plains of the Upper and Middle Cheliff, which include the aquifers of the same regions, have a population of about 817,227 inhabitants. These residents produce approximately 3,827,074 m^3 per day of domestic wastewater. This wastewater is treated by two plants located in the Ain Defla and Chlef municipalities, which have a combined treatment capacity of roughly 48,000 m^3 per day, or 4.38 million m^3 per year. Notably, the discharge point for these plants is the Wadi Cheliff. Nevertheless, because treatment facilities are lacking in other localities, it is estimated that roughly 2,983 tons of nitrogen are discharged annually through domestic wastewater. This poses a significant threat of groundwater pollution, as indicated by Madene et al. [39]. In addition, more than 616 tons of household waste are deposited daily at two Technical Landfill Centers (TLC) located in the municipalities of Ain Defla and Chlef. These centers are particularly intended for municipal solid waste from the city of Ain Defla. Waste is also deposited at other unregulated sites. While the number of landfill sites is relatively limited and the volume of waste is fairly average, there still exists a risk of pollution. This is because the waste contains high concentrations of organic matter.

It has been estimated that agri-food activities generate the largest pollutant load in this region, measured at 12,191,796 inhabitant-equivalent. This is followed by industrial manufacturing activities, which contribute a load of 1,564,975 inhabitant-equivalent. Breeding is generally a practice reserved for private individuals in this area. Farmers commonly use livestock discharges as fertilizers, producing approximately 611 tons of nitrogen annually. This results in a significant pollutant load, particularly in the municipalities of Chlef, Ouled Fares (Middle Cheliff), and Djelida (Upper Cheliff). Other affected areas include Djendel, Bir Ouled Khelifa, Ain Soltane, and Ain Defla (Upper Cheliff) [39].

4.1 Cheliff Plain

The Cheliff plain predominantly relies on groundwater from the alluvial aquifer for irrigation purposes. This has led to a high demand for water and a consequent significant decline in groundwater quantity and quality [55]. Excessive mineralization of this water often results in soil salinization, which severely limits crop yields as the water does not always meet irrigation standards (Fig. 9).

According to a study by Abdelbaki et al. [56] on the application of GIS in evaluating groundwater quality in the Lower Cheliff plain, they discovered several key findings. First, the coarse alluvium of the Lower Cheliff plain constitutes a significant aquifer, which is accessed through multiple boreholes. Second, while the aquifers contain fresh water, the surface layers of very fine formations – such as clays and silts – contain salty water at shallow depths. Lastly, they revealed that the groundwater in the Lower Cheliff plain is largely non-drinkable. The composition of this water is influenced primarily by the geological features of the region. Additionally, agricultural and industrial activities and wastewater discharged directly into the wadis contribute significantly to the chemical composition of the groundwater in this area.

In a study examining the evolution of hydrochemical quality of groundwater in the Upper Cheliff plain of Algeria, Achour et al. [57] discovered that nitrate concentrations range from 0 mg/l to 245 mg/l. These levels are significantly higher than the World Health Organization [58] standards, which are set at 50 mg/l.

The increase in nitrate concentrations has been progressive from the end of 1993 to the beginning of 2003, mainly affecting the front and center of the plain. Starting in the latter part of the year 2008, however, the nitrate contamination became more

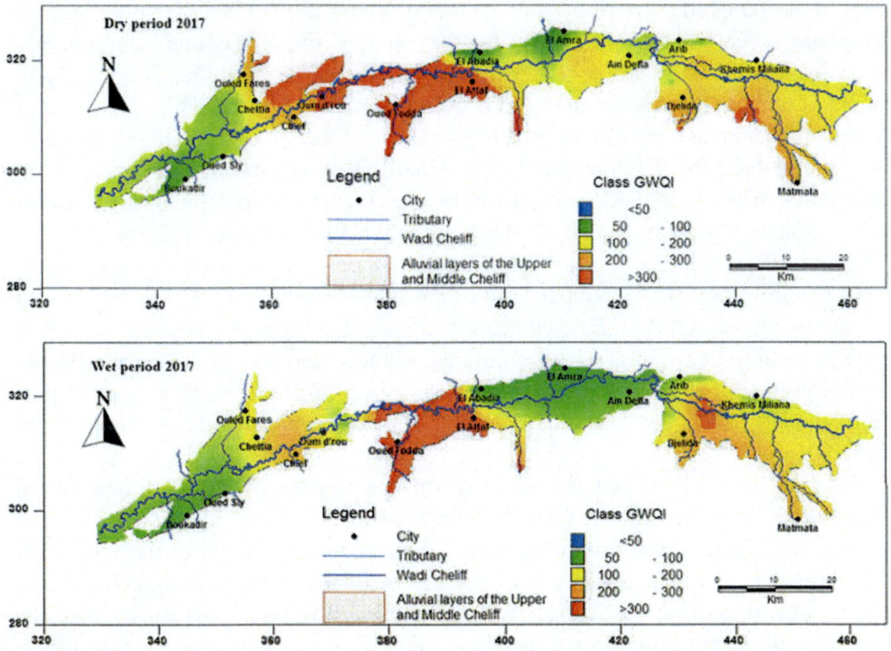

Fig. 9 Index map of the groundwater quality (GWQI) in the Upper and Middle Cheliff alluvial aquifers in 2017 ([40]; Open access)

significant and spread across the entire plain. Several factors could contribute to this excess. These may include the nitrification of organic nitrogen, leaching of nitrates from affected soils, use of chemical fertilizers, or discharges from local communities.

Bouchenouk [55] demonstrated that the chemical quality of water in the Boukadir alluvial aquifer, located in the western part of the Middle Cheliff basin, was of poor quality. This is primarily due to high salinity, which is influenced by both the existing lithology of the aquifer and prevailing climatic conditions. It is also noted that the evolution of nitrate concentrations for the period 1990–2010 of the alluvial aquifer of the Boukadir plain reflects a progressive temporal degradation of the quality of this water intended for consumption and/or irrigation. Regarding the environmental aspect, investigations into pollution sources in the Boukadir plain have revealed high levels of nitrate contamination, which are primarily of agricultural origin.

In a study on the geochemical characteristics of groundwater used for irrigation in the Cheliff plains, Bouzada [59] found that over 50% of the boreholes in all three plains had an electrical conductivity (EC) exceeding 2.25 dS/m, making them unsuitable for irrigation. However, only five boreholes – three in Lower Cheliff, one in Middle Cheliff, and one in Upper Cheliff – had sodium adsorption ratios

(SAR) that exceeded the threshold of 15 meq/l. Water in the Cheliff region tends to precipitate as calcite-gypsum, affecting the calcium ion levels and subsequently increasing SAR. When the waters from the Middle and Upper Cheliff are diluted, instead of making them safer, they can become more problematic. This is because of the potential increase in residual alkalinity. If these waters accumulate in the soil, they can be harmful. In contrast, Lower Cheliff waters present the highest risk, particularly in terms of SAR values. Understanding the geochemical evolution of these irrigation waters is crucial for predicting their impact on soil quality.

Gharbi [60] conducted a study assessing and mapping groundwater quality in the Upper Cheliff plain. He found that most of the study area is characterized by poor to medium quality water. A significant decline in water quality was observed, attributed to the gradual increase in levels of chlorides, nitrates, and sulfates. Gharbi's results also highlighted that groundwater quality deteriorates from the north to the south of the plain. However, the western part of the area exhibits relatively good water quality.

Meddi et al. [24] conducted a study on the impact of reduced rainfall on groundwater resources in the Cheliff-Zahrez basin. Meddi et al. [24] found that over the last four decades, diminished rainfall has led to a decrease in the volume of water stored in dams. Consequently, this has resulted in the over-exploitation of groundwater resources, particularly for agricultural purposes. This shows that climate variability and groundwater levels are closely interconnected, mainly due to fluctuations in precipitation. These variations have had a negative impact on the management of groundwater resources and have led to declining groundwater levels globally.

A study by Touhari et al. [61] investigated the evolution of nitrate ions from 2002 to 2008 in the Upper Cheliff aquifer. The study found that nitrate levels were consistently higher during the high-water season compared to the low water season. Notably, a spike in nitrate levels was observed during heavy rainfall in April, which is likely due to leaching from agricultural lands. The campaign in 2004 recorded maximum nitrate levels of 280 and 260 mg/l, significantly exceeding both WHO and Algerian standards of 50 mg/l. The researchers concluded that the high levels of nitrate pollution in the groundwater can primarily be attributed to various sources, including agriculture, livestock, and urban practices like the disposal of domestic and industrial waste.

Bouderbala [62] evaluated the suitability of groundwater for drinking and agricultural uses in the plain of Upper Cheliff. The study found that the concentration of most chemical constituents exceeds WHO standards, largely due to various sources of pollution. As a result, the groundwater is severely contaminated and deemed unsuitable for drinking purposes. However, the study indicates that groundwater is generally suitable for irrigation in most parts of the plain. Factors affecting the water's mineralization include the lithology of the aquifer, anthropogenic influences such as urban sewage discharge and fertilizer use, and natural processes like evaporation due to the semi-arid climate.

Further, Madene et al. [40] conducted a study to evaluate the groundwater quality in the plains of Upper and Middle Cheliff, located in Northwestern Algeria, and its suitability for both irrigation and consumption. The study found that the water quality poses a health hazard for humans, necessitating appropriate treatment before use. However, the water was generally found to be safe for irrigation, with exceptions for a few sensitive crops such as garlic, onions, beans, and strawberries.

4.2 Mitidja Plain

The alluvial aquifer of the Mitidja in Northern Algeria is a critically important underground reservoir that has experienced significant deterioration in recent years in terms of quantity and quality. Mimouni et al. [63] highlighted the pollution of this aquifer by nitrates. This pollution has been confirmed by subsequent studies conducted afterward by Hadjoudj [54], Ait Ouali [64], Yahiaoui [65], and Djoudar-Hallal [66]. Additionally, Khouli and Djabri [67] conducted an in-depth study over several years, attributing the presence of nitrates in the waters of the alluvial aquifer of the Mitidja to the excessive use of nitrogen fertilizers in agriculture [68].

In addition, Hadjoudj et al. [54] conducted a study on the physico-chemical characterization of groundwater in the Mitidja plain for the period from 2004 to 2007. They found that the concentrations of most physico-chemical characteristics met Algerian water consumption standards. However, nitrate concentrations were significantly higher than the threshold (50 mg/l). Based on these findings, they identified three potential sources of nitrate contamination:

1. Agricultural origin: The Mitidja is primarily an agricultural region. In its western area, the fertilization rate has reached nearly 400 kg of nitrogen per hectare.
2. Urban origin: Urban discharges channeled by the public sewerage system are not fully treated. Consequently, they are released into the wadis that traverse the plain. Since the aquifers in the Mitidja are alluvial, part of this polluted water infiltrates the underlying aquifers.
3. Industrial origin: The eastern region of the Mitidja has substantial industrial activity. Industrial effluents are discharged directly into the wadis without prior treatment. This has led to high levels of nitrate pollution, with concentrations averaging 120 mg/l. This is largely due to the intense socio-economic pressure exerted on this part of the Mitidja by industrial, agricultural, and urban activities.

Sengouga et al. [69] found that groundwater pollution and its spatial distribution in the Mitidja aquifer coincided with agricultural zones in most contaminated areas. The research indicates that nitrate levels in the aquifer have risen due to the intensification of agriculture in the region. Without preventive measures, this could lead to further deterioration in groundwater quality and results in environmental consequences. According to the final report from the Deutsche Gesellschaft für Internationale Zusammenarbeit (GIZ), or the German Agency for International Cooperation, published in 2016, an assessment of groundwater quality revealed an enrichment of nitrates in some regions, including the Mitidja aquifer. The main causes are the leaching of nitrates from soils, which originate from organic and mineral fertilizers not utilized by plants.

Furthermore, Zamiche et al. [70] demonstrated that the groundwater in the Mitidja plain exhibits high salinity. This can be explained by the EC values, which remain below the threshold of 3 dS/m, and the average SAR value, which stays below 5 meq/l. Overall, the high salinity is primarily due to the dissolution of geological formations and the evaporation of the groundwater. A variable influence of marine invasion is also noted, particularly in eastern Mitidja.

In summary, several studies have highlighted the critical issues facing the Mitidja aquifer in North Algeria:

Nitrate contamination: Zamiche et al. [68] found that intensive fertilizer use and poor soil retention have made the central part of the Mitidja aquifer highly vulnerable to nitrate contamination. This contamination has exceeded safety guidelines set by the World Health Organization.

Hydro-geochemical factors: Zamiche et al. [23] identified seawater intrusion and nitrate contamination as key factors affecting the aquifer's hydrochemical evolution, indicating a combination of natural processes and human activities contributing to groundwater degradation.

Groundwater depletion: Bouderbala [71] reported a significant drop in groundwater levels between 1974 and 2010, primarily attributed to reduced rainfall and over-exploitation. Coastal areas also suffer from high salinity due to seawater intrusion. Also, there are elevated nitrate concentrations attributed to agriculture and urban pollution.

Widespread nitrate pollution: Khouli et al. [25] noted disparities in nitrate levels across the entire aquifer. Nitrate concentrations rise during recharge and irrigation periods due to leaching from agricultural soils and the use of nitrogenous fertilizers, particularly in market gardening.

Growing threats: Algérie Presse Service [72] reported that the Mitidja aquifer faces severe threats from pollution and increased water withdrawals, driven by demographic and economic growth. Over-exploitation, especially for agriculture and drinking water supply, has led to surface aquifer drawdowns and marine intrusion in certain areas.

Pollution sources: Integrated Water Resources Management (AGIRE) found that nitrate concentrations in the groundwater significantly exceed standard limits and are primarily attributed to agricultural activities, urban discharges, and industrial activity. The eastern Mitidja region is particularly affected, with direct effluent discharges into wadis without treatment.

Quantitative and qualitative pressures: AGIRE's analysis highlighted that the aquifer faces both quantitative and qualitative pressures. Water withdrawals for various purposes are lowering the aquifer level, leading to marine intrusion, which jeopardizes water quality for drinking and irrigation.

It is evident that the Mitidja aquifer faces a complex set of challenges, including pollution, over-extraction, and seawater intrusion, which collectively threaten the region's groundwater quality and availability. Addressing these issues will require a comprehensive approach involving regulation, sustainable agricultural practices, and improved wastewater management to safeguard this vital resource.

4.2.1 Marine Water Intrusion in Coastal Aquifers (Mitidja Plain)

Over the past two decades in Algeria, there has been a marked increase in saltwater intrusion from the sea, predominantly due to the prolonged droughts plaguing the north. This led to excessive and unchecked water extraction to cater to rising demands. At present, every region spanning Algeria's extensive 1,200 km coastline faces the threat of this saline invasion. Multiple sites along the coast have reported groundwater contamination, with the central areas, including the plains of Wadi Nador, Wadi Mazafran, and the Bord El Bahri region, being significantly impacted [73].

Research conducted by Mania et al. [74] between 1981 and 1984 revealed a concerning trend of marine saltwater intrusion into the aquifer. This issue was attributed to the large-scale pumping primarily set up to address the water requirements of Algiers East. This included fulfilling the potable water needs of its residents and catering to several smaller industries. In addition, the aquifer was tapped to irrigate surrounding horticultural and tree-farming areas. Their findings indicated polluted regions at the Hamiz wadi mouth (Stambul) and around Bordj-El-Kiffan-Verte Rive. The saline spread could stretch up to 1 km inland, fluctuating with the aquifer's hydrodynamic conditions, notably during water-scarce periods.

Several hydraulic issues could be mitigated through the use of artificial groundwater recharge. Over-exploitation has dangerously lowered the water level of the Mitidja aquifer, even causing several instances of land subsidence in the region. Moreover, intensive pumping has led to the intrusion of saltwater into the coastal aquifers of the Wadi Nador plain, causing the saltwater wedge to continually expand within the aquifer. Now is a relevant time for Algeria to start using this water storage method, especially since it does not require significant material resources. Artificial groundwater replenishment could serve as an alternative solution to harness the billions of cubic meters of water that currently flow into the sea and cannot be captured through dam construction. This approach could also mitigate the drawdown of groundwater levels, which occurs due to the insufficient mobilization of surface resources.

Bouderbala et al. [75] conducted a study to understand the salinity levels of groundwater in the Nador coastal aquifer in Algeria. By analyzing high and low groundwater samples from 2013, they aimed to track the changes in groundwater hydrochemistry from the point of recharge to the coastal area. Their findings corroborated that the increased salinity stemmed from marine intrusion into the aquifer, extending approximately 2 km inland.

The findings by Khoualed and Remili [30] and Bechkit et al. [76] regarding marine intrusion along Algeria's coast are alarming. The evidence of lowering water tables, shifts in groundwater flow, and increasing mineralization toward the sea highlights a significant threat to the region's coastal aquifers. Bechkit et al. [76] discovery of clay's protective role against saltwater intrusion is intriguing and could inform future strategies for managing this issue.

The overall situation of coastal groundwater reserves in Algeria is deeply concerning. The existing shortfalls in water supply compared to demand are likely to worsen due to saltwater intrusion and pollution from various human activities. Despite dam construction and desalination efforts, the projected 1 billion cubic meter water shortfall by 2025 is alarming, especially for regions like Cheliff-Zahrez and Algiers-Soummam-Hodna.

Certainly, urgent and comprehensive measures are needed to address these challenges. This includes improving water resource management, reducing pollution, and exploring innovative solutions to mitigate the impact of marine intrusion on Algeria's coastal aquifers. Solving these issues is crucial for ensuring a sustainable water supply and protecting the environment and the livelihoods of those dependent on these vital water resources.

Finally, even if in the medium term, demand seems to be satisfied, the fact remains that a rigorous demand management policy is necessary. This can be achieved through several key measures:

1. **Construction of wastewater treatment plants and adherence to discharge standards**: To safeguard water quality, investing in the construction of wastewater treatment plants is imperative. Strict adherence to standards for the discharge of treated water into the receiving environment is essential. This ensures that the water returned to natural sources is safe and clean. Moreover, regulating agricultural activity in the vicinity of boreholes intended for human consumption is crucial to prevent contamination.
2. **Artificial groundwater recharge**: Utilizing artificial groundwater recharge can address various hydraulic problems. This approach offers a sustainable alternative to prevent the loss of billions of cubic meters of water that currently flows into the sea. By recharging groundwater artificially, we can bolster water resources and reduce wastage.
3. **Immediate closure of heavily contaminated boreholes and wells**: Boreholes and heavily contaminated wells should be promptly shut down to prevent further groundwater pollution. This action is critical for protecting the health of those dependent on these water sources.
4. **Cessation of operations in vulnerable areas**: Ceasing water extraction operations in areas vulnerable to intrusion, such as regions susceptible to saltwater intrusion, is essential. This proactive step helps maintain the integrity of freshwater aquifers.
5. **Promotion of micro-irrigation**: The widespread adoption of micro-irrigation techniques can significantly reduce water consumption in agriculture. This efficient method optimizes water use while sustaining crop yields.

6. **Piezometric measurement campaigns**: Regular piezometric measurement campaigns are essential to monitor groundwater level fluctuations. These measurements provide valuable data for assessing the health of aquifers and the impact of water extraction.
7. **Chemical and geophysical analyses**: Conducting chemical and geophysical analyses is vital for monitoring and locating the freshwater–saltwater interface within aquifers. This information guides effective management strategies to combat saltwater intrusion.
8. **Study and modeling of saltwater intrusion**: In-depth studies and modeling efforts are necessary to understand and predict the spread of saltwater intrusion into freshwater aquifers. This knowledge enables the development of targeted strategies to mitigate this critical issue.

In summary, a comprehensive approach that encompasses these measures is imperative for sustainable groundwater management. By implementing these strategies, we can ensure the availability of clean and secure groundwater resources for future generations while addressing the challenges posed by increasing demand and environmental pressures.

5 Conclusion

The water deficit of the last few decades has negatively affected agricultural production as well as surface and groundwater resources in the Cheliff and the Mitidja plains of Algeria. The alluvial aquifers of the Mitidja and Cheliff are very important groundwater reservoirs in the country that have suffered a great deal of deterioration in recent years both quantitatively and qualitatively.

Furthermore, it has been observed that nitrate concentrations consistently exceed established norms. This pollution primarily arises from agricultural activities and urban discharges channeled through the public sewage system and ultimately discharged into the wadis traversing the plain. At the same time, the eastern region of the Mitidja has a very high level of industrial activity whose effluents are discharged directly into the wadis without prior treatment, stressing that this region is the most affected by pollution where nitrate concentrations reach an average 120 mg/l.

Furthermore, the Cheliff plain, primarily dedicated to agriculture, heavily depends on groundwater for irrigating a diverse range of crops, including market gardening, orchards, and even supplementary irrigation for cereals. Besides the challenge of ensuring an adequate water supply, the quality of this groundwater does not consistently meet irrigation standards. Excessive mineralization of the water frequently results in soil salinity issues, significantly hampering crop yields.

6 Recommendations

To address the challenges of water resources in Algeria, a multifaceted approach is needed:

- **Water resource management**: Implementing sustainable water resource management practices is essential. This includes efficient use of available water resources, adopting modern irrigation techniques, and regulating groundwater extraction to avoid further depletion.
- **Pollution control**: Stringent measures should be taken to control pollution sources, particularly agricultural runoff and untreated industrial discharges. Implementing proper wastewater treatment before discharge is crucial.
- **Agricultural best practices**: Encouraging sustainable and eco-friendly agricultural practices can play a pivotal role in diminishing the influx of pollutants into the water system. This encompasses judicious application of fertilizers and chemicals, effective irrigation management, and the adoption of soil conservation techniques.
- **Water quality monitoring**: Regular monitoring of water quality is essential to track changes and trends. This information can guide decision-making and help identify areas where intervention is most urgently needed.
- **Public awareness**: Raising awareness among the public, industries, and agricultural communities about the importance of water conservation and pollution prevention can encourage everyone to take responsible actions.
- **Legislation and enforcement**: Enforcing water quality regulations and holding industries and individuals accountable for pollution is crucial. This may involve setting and enforcing limits on pollutants in water sources and imposing penalties for non-compliance.
- **Research and innovation**: Investing in research and innovative technologies for water treatment, pollution prevention, and efficient water use can contribute to more sustainable water management solutions.

Addressing these challenges requires collaboration between government agencies, industries, agricultural sectors, researchers, and communities to ensure the long-term health of the Cheliff and Mitidja plains' water resources and agricultural productivity.

Additionally, to bridge the whole research gap in the topic, the future research should focus on:

1. Identify the specific pollutants in the aquifers
2. Determine the sources of these pollutants
3. Track the temporal trends of pollutant concentrations
4. Assess the health and environmental risks posed by the polluted aquifers
5. Propose and test pollution mitigation strategies
6. Evaluate the sustainability of water management considering population and industrial growth
7. Engage local communities for better insights and sustainable practices

Acknowledgments This study was carried out by the authors in the GEE scientific research laboratory of the Higher National School of Hydraulics. The study is carried out within the framework of the SWATCH project (Prima project) funded by the DGRSDT, Algeria. We thank editors for their critical constructive comments on this chapter.

References

1. Zhang T, Wang P, He J, Liu D, Wang M, Wang M, Xia S (2023) Hydrochemical characteristics, water quality, and evolution of groundwater in Northeast China. Water 15(14):2669. https://doi.org/10.3390/w15142669
2. Bose S, Mazumdar A, Basu S (2023) Evolution of groundwater quality assessment on urban area – a bibliometric analysis, groundwater for sustainable development. 20:100894, ISSN 2352-801X. https://doi.org/10.1016/j.gsd.2022.100894
3. Quiggin J, Tan P-L (2004). Sustainable management of the Great Artesian Basin: an analysis based on environmental economics and law. Risk and Sustainable Management Group Working Papers 149839, University of Queensland, School of Economics
4. Hallouz F, Meddi M, Ali Rahmani S (2022) Multivariate analysis to assess the quality of irrigation water in a semi-arid region of north west of Algeria: case of Ghrib reservoir. Environ Earth Sci 81:158. https://doi.org/10.1007/s12665-022-10272-5
5. Hanak E, Escriva-Bou A, Gray B, Green S, Harter T, Jezdimirovic J, et al. (2019) Water and the future of the San Joaquin Valley. Public Policy Institute of California. https://doi.org/10.13140/RG.2.2.24360.83208
6. Koshy J, 2023. Groundwater exploitation is silently sinking the ground beneath India's feet. Subduction in parts of Haryana, Punjab, and Delhi is as high as 7–12 cm a year, but a reversal is possible if aquifers are left to charge. May 14, 2023, - New Delhi. The Hindu ePaper. https://www.thehindu.com/sci-tech/energy-andenvironment/groundwater-exploitation-is-silently-sinking-the-ground-beneath-indias-feet/article66847379.ece
7. Besser H, Dhaouadi L (2022) An overview of groundwater resources evolution in North Africa: sustainability assessment of the CI aquifer under natural and anthropogenic constraints. Meteorol Hydrol Water Manag 10(1):73–94. https://doi.org/10.26491/mhwm/150572
8. Drouiche A, Nezzal F, Djema M (2019) Interannual variability of precipitation in the Mitidja plain in northern Algeria. Revue des sciences de l'eau / Journal of Water. Science 32(2): 165–177. (In French). https://doi.org/10.7202/1065205ar
9. Hallouz F, Meddi M, Mahe G (2013) Changes of the hydroclimatic regime in the basin of the Wadi Mina (Northwest Algeria). J Water Sci 26(1):33–38. https://doi.org/10.7202/1014917ar. In French
10. Hallouz F, Meddi M, Mahé G, Ali Rahmani S, Karahacane H, Brahimi S (2020) Analysis of meteorological drought sequences at various timescales in semi-arid climate: case of the Cheliff watershed (northwest of Algeria). Arab J Geosci 13:280. https://doi.org/10.1007/s12517-020-5256-5
11. Meddi M, Hubert P (2003) Impact of the modification of the rainfall regime on water resources in northwestern Algeria. Hydrology of the Mediterranean and semi-arid regions. IAHS 278:1–7. https://api.semanticscholar.org/CorpusID:162720635 (In French)
12. Meddi H, Meddi M (2007) Spatial and temporal variability of precipitation in northwestern Algeria. Geogr Tech 2:49–55. ISSN 1842-5135 (Printed version. (In French))
13. Meddi H, Meddi M (2009) Variability of annual precipitation in northwestern Algeria. Sécheresse 20(1):173–184. https://doi.org/10.1684/sec.2009.0169. In French
14. Meddi M, Talia A, Martin C (2009) Recent evolution of climatic conditions and runoff in the Macta watershed (northwestern Algeria). Physio-Géo [online], vol 3 | 2009, posted on January 1, 2009. Accessed 18 Aug 2023. http://journals.openedition.org/physio-geo/686; https://doi.org/10.4000/physio-geo.686. In French

15. Merabti A, Meddi M, Martins DS, Pereira LS (2018) Comparing SPI and RDI applied at local scale as influenced by climate. Water Resour Manag 32:1071–1085. https://doi.org/10.1007/s11269-017-1855-7
16. Taibi S, Meddi M, Souag D, Mahé G (2013) Evolution and regionalization of precipitation in northern Algeria (1936-2009). In: Taibi S (ed) Considering hydrological change in reservoir planning and management. Proceedings of IAHS-IAPSO-IASPEI assembly, Gothenburg, Sweden, July 2013, IAHS Publ, p 362. http://iahs-iapso-iaspei2013.com/. (In French)
17. Ladjal R (2013) Issues of mobilization and preservation of water resources in Sersou (upper Cheliff Basin, Boughzoul). Master's thesis in hydraulic engineering. Abou Bakr Belkaid University, Tlemcen. 145 p. In French
18. Meddi M, Boucefiane A, Sadeuk Belabbes A (2010) Impact of climate change on streamflow in the Chellif Basin, Algeria. In: Global change: facing risks and threats to water resources (Proceedings of the sixth world FRIEND conference, fez, Morocco, October 2010), IAHS Publ. 340, pp 95–102. (In French)
19. Ghenim A, Megnounif A (2013) Ampleur de la sécheresse dans le bassin d'alimentation du barrage Meffrouche (Nord-Ouest de l'Algérie). Géographie Physique et Environnement 7:35–49. Disponible sur: http://physio-geo.revues.org/3173
20. Hallouz F, Meddi M, Mahé G, Karahacane H, Ali Rahmani SE (2019) Tendance des précipitations et évolution des écoulements dans un cadre de changement climatique: bassin versant de l'oued Mina en Algérie. Revue des sciences de l'eau/J Water Sci 32(2):83–114. https://doi.org/10.7202/1065202ar
21. Boufekane A, Yahiaoui S, Meddi H, Meddi M, Busico G (2021) Modified DRASTIC index model for groundwater vulnerability mapping using geostatistic methods and GIS in the Mitidja plain area (Algeria). Environ Forensics 23(5–6):539–556. https://doi.org/10.1080/15275922.2021.1913674
22. El Meddahi Y (2016) Climate change and its impacts on water resources: the case of the Cheliff basin. Doctoral thesis, Hassiba Benbouali University of Chlef, Algeria, 99 p, (In French)
23. Zamiche S, Hamaidi-Chergui F, Demiai A, Belaidi M (2018b) Identification of factors controlling the quality of groundwater in Mitidja plain (North Algeria) using indexing method and statistical analysis. J Fundam Appl Sci 10(1):248–267. http://www.jfas.info
24. Meddi M, Boufekane A, Meddi H, (2015) Artificial recharge of the Mitidja aquifer. European University Editions (November 9, 2015), ISBN-10: 3841678351, ISBN-13: 978-3841678355, 76 p, (In French)
25. Khouli MR, Haouchine A, Banton (2021) Deterioration of the quality of ground water in agricultural region. Case Mitidja (Algeria). J Fundam Appl Sci 13(1):172–184. ISSN 1112-9867. http://www.jfas.info/. https://doi.org/10.4314/jfas.v13i1.10
26. Namane L (2009) Irrigation monitoring on a farm in the western Mitidja region, Mouzaia municipality. State Engineer's thesis in agronomy, specializing in hydraulics, ENSA of El Harrach, Algiers, 66 p. In French
27. Boufekane A, Meddi H, Meddi M (2020) Delineation of groundwater recharge zones in the Mitidja plain, North Algeria, using multi-criteria analysis. J Hydroinf. https://doi.org/10.2166/hydro.2020.082
28. Meddi H, Boufekane A, Meddi M (2014) Impact of climate change on groundwater (the Mitidja plain). 41st IAH international congress "Groundwater: Challenges and Strategies" Marrakech, September, 15–19. https://doi.org/10.13140/2.1.4210.0000
29. ABH (2000) River Basin Agency report
30. Khoualed R, Remili S (2017) Detection of marine intrusion into coastal aquifers: case study of the eastern Mitidja plain. Master's thesis, University of Ouargla, 178 p
31. Haouchine A, Haouchine FZ, Labadi A (2015) Changements climatiques et activités anthropiques: impacts sur les aquifères côtiers en Algérie. Larhyss J., ISSN 1112-3680 24:227–241
32. Gomer D (1994) Flow and erosion in small watersheds with marly soils under semi-mediterranean climate. Algerian-German Technical Cooperation. GTZ-ANRH, Germany. (In French)

33. Ahmed Rahmani R (2018) Optimization of water resources in a semi-arid to arid zone (case of the eastern middle Cheliff Basin). Master's thesis in Earth Sciences University of Oran 2. 218 p. (In French)
34. SCET – ARGI (1985) Assessment of water resources, study of the reorganization and extension of the Middle Cheliff Area Report A1.1. Published by the Ministry of Hydraulics, pp 4–28. (In French)
35. Seltzer P (1946) The climate of Algeria. Works of the Institute of Meteorology and Geophysics of Algeria, special issue, University of Algiers. Printed by La Typo-Litho and J Carbonel, 219 p. (In French)
36. Habibi B, Meddi M, Torfs PJJF, Remaoun M, Van Lanen HAJ (2018) Characterisation and prediction of meteorological drought using stochastic models in the semi-arid Chéliff–Zahrez basin (Algeria). J Hydrol Reg Stud 16:15–31., ISSN 2214-5818. https://doi.org/10.1016/j.ejrh.2018.02.005
37. Khedimallah A (2021) Impact of climate and anthropogenic changes on water resources in the Cheliff and Medjerda basins. Doctoral thesis from the National School of Hydraulics, 279 p
38. Rodier J, Colombani J, Claude J, Kallel R (1981) Hydrological monographs of the Medjerda basin; ORSTOM, N°6. Paris. 472 p, ISBN 2-7099-0595-7. http://www.hydrosciences.fr/sierem/Bibliotheque/biblio/Monographies/N%C2%B06-Monographie%20de%20la%20Mejerdah.pdf. (In French)
39. Madene EL, Meddi H, Boufekane A, Meddi M (2020) Contribution of hydrogeochemical and isotopic tools to the management of the upper and middle Cheliff aquifers. J Earth Sci 31(5): 993–1006. https://doi.org/10.1007/s12583-020-1293-y
40. Madene EL, Boufekane A, Meddi M, Gianluigi B, Zghibi A (2022) Spatial analysis and mapping of the groundwater quality index for drinking and irrigation purpose in the alluvial aquifers of upper and middle Cheliff basin (north-West Algeria). Water Suppl 22(4): 4422–4444. https://doi.org/10.2166/ws.2022.107
41. Zerarka S (2018) Water resources in the middle Cheliff plain. Bul Soc Géo D'Egypte Tome XCI:89–99. Print ISSN1110-5232, Online ISSN2735-3036. (In French)
42. Nichane M, Khelil MA (2015) Climate change and water resources in Algeria: vulnerability, impact, and adaptation strategy. Larhyss J 21:15–23. ISSN 1112-3680, (In French)
43. Progress (Engineering Office) (2016) Study of Water Sector Assessment in Algeria, Current Situation, Final Version. Final Study Report (Deutsche Gesellschaft für Internationale Zusammenarbeit (GIZ)), Algeria, December, 2016, PROGRESS Engineering Office (Algeria), 102 p. http://projet.oss-online.org/maghreb-eau/sites/default/files/2020-01/Rapport-GIZ-Final_ALGERIE_0%20%283%29.pdf. (In French)
44. Del Vecchio K, Kuper M (2021) Making Groundwater Visible in Morocco: A Process Historically Linked to Irrigation Development Policies. "Sustainable Development and Territories", [Online], Vol. 12, No. 3 | December 2021, published on March 22, 2022. Accessed 25 Mar 2022. http://journals.openedition.org/developpementdurable/19675. (In French)
45. Rutulis M (1989) Groundwater drought sensitivity of southern Manitoba. Can Water Res J 14(1):18–33. https://doi.org/10.4296/cwrj1401018
46. Hafmann N, Mortsch L, Donner S, Dunacan K, Kreutzwiser R, Kulshreshtha S, Piggott A, Shellenberg S, Schertzer B, Slivizky M (2000) Climate change and variability: impacts on Canadian water. Environmental Adaptation Research Group, Environment Canada, University of Waterloo, 120 p. https://citeseerx.ist.psu.edu/document?repid=rep1&type=pdf&doi=2ea17a36dab1c47e91020893b3eb44285bcb74da#page=29
47. Koudjega HK, Kodja DJ, Vissin WEE (2020) Evaluation of the recharge of deep groundwater in the Upper Paleocene aquifer in the Mono-Couffo Basin in Benin. Int J Progr Sci Technol (IJPSAT). ISSN: 2509-0119. Int J Sci High Technol 20(2):145–155. http://ijpsat.ijsht-journals.org. (In French)
48. Mohammed A, Benchaben H (2016) Réflexions Sur Les Variations Pluviométriques De La Région De Tiaret (Algérie Occidentale) Durant La Période 1984 – 2015. Eur Sci J 12(11):498. https://doi.org/10.19044/esj.2016.v12n11p498

49. Bah OA, Kone T, Yaffa S, Ndiaye ML (2019) Land use and land cover dynamics in Central River Region of the Gambia, West Africa from 1984 to 2017. Am J Mod Energy 5(2):5–18. https://doi.org/10.11648/j.ajme.20190502.11
50. Ouis S (2012) Impact of climatic fluctuations on the quantity and quality of groundwater in a semi-arid region: case of the Ghrisss Plain (Northwest Algeria). Larhyss J 11:119–131. (In English)
51. Bellal S-A, Baiche A, Dari O (2020). Drought and fluctuations of groundwater resources: the case of the Mostaganem Plateau (Western Algeria). The "Geographical Notebooks of the West" (CGO). Double issues 14 and 15. (In English)
52. Benlecheheb MW, Bouzid-Lagha S (2014) Cartographic approach to the study of water pollution in the alluvial aquifer of eastern Mitidja (Algeria). TSM 9:20–30. https://doi.org/10.1051/tsm/201509018. In French
53. Gorine M, Benkhelifa M, Gacem F, Bellague D (2019) Evaluation of the quality of groundwater used for irrigation in the Mina plain, Algeria. J Water Sci 32(3):223–234. https://doi.org/10.7202/1067306ar. In French
54. Hadjoudj O, Bensemmane R, Saoud Z, Reggabi M (2014) Groundwater pollution by nitrates in Mitidja: current situation and corrective measures. Eur J Water Qual 45:57–68. https://doi.org/10.1051/wqual/20140010. In French
55. Bouchenouk I (2013) Enrichment process of nitrates in groundwater in semi-arid areas, case of the Boukadir plain (northwestern Algeria). Master's thesis in Geology, University of Oran, 128 p. (In French)
56. Abdelbaki C, Rahmani B, Kassoul L (2004) Application of geographic information systems in the assessment of groundwater quality – case of the lower Cheliff. Scientific and technical semi-annual review. Water Environ J:31–43. https://www.asjp.cerist.dz/en/downArticle/243/3/5/37807. (In French)
57. Achour K, Kahila A, Sadeuk Ben Abbes A (2013) Evolution of the hydrochemical quality of groundwater in the upper Cheliff plain, Algeria. The International Seminar on Hydrogeology and Environment, November 5th - 7th, 2013. Ouargla (Algeria). (In French)
58. WHO (2017) Guidelines for drinking water quality: 4th edn. incorporating first addendum Switzerland: Geneva. 564 p, ISBN 978-92-4-254995-9 file:///C:/Users/FAIZA/Downloads/9789242549959-fre.pdf (In French)
59. Bouzada N (2013) Geochemical characterization of groundwater used for irrigation in the Cheliff plains. Master's thesis, Hassiba Ben Bouali University of Chlef, 128 p. http://dspace.univ-chlef.dz:8080/jspui/handle/123456789/116. (In French)
60. Gharbi B (2013) Mapping of groundwater quality using geographic information systems (GIS), 3D geomodeling, and spatial analysis: case study of the alluvial plain of Upper Cheliff, Algeria. Int Sem Hydrogeol Environ. November 5th–7th, Ouargla (Algeria), http://dspace.univ-ouargla.dz/jspui/handle/123456789/12085. In French
61. Touhari F, Meddi M, Mehaiguene M, Razack M (2015) Hydrogeochemical assessment of the upper Cheliff groundwater (north West Algeria). Environ Earth Sci. https://doi.org/10.1007/s12665-014-3598-6. ISSN 1866-6280
62. Bouderbala A (2017) Assessment of water quality index for the groundwater in the upper Cheliff plain, Algeria. J Geol Soc India 90:347–356. September 2017, 0016-7622/2017-90-3-347/$ 1.00 © Geol. Soc. India. https://doi.org/10.1007/s12594-017-0723-7
63. Mimouni O, Gaid A, Chibane B, Akli A (1989) Pollution by nitrates of groundwater in the Mitidja plain. J Water Soils Algeria 3. ANRH, Algiers. ISSN 2335-2124, In French
64. Ait-Ouali A (2007) Hydrogeological synthesis and vulnerability to pollution of the quaternary aquifer system of Mitidja. Master's thesis, USTHB (University of Science and Technology Houari Boumediene), 123 p. (In French)
65. Yahiaoui S (2011) Evaluation of groundwater quality in Mitidja and its suitability for irrigation. Master's thesis Water Engineering ENSH. (In French)
66. Djoudar-Hallal D (2014) Methodological approach to the vulnerability of groundwater resources in highly urbanized environments: the example of the coastal plains (Mitidja) in

Algeria. Doctoral thesis, USTHB (University of Science and Technology Houari Boumediene), 2014., 157 pages., (In French)

67. Khouli MR, Djabri L (2011) Impact of use of agricultural inputs on the quality of groundwater case of Mitidja plain (Algeria). Geogr Tech 2:35–44. ISSN 1842–5135 (Printed version)
68. Zamiche S, Hamaidi-Chergui F, Demiai A (2018a) Pollution of the quaternary aquifer of Mitidja (Algeria) by nitrates: origins and impacts of the quality for human consumption. J Fundam Appl Sci 10(1):113–131. ISSN 1112-9867. Available online at http://www.jfas.info
69. Sengouga A, Semar A, Bentala B (2015) Spatio-temporal variation of nitrate pollution in the western Mitidja aquifer. Conference: First National Biodiversity Seminar, Environment and Food Security in Algeria, October 2015. https://www.researchgate.net/publication/330514900. (In French)
70. Zamiche S, Hamaidi-Chergui F, Demiai A, Belaidi M (2017) Origins and mechanisms of salinity acquisition in the quaternary aquifer waters of Mitidja (northern Algeria). Rev Agrobiol 7(2):644–657. www.agrobiologia.net ISSN (print): 2170-1652. e-ISSN (online): 2507–7627
71. Bouderbala A (2019) The impact of climate change on groundwater resources in coastal aquifers: case of the alluvial aquifer of Mitidja in Algeria. Environ Earth Sci. https://doi.org/10.1007/s12665-019-8702-5
72. Algeria Press Service (2022) Water resources: The Mitidja aquifer threatened by pollution. Published on: Tuesday, March 23, 2021, 6:30 PM. Water resources: the Mitidja aquifer threatened by pollution (aps.dz). (In French)
73. Remini B (2010) The issue of water in Northern Algeria. Larhyss J 8:27–46. ISSN 1112-3680, In French
74. Mania J, Imerzoukene S, Braillon JM (1985) Saline pollution of the coastal aquifer to the east of Algiers. Hydrogéologie 3:213–226. 13. (In French)
75. Bouderbala A, Remini B, Saaed Hamoudi A, Pulido-Bosch A (2016) Application of multivariate statistical techniques for characterization of groundwater quality in the coastal aquifer of Nador, Tipaza (Algeria). Acta Geophys 64(3):670–693. https://doi.org/10.1515/cage-2016-0027
76. Bechkit MA, Bourouis S, Mouhoub FC, Aichaoui M, Arrache K (2022) Characterization of a marine intrusion using electrical sounding and seismic refraction: case of the Wadi Nador plain Tipaza, Algeria. https://doi.org/10.21203/rs.3.rs-1327688/v1

Groundwater Pollution Sources and Its Quality in the Kingdom of Saudi Arabia: State of the Art

Mustafa El-Rawy and Heba Fathi

Contents

1 Introduction 216
2 Materials and Methods 219
2.1 Study Area Description 219
2.2 Aquifers in Saudi Arabia 220
2.3 Pollution Sources 223
3 Groundwater Quality Assessments: Methods and Findings 226
4 Discussions 228
5 Conclusions 230
6 Recommendations 231
References 231

Abstract The Kingdom of Saudi Arabia (KSA) is noted for being one of the driest places in the world, with few drinking water resources. The use of fertilizers in agriculture, home sewage in the area, industrial effluent discharge along the Red Sea coast, and petroleum industries are all contributing to the pollution of the shallow groundwater in KSA's major towns. This chapter concentrates on groundwater quality issues caused by anthropogenic and/or geogenic sources in KSA. To evaluate groundwater quality and assess aquifer sensitivity to pollutants, we investigated several water quality factors in this study. This study comprehensively provides the various aquifer layers and the groundwater quality issue. Although most

M. El-Rawy (✉)
Civil Engineering Department, Faculty of Engineering, Minia University, Minia, Egypt

Civil Engineering Department, College of Engineering, Shaqra University, Dawadmi, Saudi Arabia
e-mail: mustafa.elrawy@mu.edu.eg; mustafa.elrawy@su.edu.sa

H. Fathi
College of Design and Architecture, Jazan University, Jazan, Saudi Arabia
e-mail: hamansour@jazanu.edu.sa

Shakir Ali and Abdelazim Negm (eds.), *Groundwater Quality and Geochemistry in Arid and Semi-Arid Regions*, Hdb Env Chem (2024) 126: 215–236, DOI 10.1007/698_2023_1050,

Published online: 9 December 2023

groundwater contaminants are of natural origin, the area also experiences extensive wastewater effluent discharge, industrial, mining, and agricultural activities, all of which contribute to the heavy metal pollution of aquifers. Previous studies, carried out in various regions of the kingdom, revealed that a significant portion of the groundwater quality exceeded the allowable levels, and the majority of it is considered unfit for drinking. However, taking into account the high salinity levels that are ideal for some crops, it may be suitable for irrigation uses. Therefore, this chapter provides precise updated information that aids decision-makers and national planners in making the best decisions that balance society's needs for water resources and satisfactory environmental impacts.

Keywords Aquifers in Kingdom of Saudi Arabia, Arabian shield, Groundwater quality in the middle east, Water quality index, Water resources

1 Introduction

Water is one of the most important factors of economic and social development, as it is essential to meet the needs of humanity, provide environmental management, and ensure the sustainability of economic development [1, 2]. Despite the importance of water, the Kingdom of Saudi Arabia (KSA) faces great challenges due to the unsustainable use of water resources [3], as there is a significant demand-supply mismatch, and climate change has added to the strain [4].

According to the limited water reserves of nonrenewable groundwater, which is rapidly depleting, and under arid climatic conditions, water is a renewable resource that is extremely scarce [5, 6]. In addition, the high demand for water in the agricultural sector exacerbates the problem of water scarcity in the kingdom, and the government bears a high cost of water production and drainage services in the urban sector [7].

According to Fazel et al. [8], KSA has the third highest consumption of freshwater in the world, which grew from 227 L/capita/day (L/c/d) in 2009 to 278 L/c/day in 2018. Long-standing challenges caused by the increasing human population, rising temperatures, increased evaporation, and increased groundwater abstraction might be exacerbated or accelerated by climate change in arid regions. Moreover, in arid regions aquifer recharge rates are limited and groundwater use for agriculture is more prevalent [1].

A comparison of previous studies revealed a consensus on the need to rationalize demand on water, particularly in the agricultural sector, besides the importance of adopting an integrated approach to develop, direct, and diversify national and local water supply sources [9]. According to MEWA [10] in the report of National Water Strategy 2030, the water requirements in the kingdom (estimated in 2015 at about 24.8 billion cubic meters) are increasing at a constant annual rate of 7%, bearing in mind that the agricultural sector is the largest consumer of water in the kingdom, and accounts for 84% of the total water demand. Thus, reflecting water use in the agricultural sector is an environmental challenge due to its dependence on nonrenewable resources, which account for 90% of the total water supplied to the sector. Also, El-Rawy et al. [11] came to a conclusion that, in considering the anticipated rise in temperature values (particularly in Al Quassim, Al Jawf, Ha'il, Al Madinah, the Northern region, Najran, Baha, and Tabuk, which have the highest reference evapotranspiration (ETo) values in KSA), more water resources will be needed in the future to meet the increased irrigation water requirements.

Groundwater is a vital component of the hydrologic cycle and is therefore extremely sensitive to climate change; these quick changes are anticipated to have significant effects on KSA's water supplies [1]. Groundwater quality is significantly impacted by anthropogenic, including industrial waste discharge, usage of pesticides, high fertilizer use, petroleum and natural gas spills, landfills, and waste from mining. When groundwater becomes polluted, cleanup is challenging and expensive [12]. To develop policies to safeguard and manage groundwater quality, it is crucial to understand the condition of groundwater quality, any associated health issues, and the variables that affect it.

Lately, a significant and developing set of studies focused on assessing and comprehending hydrochemical characteristics, pollutant sources, and groundwater quality using a number of efficient analysis approaches including:

- Geographic information system (GIS) methods
- Remote sensing (RS)
- Statistical procedures
- Multivariate modeling

while common analyses, such as:

- The Piper trilinear diagram
- Gibbs diagram
- Hydrochemical facies evolution diagram (HFE)
- Chadha's diagram

are appropriate and regularly used for determining the major geochemical factors precisely [13]. Additionally, multivariate analyses such as:

- Discriminant analysis (DA)
- Factor analysis (FA)
- Principal component analysis (PCA)
- Cluster analysis (CA)

are efficient methods to recognize the physicochemical factors in groundwater [14]. The development of groundwater is viewed as a crucial alternative for filling in action plan gaps and offering flexibility to deal with the effects of climate change [15]. As a result of documented climate change and anthropogenic activities, grave issues with rising groundwater depletion and groundwater quality degradation are already beginning to arise in Saudi Arabia [16]. Many researchers and environmental institutions have been interested in studying the status and quality of groundwater in countries with limited surface water resources in general, and in the KSA in particular, due to its previously mentioned importance. The following is a summary of the most recent studies and their conclusions.

Rajmohan et al. [17] developed an integrated approach ($n = 50$) using groundwater suitability zone (GWSZ) maps, drinking water quality index (DWQI), irrigation water quality index (IWQI), chronic daily index (CDI), and hazard quotient (HQ); these approaches are used to assess groundwater quality in the arid environment of the Rabigh basin, Western Saudi Arabia. According to the DWQI, more than half of the groundwater samples were unfit for drinking. According to IWQI readings, just 6% of wells were highly suited for irrigation, while the groundwater quality index (GWQI) and multivariate statistical techniques were used to examine nearly a hundred groundwater samples in order to assess the heavy metals pollution and groundwater quality in southern Saudi Arabia by Alfaifi et al. [18]. According to pollution indices, 20–52% of the groundwater samples might be used for household and agricultural uses. According to multivariate statistical studies, the primary cations, anions, and heavy metals in the research region may have originated from the dissolution of halite and gypsum (in sabkha deposits), carbonates, and agricultural activities. Alharbi et al. [19] used statistical analyses, to assess the heavy metals pollution and groundwater quality in central Saudi Arabia. The findings demonstrated that the average concentrations of most of the major ions including salinity exceeded the WHO permissible limits for drinking water. The variables impacting groundwater chemistry included soil leaching, gypsum dissolution/precipitation, heavy use of fertilizers and pesticides, carbonates, and silicates. On the other side, Alqarawy et al. [20] assessed the viability of huge groundwater samples for drinking and its controlling factors in Makkah Al-Mukarramah Province using hyperspectral reflectance and water quality indices. The overall quality of groundwater samples in the examined sites ranged widely, according to the DWQI evaluation, from good 2.5% to unsuitable for drinking 30%. Despite the fact that the heavy metal index (HPI), contamination degree (C_d), and pollution index (PI) showed that all groundwater samples had a low level of pollution and were often unaffected by metals, Alramthi et al. [21] also evaluated the quality of groundwater in southern KSA and compared it with the standards of the Saudi Arabian Standards Organization (SASO) and WHO guideline. However, Alramthi et al.'s findings varied somewhat between different site locations, turbidity, and total coliform remained within the acceptable limit of international criteria. As a result of previously provided information and facts, efforts are made in this chapter to review the quality and current state of groundwater resources in the kingdom.

The chapter lists issues and difficulties and makes recommendations to increase the sustainability of the groundwater resources, and the chapter's major goals may summarize as:

1. Evaluate and summarize the most significant studies and research projects that were carried out in various KSA areas to assess the quality of the groundwater.

2. Review the most recent procedures and techniques that demonstrated excellent accuracy in assessing the appropriateness of groundwater in the kingdom for a variety of uses, including drinking and irrigation.
3. Study the sources of pollution and their impacts on the aquifers, focusing on discussing the proposed solutions.

2 Materials and Methods

The above objectives are achieved through the following procedures shown in Fig. 1 with a brief description in the subsections:

2.1 Study Area Description

Saudi Arabia is located in the Arabian Desert, extends from northern Africa into Asia, and comprises more than 70% of the Arabian Peninsula; covering an area of

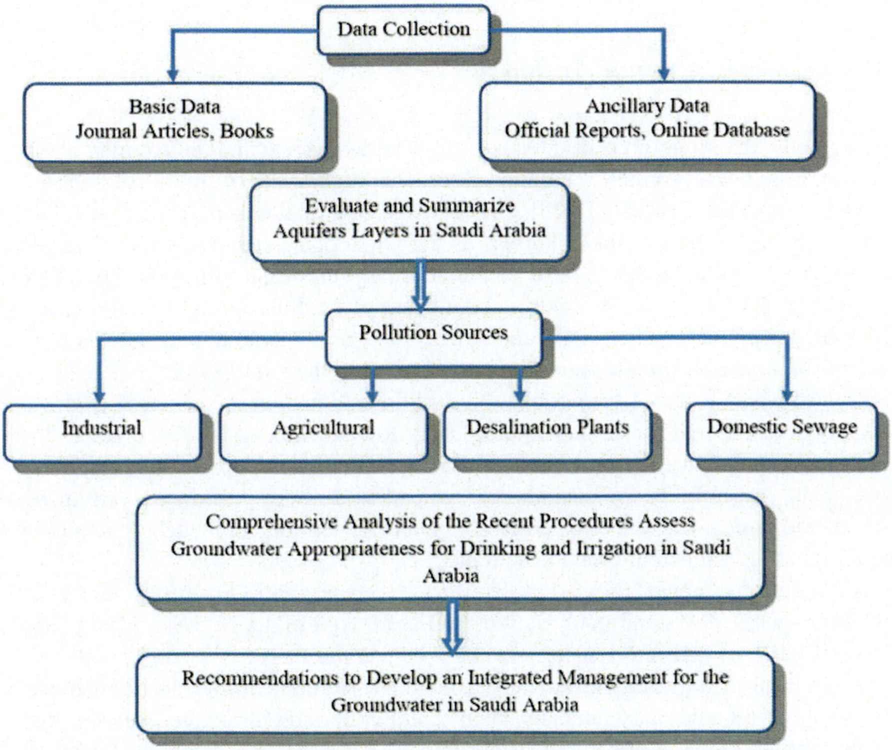

Fig. 1 Detailed flowchart showing methodology adopted for this study

about 2.25 million km^2. The country's hot, arid climate is caused by the Red Sea and the adjacent mountain chains of Asir, As Shifa, and Hejaz [22]. The weather is harsh, with limited precipitation and a high rate of evapotranspiration [23]. Rainfall is limited and inconsistent, and the annual precipitation ranges from less than 2 cm in the north to over 30 cm in the southern top of the Asir Mountain in the southwest, with an average of 7–13 cm. The Kingdom of Saudi Arabia's distinguished geographical location, as well as the presence of the Red Sea on the western side and the Arabian Gulf on the eastern side, drew the attention of researchers over the years [24]. The country has borders with Kuwait, Jordan, and Iraq in the north, Qatar in the southeast, and Oman, the United Arab Emirates, and Yemen in the south, as shown in Fig. 2. Water availability is an important issue in KSA because of the limited available surface water. In KSA, desalinated seawater, groundwater, and surface water are the primary sources of water that is clear in Fig. 3 according to the observation of MEWA data.

It is regrettable to remark that groundwater in certain areas has become contaminated in recent years; particularly shallow aquifers are highly vulnerable. Geological processes and human activity are to blame. As previously mentioned, numerous pollutants of both organic and inorganic type have been found in groundwater at various locations around the kingdom.

2.2 *Aquifers in Saudi Arabia*

Geologically, the kingdom is divided into the Arabian Shield and the Arabian Shelf. Arabian Shield is originally composed of the highlands of the Precambrian-Cambrian basement rocks [1, 25] and igneous and metamorphic complexes in some areas with volcanic flows known as the Quaternary sediments for all recent ages, and this shield covers a third of the Arabian Peninsula within the Dammam region (Fig. 4). The Arabian Shield is composed of an outcropping of solid rocks, which begins in the western part of the Kingdom of Saudi Arabia, and extends from the Gulf of Aqaba in the north to the Gulf of Aden in the south [26].

The Arabian Shelf is a formation of a pseudo-sedimentary sequence disproportionately above shield rocks and dipping away toward the Persian Gulf, where the sedimentary sequence begins with sediment stimulation up to the Cambro-Ordovician assemblage. Precambrian volcanic activity produced complex, interlayered volcanically delivered and epiclastic sedimentary rocks, including volcano-clastics and subordinate flow rocks [27, 28].

The region has newer rocks that date to the Tertiary and Quaternary that were formed by basalt flows and gabbro dikes and are linked to the Red Sea rifting. The gabbro dikes intruded into tension cracks, whereas the basalt is a component of a wide region of flow rocks and volcanic cones that resulted from volcanic activity. Unconsolidated Quaternary deposits, such as alluvium, conglomerate from the Red Sea escarpment, terrace gravels, coastal-plain silt, and Aeolian sand, are found on top of the bedrock [29].

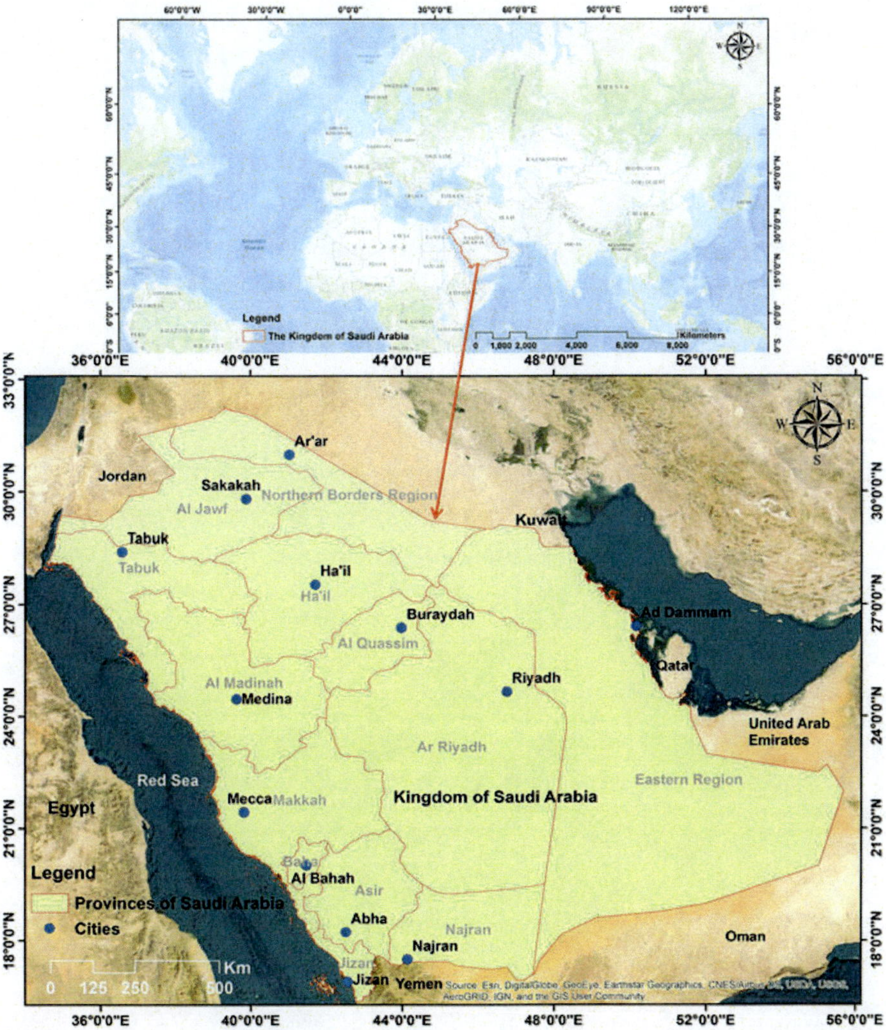

Fig. 2 Location map of the study area

These formed during a period of vigorous erosion that followed the elevation of the region and the entrance of the Red Sea, leading to the creation of a Wadi system that drains to the east and west as well as the erosional retreat of the Red Sea escarpment.

The productive aquifer is located under the Precambrian basement in the sedimentary layers and porous volcanic rocks. Based on earlier investigations, the lithological succession may be divided into eight primary aquifers, as indicated in Table 1. Based on its hydrologic characteristics and geographic range, major and secondary aquifers are distinguished from one another [9]. The main sandstone

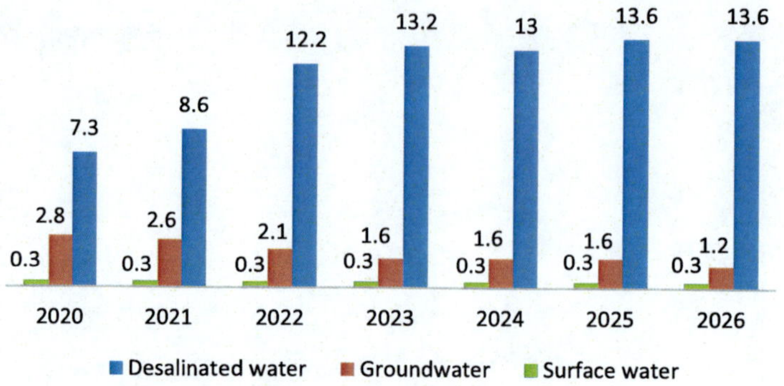

Fig. 3 Total water supply in KSA (Million m^3/day). Source: [10]

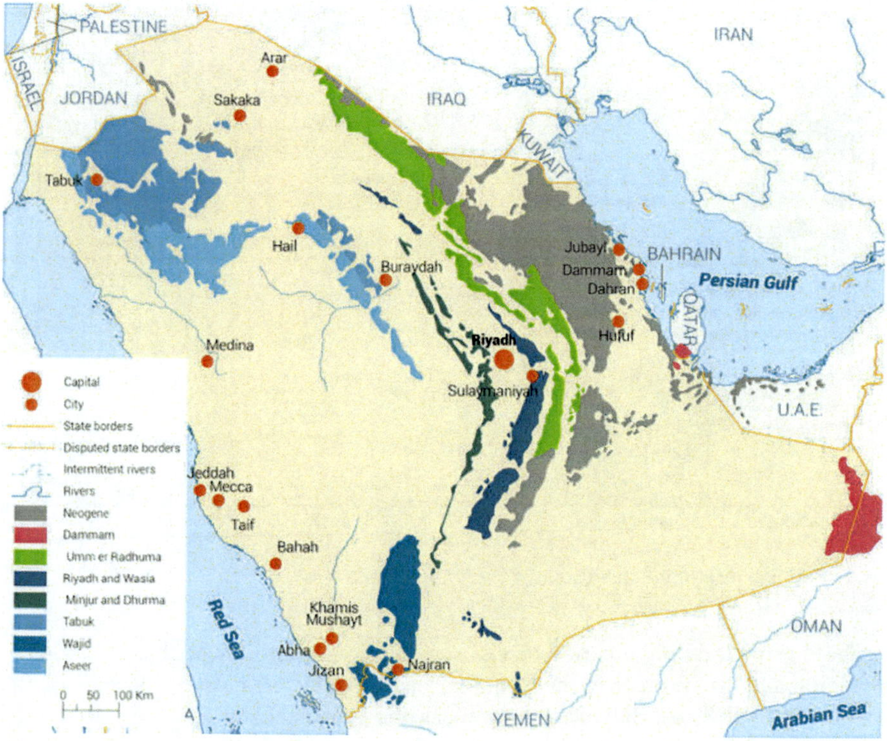

Fig. 4 Locations of the potential aquifers in Saudi Arabia (Ministry of Environment, Water and Agriculture, 2018). National Water Strategy 2030

Table 1 Lithological sequence and major aquifers in Saudi Arabia; Source: modified [9]

Lithologic sequence	• Quaternary and tertiary • Pliocene and miocene clastic rocks • Eocene carbonate to upper cretaceous rocks. • Middle and lower cretaceous clastic rocks. • Lower and upper Jurassic cretaceous carbonate • Middle and lower Jurassic clastic and carbonate rocks • Jurassic, Triassic, and Permian clastic rocks • Lower Paleozoic clastic rocks
Principal aquifers	• Neogene • Damman • Ummer Radhuma • Wasia Biyadh • Minjur/Dhrumma • Tabuk, Wajid, Saq
Secondary aquifers	• Alluvium • Basalt • Aruma • Sakaka • Buwaib, Yamama, Sulay, Arab, Juballa, Hanifah • Dhruma • Jilh • Jauf

aquifers are widespread throughout the southeast and have strong local water-bearing capabilities. Aquifers with a high potential production are found in water-bearing Mesozoic sandstone and limestone strata like Wajid and the Minjur/Dhruma aquifer in the southeast of the Asir and Najran provinces. The secondary aquifers consist of sandstone interspersed with less permeable layers, which operate as confining beds, generating less and storing less water. These aquifers are dispersed throughout the area and serve as limited water sources [30].

Some aquifers have strong potential yields and are hydraulically linked to the major aquifers underneath them. In Saudi Arabia's center and Northern provinces, a reliable supply of groundwater is provided by the vast majority of primary deep aquifers. Renewable groundwater for 2018 is estimated at 2.8 BMC in the Arabian Shield region [6]. Although there is a significant volume of water stored in these deep aquifers in Saudi Arabia's Saq, Tabuk, and Wajid (Table 1), the water is unfit for drinking.

2.3 Pollution Sources

Finding out the origins of groundwater pollution is crucial, with a focus on the level of contamination, pollutant release rates, and quantity in relation to time. The size of the aquifer, the geological structure of layers, and the geological activities all generally affect the quality of the groundwater reservoir. It has been noted that the pollutants released from domestic, industrial, mining, and agricultural activities are the primary causes of groundwater pollution [25]. As compared to the other industrial sites along the Red Sea coast, Jeddah has the most heavily polluted sediments on heavy metals in Red Sea sediments near Jeddah [31].

The harm produced by petroleum hydrocarbons in aquatic organisms can offer immediate availability of trustworthy information on water quality and early warning of pollution. Halawani et al. [31] confirmed that multiple geographical and temporal scales allow sediments to interact directly with other environmental media. As a result of their susceptibility to anthropogenic and industrial activity, Saudi Arabia's coastal sediments are appealing to researchers interested in studying heavy metal enrichment. Both natural and human-made processes, including oil spills, wastewater discharge, desalination plant effluents, building operations, and maritime traffic, can contribute heavy metals to the local marine ecosystem [32, 33].

Sediment concentrations of heavy metals are related to aquatic concentrations by a complicated sequence of mechanisms, as demonstrated by their interaction with pH, conductivity, salinity, and the availability of organic matter [34]. Metals in sediment and aqueous solutions can be bioavailable. Lead, cadmium, mercury, chromium, and arsenic are among the heavy metals that have been shown to be hazardous to living things, whereas copper, manganese, sodium, iron, and zinc are necessary metals but can be poisonous if their concentrations are higher than what is considered safe [35]. According to several studies, mining, mineral leaching, and metal plating release copper and zinc into the water, while agricultural runoff and household sewage, which contains human and animal waste as well as phosphate fertilizers and pesticides, are likely to cause lead pollution in the groundwater [36, 37]. The shallow aquifer is in hydraulic contact with both the irrigation system and the surface water, making it susceptible to contamination from surface water sources. The groundwater aquifer loses water through seepage to the surface water, abstraction, and evaporation from the surface [38–43]. The sources of groundwater pollution in the kingdom can be summarized as follows:

2.3.1 Domestic Sewage

There is a substantial risk of groundwater pollution as a result of the wastewater treatment due to seepage from drying beds and sludge storage [39, 41]. Sewage water from the Jazan Region sewage treatment plant in the south of the Kingdom of Saudi Arabia was collected and tested [44]. The results showed that the potentially toxic heavy metals, including Cd, Cu, Mn, and Zn, were released into the aquatic environment, influencing the pH, COD, and other factors, which would have a direct effect on the Red Sea marine ecology and in the adjoining aquifers.

The Ministry of Environment Water and Agriculture [6] predicted that the kingdom's population will increase to 40.1 million in 2030, compared to 33.6 million in 2018, which implies an increase in the total wastewater quantities from 7.2 million cubic meters/day in 2018 to 10.8 million cubic meters/day in 2030. In addition to this, the average annual growth in population will range between 1.28%

and approximately 2.05%, and the average water consumption per person will reach about 250 L per day in 2030 [22]. It is also expected that the same year will witness an increase in the demand for water from pilgrims and Umrah performers to about 2.1 million cubic meters/day compared to 1.1 million cubic meters/day in 2018, and the amount of wastewater generated by each person will reach about 212.5 L per day. In addition, it is estimated that groundwater infiltration rates and unaccounted-for rainwater flows into sewage networks in coastal and non-coastal cities will be as follows [6]:

- Coastal cities: 20% in 2018 AD (the number is higher in some cities such as Dammam and Jeddah) and 10% in 2030 AD
- Landlocked cities: 10% in 2018 AD and 5% in 2030 AD

It should be noted that the current aggregate capacity of the sewage network is only about 50% of the total wastewater generated, and it is intended to reach this ratio to 95–100 in the year 2030 AD. Therefore, the amount of wastewater collected as a result of the planned expansion projects will reach 10.3 million cubic meters/day in 2030, which requires the provision of treatment capacities of 11.1 million cubic meters/day, with an excess of 5–10% in treatment capacities to ensure appropriate treatment of increase in unpredictable wastewater and avoiding any environmental impact associated with untreated wastewater [6].

2.3.2 Desalination Plants

Saudi Arabia currently has 35 desalination facilities. The overall daily capacity of desalination reached 6.28 million cubic meters in 2015 [10]. Every year, the amount of desalinated water consumed rises by around 14% [22]; this is six times the pace of population increase and double the growth in total residential water demand.

About half of the drinking water in the kingdom is actually desalinated water. Desalination plants employ three types of techniques: (1) multi-stage flash distillation, (2) reverse osmosis, and (3) multiple distillations, but there are a number of drawbacks to these technologies, especially their environmental impact, as desalination plants dispose of brine, a liquid with a high salt concentration, through wastewater and then it is transferred to water bodies, causing incalculable environmental damage, quoting the Minister of Environment, Water and Agriculture report [10]. As a result, it is important to select the best intake outfall design, reduce recirculation, and research how brine discharge from outfalls along the Saudi coastline is dispersed [45].

2.3.3 Agricultural

The wrong use of mineral fertilizers to fertilize crops at rates higher than the optimum rates needed by the plant leads in the long run to a deterioration of soil structure and a deterioration of the natural and chemical properties of groundwater.

The fertilizers' effect on water salinity, pH number, and the activity of microorganisms leads to negative effects on groundwater [35, 46]. This was demonstrated by some studies that evaluated the quality of groundwater in agricultural areas in Saudi Arabia. Alsalah et al. [47] studied the groundwater quality nearby agricultural fields. According to the study, anthropogenic pollution caused a change in the dominating bacterial phyla in the groundwater microbial communities. In addition, the majority of the groundwater samples had total nitrogen concentrations that were higher than the 15 mg/L allowed limit for agricultural irrigation. Findings showed that the study area in Western Saudi Arabia was not of the highest quality, and better agricultural management techniques are required in addition to groundwater treatment methods. The findings of indicators used to evaluate the quality of the groundwater in central Saudi Arabia by Alharbi and El-Sorogy [48] showed that more than half of the collected samples were considered harmless for human consumption, and the concentration of contamination was traced back to farming activities.

The presence of nitrates in irrigation water is one of the problems that affect public health, as nitrates lead to pollution of drinking water, whether ground or surface water, through deep drainage to groundwater or surface runoff to surface water channels. The percentage of nitrates in drinking water should not exceed 50 mg/L, and to avoid this problem, fertilizer doses must be added to the plant in a good way, and these doses should be based on the analysis of both the plant and the soil, in order to avoid leaching of nitrates into freshwater streams or groundwater [49, 50].

2.3.4 Industrial

The industry works to pollute the waterways by dumping its waste and by-products into it, whether from ships, factories, or hot springs. Thermal pollution leads to contamination and deterioration of ecosystems [51]. An increase in the levels of pollutants resulting from industrial activity and petroleum industries has been observed in the coastal areas extending to the east and west of the kingdom, especially the port areas such as the cities of Jeddah and Jizan [24, 52].

Along the coastal plain of Jizan region, southwest Saudi Arabia, the alluvial deposits of the long-buried wadi (dry valley) channels create a hydraulic connection between the aquifer and the seawater and offer possible routes for saltwater intrusion [53].

3 Groundwater Quality Assessments: Methods and Findings

Due to the scarcity and poor quality of surface water in the later part of the twentieth century, groundwater consumption increased dramatically in the majority of the world's countries including KSA. Groundwater quality offers significant advantages over surface water if it is free of viruses and hazardous microorganisms [8].

Environmental indicators, such as those measuring the quality of groundwater, deteriorate when pollutants, particularly heavy metals, are present in larger concentrations. Therefore, it is necessary to use thorough indices for classifying the contamination level in order to determine the condition of heavy metals and their distribution in ground layers in order to guide future management decisions [33]. The data and analysis presented here will provide useful benchmarks for the local area as well as for similar dry regions worldwide and for the region as a whole.

According to Elshemy and Meon [54], there are four main methods that may be used to evaluate water quality:

1. The water quality index method
2. The statistical analysis method
3. The trophic status index method
4. The biological analysis method

The water quality index is the most popular and widely used for a variety of reasons, including the fact that it is a superior method for comprehending and summarizing enormous amounts of data on water quality by integrating complex information and expressions, which represent a composite impact of involved variables on the water quality in every watering cycle [55].

In this context, the MEWA has undertaken the task of developing a unified reference framework for the water sector. This includes a comprehensive water strategy and works to integrate directions, policies, legislation, and practices in the water sector on a global scale with the main objective of setting environmental standards in line with international standards of the World Health Organization [10], as shown in Table 2.

Table 2 Drinking water standards and the acceptable range of various physical parameters according to Saudi Arabian Standards Organization ([56] and World Health Organization standards. Source:[56, 57]

Parameters	SASO [56] standards	WHO [57] allowed limits
pH	6.5–8.5	6.5–8.5
TDS (mg/L)	500	1,000
Turbidity (NTU)	5	5
Total hardness (mg/L)	500	500
Chloride (mg/L)	250	250
Bicarbonate (mg/L)	125	100
Sulfate (mg/L)	250	250
Nitrate (mg/L)	10	50
Fluoride (mg/L)	1.5	1.5
Sodium (mg/L)	200	200
Potassium (mg/L)	12	30
Calcium (mg/L)	100	75
Magnesium (mg/L)	50	30
Iron (mg/L)	0.3	0.3
Manganese (μg/L or ppb)	100	80

This study took into account a total of 10 researches that were carried out during the last 12 years in order to better understand the issues related to groundwater quality in Saudi Arabia. Table 3 summarizes the approaches and findings on groundwater assessment for the recently published studies during the period from 2011 to 2022.

The aforementioned studies confirmed that the presence of anthropogenic activities, such as excessive groundwater withdrawal, resulted in increased urbanization and industrial activity, use of extensive fertilizers and pesticides in agriculture, dumping of untreated wastewater and sewage discharge, leakage of septic tanks, and landfills. Previous studies conducted in various places in the kingdom showed a high percentage of samples exceeds the permissible levels, and most of it is considered unfit for drinking, and may be suitable for irrigation uses, taking into account the high salinity levels that suit certain crops.

4 Discussions

Rapid population development, economic devolution, and unsuitable water management placed demand on the freshwater resources in the Kingdom of Saudi Arabia's (KSA) surrounding territories as the Middle East and North Africa (MENA) [62, 63]. Rateb et al. [64] confirmed that water quality varies across MENA, ranging from very contaminated in the Gulf Cooperation countries (GCC), like Jordan, Yemen, and Libya, to less so in countries with substantial groundwater supplies (e.g., Lebanon and Iran), or big rivers (e.g., Iraq). According to various studies conducted in different regions in Egypt, high salinity is the most serious concern influencing groundwater quality. This is due to the country's arid climate, which causes substantial evaporation rates and salt deposition from fertilizer use in the soil and groundwater [38–43, 65–67].

According to a database prepared from a literature review, the MENA nations are mostly polluted by industrial effluents, agricultural activities, and the discharge or reuse of treated wastewater, while the Arabian Peninsula suffering from petroleum waste. According to Halawani et al. [31], Jamoussi et al. [24], and Zanaty et al. [33], the coastal strip along the Red Sea shoreline in the western border of KSA contains several indicators of oil pollution and heavy metals. Otherwise, in the eastern region, the principal sources of groundwater deterioration in the Al-Kharj area are large agricultural investment enterprises that overexploit groundwater – in addition to exceeding the maximum permitted level of TDS content in the provinces of Qassim and Riyadh [9]. Several researchers investigated the water quality in various locations of the KSA and their findings revealed that toxic metal ions are contaminating groundwater due to direct contact of existing geological rocks with groundwater [25].

Table 3 Assessments of groundwater-quality studies present in the groundwater of Saudi Arabia

Approach	Results	No. of samples	Region	Studies
Hyperspectral reflectance and water quality indices	According to the DWQI evaluation, quality ranges from good 2.5% to unsuitable for drinking 30% despite the fact that the HPI, Cd, and PI readings showed that all groundwater samples had a low level of pollution and were often unaffected by metals	173	Makkah Al-Mukarramah Province	[20]
An integrated approach, including groundwater suitability zone (GWSZ) maps for drinking water quality index (DWQI), irrigation water quality index (IWQI), chronic daily index (CDI), and hazard quotient (HQ)	According to the DWQI, more than half of groundwater samples were unfit for drinking. According to IWQI readings, just 6% of wells were highly suited for irrigation	50	Rabigh basin	[17]
Multivariate statistical analyses	The average concentrations of TDS, Ca +, Na +, K +, Cl +, SO $_4$ 2 +, and F + exceeded the WHO guidelines' permitted limits for drinking water	30	Central Saudi Arabia	[19]
Groundwater quality index (GWQI) and multivariate statistical	According to pollution indices, 20–52% of the groundwater samples might be used for household and agricultural uses	105	Southern Saudi Arabia	[18]
Groundwater suitability maps using GIS	The groundwater in the coastal region is unsuitable due to its high salinity, high TH, and high concentrations of main ions. However, the wells located in the upstream region are suitable for drinking and irrigation despite their high TDS and total hardness (TH) values	182	Wadi Baysh Basin, Western Saudi Arabia	[58]
Descriptive statistics, histograms, and normal quantile plots	Ionic interactions show significant rock–water interaction in a variety of ways	295	Saq aquifer in the northwestern part of Saudi Arabia	[59]

(continued)

Table 3 (continued)

Approach	Results	No. of samples	Region	Studies
Multivariate analysis	It was discovered that the groundwater collected at this location was polluted with excessive levels of nutrients and human wastes	11	Makkah Al-Mukarramah Province	[47]
Hydrochemical analyses	The groundwater was found to be unsuitable for drinking but suitable for irrigation purpose	8	Dammam, eastern part of Saudi Arabia	[15]
Hydrochemical analyses	In most of the study area, groundwater samples are generally unsuitable for domestic and irrigation purpose	31	Wadi Jazan, southwest Saudi Arabia	[60]
Hydrochemical analyses with GIS	The aquifer's water is considered brackish, and to some extent, it might be used in agriculture. Because of the influence of seawater intrusion, the aquifer's downstream region is extremely salty	34	Wadi Rabigh, Western Saudi Arabia	[61]

5 Conclusions

The Kingdom of Saudi Arabia (KSA) is the third highest consumer of freshwater in the world. However, the kingdom faces great challenges due to the limited water reserves of nonrenewable groundwater, which is rapidly depleting, under arid climatic conditions. The data provided above reflect water use in the agricultural sector and the corresponding environmental challenge due to its dependence on nonrenewable water resources, so more water resources will be needed in the future to meet the increased irrigation water requirements. In Saudi Arabia's center and Northern provinces, a reliable supply of groundwater is provided by the vast majority of primary deep aquifers. Although there is a significant volume of water stored in these deep aquifers in Saudi Arabia's Saq, Tabuk, and Wajid, the water is unfit for drinking due to quality-related problems.

On the other side, every year, the amount of desalinated water consumed rises by around 14%. This is six times the pace of population increase and double the growth in total residential water demand. About half of the drinking water in the kingdom is actually desalinated water. Desalination plants dispose of brine, a liquid with a high salt concentration, through wastewater, and then it is transferred to water bodies, causing incalculable environmental damage.

Lately, a significant and developing set of studies have focused on assessing and comprehending hydrochemical characteristics, pollutant sources, and groundwater quality using a number of efficient analysis approaches. Regrettably, numerous pollutants of both organic and inorganic type have been found in groundwater at various locations around the kingdom. Most of these groundwater samples are generally unsuitable for domestic and irrigation purposes and polluted with excessive levels of nutrients and human wastes, while the groundwater in the coastal region is unsuitable due to its high salinity and high concentrations of major ions.

6 Recommendations

In order to develop an integrated management for the groundwater in KSA, we drew out a number of recommendations.

- Technical procedures for groundwater quality management would be needed to implement the control on degrading water quality, using more flow model programs and calibration to predict the groundwater quality.
- Groundwater monitoring systems must be improved and expanded.
- Modern irrigation techniques should be promoted for agricultural water management, together with an estimation of crop water requirements.
- Groundwater recharge (GWR) is critical to maintain the abundance of groundwater.

References

1. Alghamdi AG, Aly AA, Majrashi MA, Ibrahim HM (2023) Impact of climate change on hydrochemical properties and quality of groundwater for domestic and irrigation purposes in arid environment: a case study of Al-Baha region, Saudi Arabia. Environ Earth Sci 82:39. https://doi.org/10.1007/s12665-022-10731-z
2. Mushtaq F, Rehman H, Ali U, Babar MS, Al-Suwaiyan MS, Yaseen ZM (2023) An investigation of recharging groundwater levels through river ponding: new strategy for water Management in Sutlej River. Sustainability 15:1047. https://doi.org/10.3390/su15021047
3. Alotaibi BA, Baig MB, Najim MM, Shah AA, Alamri YA (2023) Water scarcity management to ensure food scarcity through sustainable water resources management in Saudi Arabia. Sustainability 15:10648. https://doi.org/10.3390/su151310648
4. Sharma SK (2022) A novel approach on water resource management with multi-criteria optimization and intelligent water demand forecasting in Saudi Arabia. Environ Res 208: 112578. https://doi.org/10.1016/j.envres.2021.112578
5. Abdelkareem M, Abdalla F (2022) Revealing potential areas of water resources using integrated remote-sensing data and GIS-based analytical hierarchy process. Geocarto Int 37:8672–8696. https://doi.org/10.1080/10106049.2021.2005155
6. SWPC (2020) Seven year statement for KSA's water (Arabic language). Ministry of Environment Water & Agriculture

7. El-Rawy M, Fathi H, Zijl W, Alshehri F, Almadani S, Zaidi FK, Aldawsri M, Gabr ME (2023) Potential effects of climate change on agricultural water resources in Riyadh region, Saudi Arabia. Sustainability 15:9513. https://doi.org/10.3390/su15129513
8. Fazel HK, Abdo SM, Althaqafi A, Eldosari SH, Zhu B-K, Safaa HM (2022) View of Saudi Arabia strategy for water resources management at Bishah, Aseer Southern Region. Water Assess Sustain 14:4198. https://doi.org/10.3390/su14074198
9. Mallick J, Singh CK, AlMesfer MK, Singh VP, Alsubih M (2021) Groundwater quality studies in the Kingdom of Saudi Arabia: prevalent research and management dimensions. Water 13: 1266. https://doi.org/10.3390/w13091266
10. MEWA (2018) National Water Strategy 2030 (Arabic language), Ministry of Environment, Water and Agriculture
11. El-Rawy M, Batelaan O, Al-Arifi N, Alotaibi A, Abdalla F, Gabr ME (2023) Climate change impacts on water resources in arid and semi-arid regions: a case study in Saudi Arabia. Water 15:606. https://doi.org/10.3390/w15030606
12. Negm A (2019) Groundwater in the Nile Delta. Springer. https://doi.org/10.1007/978-3-319-94283-4
13. El Osta M, Masoud M, Alqarawy A, Elsayed S, Gad M (2022) Groundwater suitability for drinking and irrigation using water quality indices and multivariate modeling in Makkah Al-Mukarramah province, Saudi Arabia. Water 14:483. https://doi.org/10.3390/w14030483
14. El-Rawy M, Fathi H, Abdalla F, Alshehri F, Eldeeb H (2023) An integrated principal component and hierarchical cluster analysis approach for groundwater quality assessment in Jazan, Saudi Arabia. Water 15:1466. https://doi.org/10.3390/w15081466
15. Iwalewa TM, Makkawi MH, Elamin AS, Al-Shaibani AM (2013) Groundwater management case study, eastern Saudi Arabia: part II–solute transport simulation and hydrochemistry. Eur J Sci Res 109:650–667. ISSN 1450-216X/1450-202X
16. Waheed R (2022) The significance of energy factors, green economic indicators, blue economic aspects towards carbon intensity: a study of Saudi vision 2030. Sustainability 14:6893. https://doi.org/10.3390/su14116893
17. Rajmohan N, Masoud MHZ, Niyazi BAM (2021) Assessment of groundwater quality and associated health risk in the arid environment, Western Saudi Arabia. Environ Sci Pollut Res 28: 9628–9646. https://doi.org/10.1007/s11356-020-11383-x
18. Alfaifi H, El-Sorogy AS, Qaysi S, Kahal A, Almadani S, Alshehri F, Zaidi FK (2021) Evaluation of heavy metal contamination and groundwater quality along the Red Sea coast, southern Saudi Arabia. Mar Pollut Bull 163:111975. https://doi.org/10.1016/j.marpolbul.2021.111975
19. Alharbi T, El-Sorogy AS, Qaysi S, Alshehri F (2021) Evaluation of groundwater quality in Central Saudi Arabia using hydrogeochemical characteristics and pollution indices. Environ Sci Pollut Res 28:53819–53832. https://doi.org/10.1007/s11356-021-14575-1
20. Alqarawy A, El Osta M, Masoud M, Elsayed S, Gad M (2022) Use of hyperspectral reflectance and water quality indices to assess groundwater quality for drinking in arid regions, Saudi Arabia. Water 14:2311. https://doi.org/10.3390/w14152311
21. Alramthi SM, Ali GH, Shaban AM, Abdou TA, Elthagafi AM, Eldosari SH, Zhu B-K, Safaa HM (2022) Quality characterization of groundwater for drinking purposes and its network distribution to assure sustainability in southern region of Saudi Arabia. Water 14:3565. https://doi.org/10.3390/w14213565
22. Baig MB, Alotibi Y, Straquadine GS and Alataway A (2020) Water resources in the Kingdom of Saudi Arabia: challenges and strategies for improvement. Water policies in MENA countries, pp 135–160. https://doi.org/10.1007/978-3-030-29274-4_7
23. GAS (2022) Anural report (Arabic Language), general authority for statistics. https://www.stats.gov.sa.
24. Jamoussi B, Chakroun R, Al-Mur B (2022) Assessment of total petroleum hydrocarbon contamination of the Red Sea with endemic fish from Jeddah (Saudi Arabia) as bioindicator of aquatic environmental pollution. Water 14:1706. https://doi.org/10.3390/w14111706

25. Ali I, Hasan MA, Alharbi OM (2020) Toxic metal ions contamination in the groundwater, Kingdom of Saudi Arabia. J Taibah Univ Sci 14:1571–1579. https://doi.org/10.1080/16583655.2020.1847807
26. Ahmed I, Nazzal Y, Zaidi FK, Al-Arifi NS, Ghrefat H, Naeem M (2015) Hydrogeological vulnerability and pollution risk mapping of the Saq and overlying aquifers using the DRASTIC model and GIS techniques, NW Saudi Arabia. Environ Earth Sci 74:1303–1318. https://doi.org/10.1007/s12665-015-4120-5
27. Abdulkadir IT, Abdullatif OM (2013) Facies, depositional environments, reservoir potential and palaeogeography of the Cambro-Ordovician Dibsiyah formation, Wajid outcrop belt, Saudi Arabia. Arab J Sci Eng 38:1785–1806. https://doi.org/10.1007/s13369-012-0388-x
28. Ali S, Alshammari AS (2021) Genesis of gabbroic intrusions in the Arabian shield, Saudi Arabia: mineralogical, geochemical and tectonic fingerprints of the Neoproterozoic arc magmatism. Geol Mag 158:1639–1656. https://doi.org/10.1017/s0016756821000182
29. El-Sawy E, Eldougdoug A, Gobashy M (2018) Geological and geophysical investigations to delineate the subsurface extension and the geological setting of Al Ji'lani layered intrusion and its mineralization potentiality, Ad Dawadimi District, Kingdom of Saudi Arabia. Arab J Geosci 11:1–25. https://doi.org/10.1007/s12517-017-3368-3
30. Alsharhan A, Rizk Z, Nairn AEM, Bakhit D, Alhajari S (2001) Hydrogeology of an arid region: the Arabian Gulf and adjoining areas. Elsevier. https://doi.org/10.1016/b978-044450225-4/50006-3
31. Halawani RF, Wilson ME, Hamilton KM, Aloufi FA, Taleb MA, Al-Zubieri AG, Quicksall AN (2022) Spatial distribution of heavy metals in near-shore marine sediments of the Jeddah, Saudi Arabia region: enrichment and associated risk indices. J Marine Sci Eng 10:614. https://doi.org/10.3390/jmse10050614
32. Alzahrani DA, Selim E-MM, El-Sherbiny MM (2018) Ecological assessment of heavy metals in the grey mangrove (Avicennia marina) and associated sediments along the Red Sea coast of Saudi Arabia. Oceanologia 60:513–526. https://doi.org/10.1016/j.oceano.2018.04.002
33. Zanaty N, Mansour K, Fathi H (2023) Satellite-based assessment of the anthropogenic impacts on environmental sustainability in Jazan region, Red Sea. Egypt J Remote Sens Space Sci 26: 117–127. https://doi.org/10.1016/j.ejrs.2022.12.002
34. El-Zeiny AM, Abd El-Hamid HT (2022) Environmental and human risk assessment of heavy metals at northern Nile Delta region using geostatistical analyses. Egypt J Remote Sens Space Sci 25:21–35. https://doi.org/10.1016/j.ejrs.2021.12.005
35. Kahal A, El-Sorogy AS, Qaysi S, Almadani S, Kassem OM, Al-Dossari A (2020) Contamination and ecological risk assessment of the Red Sea coastal sediments, Southwest Saudi Arabia. Mar Pollut Bull 154:111125. https://doi.org/10.1016/j.marpolbul.2020.111125
36. El-Azeim A, Mohamed W, Hammam A (2016) Soil physiochemical properties in relation to heavy metals status of agricultural soils in El-Minia governorate, Egypt. J Soil Sci Agric Eng 7: 423–431. https://doi.org/10.21608/jssae.2016.39676
37. Goher ME, Ali MH, El-Sayed SM (2019) Heavy metals contents in Nasser Lake and the Nile River, Egypt: an overview. Egypt J Aquat Res 45:301–312. https://doi.org/10.1016/j.ejar.2019.12.002
38. Abdelmawgoud A, El-Rawy M, Moussa Abu Bakr A (2020) Evaluating the suitability of groundwater quality for drinking and irrigation purposes in El-Minia governorate, Egypt. J Adv Eng Trends:1003. https://doi.org/10.21608/JAET
39. El-Rawy M, Fathi H, Abdelrady A, Negm AM (2022) Environmental impacts of treated wastewater contaminates on groundwater quality in the Nile River valley, Egypt, in Sustainability of groundwater in the Nile Valley, Egypt. Springer, pp 91–120. https://doi.org/10.1007/978-3-031-12676-5_5
40. El-Rawy M, Abdalla F, Negm AM (2021) Groundwater characterization and quality assessment in Nubian sandstone aquifer, Kharga oasis, Egypt. In: Groundwater in Egypt's deserts. Springer, Cham, pp 177–199

41. El-Rawy M, Fathi H, Abdelrady A, Negm AM (2022) Environmental impacts of treated wastewater contaminates on groundwater quality in the Nile River valley, Egypt. In: Sustainability of groundwater in the Nile Valley, Egypt. Springer, Cham, pp 91–120
42. El-Rawy M, Ismail E, Abdalla O (2019) Assessment of groundwater quality using GIS, hydrogeochemistry, and factor statistical analysis in Qena governorate, Egypt. Desalination Water Treat 162:14–29. https://doi.org/10.5004/dwt.2019.24423
43. Ismail E, El-Rawy M, Mauritsch H (2022) Evaluation of groundwater potential zones using electrical resistivity and hydrogeochemistry in west Tahta region, upper Egypt. In: Sustainability of groundwater in the Nile Valley, Egypt. Springer, Cham, pp 267–291
44. Alnashiri HM (2021) Assessment of physicochemical parameters and heavy metal concentration in the effuents of sewage treatment plants in Jazan Region, Saudi Arabia. J King Saud Univ Sci 33:101600. https://doi.org/10.1016/j.jksus.2021.101600
45. Aljohani NS, Kavil YN, Shanas PR, Al-Farawati RK, Shabbaj II, Aljohani NH, Turki AJ, Abdel Salam M (2022) Environmental impacts of thermal and brine dispersion using hydrodynamic modelling for Yanbu desalination plant, on the eastern coast of the Red Sea. Sustainability 14: 4389. https://doi.org/10.3390/su14084389
46. Hossain ME, Shahrukh S, Hossain SA (2022) Chemical fertilizers and pesticides: impacts on soil degradation, groundwater, and human health in Bangladesh. In: Environmental degradation: challenges and strategies for mitigation, Springer, pp 63–92. https://doi.org/10.1007/978-3-030-95542-7_4
47. Alsalah D, Al-Jassim N, Timraz K, Hong P-Y (2015) Assessing the groundwater quality at a Saudi Arabian agricultural site and the occurrence of opportunistic pathogens on irrigated food produce. Int J Environ Res Public Health 12:12391–12411. https://doi.org/10.3390/ijerph121012391
48. Alharbi T, El-Sorogy AS (2023) Quality and groundwater contamination of Wadi Hanifa, Central Saudi Arabia. Environ Monit Assess 195:525. https://doi.org/10.1007/s10661-023-11093-0
49. Canter LW (2019) Fundamental aspects of nitrates in groundwater. Nitrates in groundwater, Routledge. https://doi.org/10.1201/9780203745793-1
50. Zhao B, Sun Z, Liu Y (2022) An overview of in-situ remediation for nitrate in groundwater. Sci Total Environ 804:149981. https://doi.org/10.1016/j.scitotenv.2021.149981
51. Chidambaram S, Prasad MBK, Prasanna MV, Manivannan R, Anandhan P (2015) Evaluation of metal pollution in groundwater in the industrialized environs in and around Dindigul, Tamilnadu, India. Water Qual Expo Health 7:307–317. https://doi.org/10.1007/s12403-014-0150-6
52. Saad M, Awni B, Eslam E, Abdulaziz A-B, Elkhedr I, Saleh Q (2011) Aquifer boundaries explored by geoelectrical measurements in the Red Sea coastal plain of Jazan area, Southwest Saudi Arabia. Int J Phys Sci 6:3688–3696. https://doi.org/10.5897/IJPS11.803
53. Mogren S (2015) Saltwater intrusion in Jizan coastal zone, Southwest Saudi Arabia, inferred from Geoelectric resistivity survey. Int J Geosci 6(3):12. https://doi.org/10.4236/ijg.2015.63022
54. Elshemy M, Meon G (2011) Climate change impacts on water quality indices of the southern part of Aswan high dam reservoir. Lake Nubia. In Fifteenth international water technology conference, IWTC-15, p 17.
55. Tiwari AK, Singh PK, Mahato MK (2014) GIS-based evaluation of water quality index of ground water resources in west Bokaro coalfield, India. Current world. Environment 9:843. https://doi.org/10.12944/cwe.9.3.35
56. SASO (2009) Saudi Arabian Standards Organization. https://saso.gov.sa/ar/pages/default.aspx
57. WHO (2011) Guidelines for drinking-water quality. WHO Chron 38:104–108
58. Masoud MH, Basahi JM, Rajmohan N (2018) Impact of flash flood recharge on groundwater quality and its suitability in the Wadi Baysh Basin, Western Saudi Arabia: an integrated approach. Environ Earth Sci 77:1–19. https://doi.org/10.1007/s12665-018-7578-0

59. Nazzal Y, Ahmed I, Al-Arifi NS, Ghrefat H, Batayneh A, Abuamarah BA, Zaidi FK (2015) A combined hydrochemical-statistical analysis of Saq aquifer, northwestern part of the Kingdom of Saudi Arabia. Geosci J 19:145–155. https://doi.org/10.1007/s12303-014-0016-8
60. Awni B, Eslam E, Saad M, Elkhedr I, Saleh Q (2012) Groundwater quality of the shallow alluvial aquifer of Wadi Jazan (Southwest Saudi Arabia) and its suitability for domestic and irrigation purpose. Sci Res Essays 7:352–364. https://doi.org/10.5897/SRE11.1194
61. El-Hames A, Al-Ahmadi M, Al-Amri N (2011) A GIS approach for the assessment of groundwater quality in Wadi Rabigh aquifer, Saudi Arabia. Environ Earth Sci 63:1319–1331. https://doi.org/10.1007/s12665-010-0803-0
62. Haddaoui I, Mateo-Sagasta J (2021) A review on occurrence of emerging pollutants in waters of the MENA region. Environ Sci Pollut Res 28:68090–68110. https://doi.org/10.1007/s11356-021-16558-8
63. Hassen W, Mehri I, Beltifa A, Giorgia Potortì A, Khellaf N, Amer R, Van Loco J, Hassen A, Di Bella G, Khdary NH, Ben Mansour H (2022) Chemical and microbiological assessment of wastewater discharged along the Mediterranean Sea. Sustainability 14:2746. https://doi.org/10.3390/su14052746
64. Rateb A, Scanlon BR, Fakhreddine S (2022) How severe is water stress in the MENA region? Insights from GRACE and GRACE-FO satellites and global hydrological modeling. In: Applications of space techniques on the natural hazards in the MENA region. Springer, pp 51–65. https://doi.org/10.1007/978-3-030-88874-9_4
65. Fathi H, El-Rawy M (2018) GIS-based evaluation of water quality index for groundwater resources nearby wastewater treatment plants. Egypt Poll Res 37:105–116
66. Ismail E, El-Rawy M (2018) Assessment of groundwater quality in West Sohag, Egypt. Desalination Water Treat 123:101–108
67. Obeid AA, Abdelrady A, El-Rawy M (2022) Hydrochemistry and hydrogeology aspects of alluvial aquifer in Aswan City, Egypt. In: Sustainability of groundwater in the Nile Valley, Egypt. Springer, Cham, pp 121–139

Isotopic and Chemical Composition of Egypt's Groundwater Resources

Mohamed Ahmed and Mahmoud M. Khalil

Contents

1 Introduction 238
2 Egypt Major Aquifer Systems 240
2.1 Alluvial Aquifer 240
2.2 Nile Valley and Nile Delta Aquifers 240
2.3 Coastal Aquifers 241
2.4 Moghra Aquifer 241
2.5 Carbonate Aquifer 241
2.6 Nubian Aquifer 241
2.7 Fractured Basement Aquifer 242
3 Data and Methods 242
4 Isotopic and Chemical Compositions of Egypt Groundwater Resources 243
4.1 Alluvial Aquifer 243
4.2 Nile Valley and Delta Aquifers 245
4.3 Coastal Aquifers 252
4.4 Moghra Aquifer 254
4.5 Carbonate Aquifer 254
4.6 Nubian Aquifer 257
4.7 Fractured Basement Aquifer 259
5 Summary and Conclusions 261
6 Recommendations 262
References 262

Abstract Aquifers in arid regions are mostly the sole sources of freshwater under the limited surface water resources. Therefore, they are affected by diverse natural

M. Ahmed (✉)
Department of Physical and Environmental Sciences, Texas A&M University-Corpus Christi, Corpus Christi, Texas, USA
e-mail: mohamed.ahmed@tamucc.edu

M. M. Khalil
Geology Department, Faculty of Science, Minia University, El Minia, Egypt
e-mail: mahmoud.khalil@mu.edu.eg

Shakir Ali and Abdelazim Negm (eds.), *Groundwater Quality and Geochemistry in Arid and Semi-Arid Regions*, Hdb Env Chem (2024) 126: 237–266, DOI 10.1007/698_2023_1046,

Published online: 19 December 2023

and anthropogenic pressures resulting in over-exploitation and quality degradation. Investigating the spatial variability in isotopic and chemical compositions of these aquifers provides a better understanding of the recharge and contamination sources and mechanisms and ultimately guides the sustainable management scenarios of these aquifers. In this chapter, we compiled all historical isotopic (hydrogen and oxygen) and chemical (salinity) compositions of all groundwater samples collected for seven aquifer systems in Egypt (alluvial, Nile Valley and Nile Delta, coastal, Moghra, carbonate, Nubian, and fractured basement). The water chemical and isotopic compositions were then used to provide an improved understanding of the country-wide groundwater resources' isotopic and chemical compositions and to investigate the recharge and contamination sources and mechanisms. This study also provides sustainable management scenarios for the seven Egyptian aquifers.

Keyword Egypt, Environmental isotopes, Groundwater, Salinity, Stable water isotope

1 Introduction

In arid and semi-arid environments, groundwater resources are vital for economic, social, and environmental sustainability. However, both natural and anthropogenic factors impact the availability and quality of groundwater resources in these regions [1]. Natural factors include climate changes, altered rainfall and temperature patterns, and extreme weather events [2]. Anthropogenic forces, on the other hand, include rapid population growth, substantial land cover and land use changes, surface water damming, groundwater over-exploitation, and pollution [3, 4]. Understanding the influences of both natural and human interventions on groundwater availability and quality is crucial for security and sustainability of groundwater resources in arid and semi-arid environments [5].

Historically, the responses of aquifer systems to natural and human influences have primarily relied on gathering and analyzing field observations such as geophysical, hydrogeological, and geochemical data [3, 4, 6–10]. Such observations are employed to determine the groundwater table's position, assess storage volumes, map areas of recharge and depletion, quantify recharge and discharge rates, and assess groundwater quality [2, 11–22].

Chemical analyses offer insights into groundwater quality, while isotopic compositions provide valuable information about groundwater age and recharge processes. Measuring groundwater salinity, for example, is crucial for water quality assessment, environmental monitoring, agricultural practices, aquifer management, and overall water resource planning (e.g., [13, 23, 24]). Stable water isotopes (hydrogen: ^{1}H, ^{2}H; oxygen: ^{16}O, ^{18}O) have proven particularly useful in mapping the spatial distribution, physical properties, and recharge mechanisms of different water types [5, 25–30].

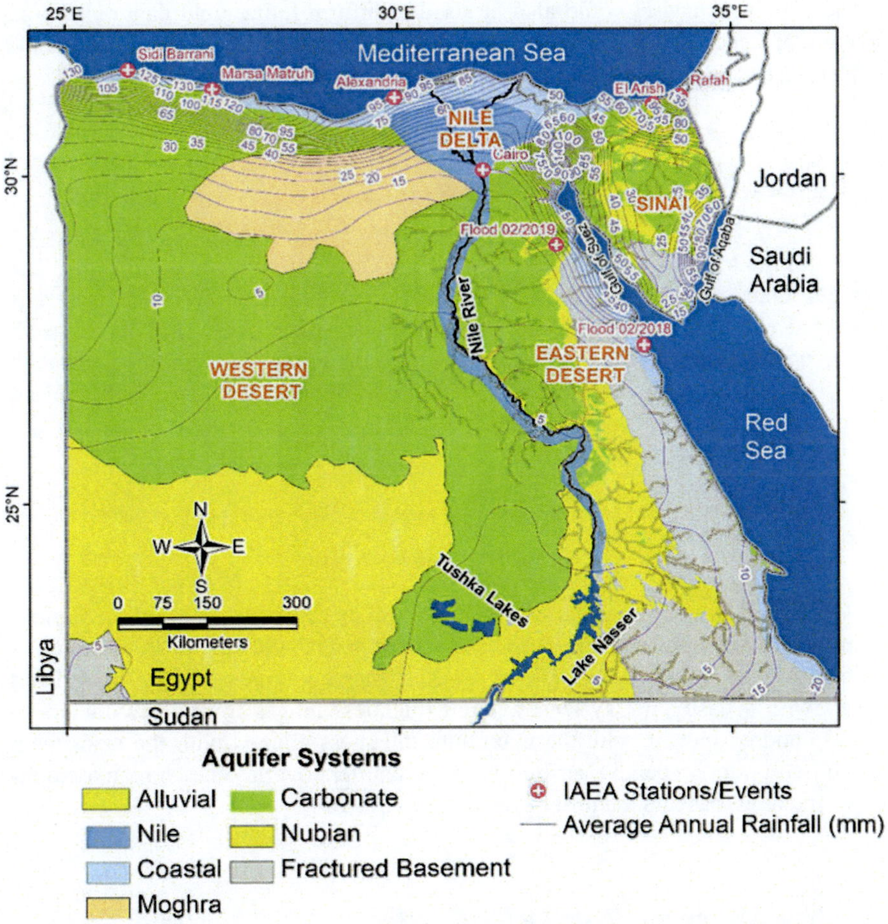

Fig. 1 Spatial distribution of the seven (alluvial, Nile Valley and Nile Delta, coastal, Moghra, carbonate, Nubian, fractured basement) major aquifer systems in Egypt. The contour lines on the map represent Egypt's average annual rainfall (in mm). The locations of rainfall stations and recent flood events, for which isotopic compositions were determined, are also indicated [5]

In this chapter, we built upon our recent publications on Egypt's major aquifer systems [5]. Specifically, we compiled all historical chemical and isotopic composition datasets for groundwater samples collected from seven major aquifer systems of Egypt. The complied datasets were then used to:

- Provide enhanced understanding of the country-wide water resources' isotopic and chemical compositions,
- Investigate sources and mechanisms of aquifer recharge, and
- Provide sustainable management scenarios for the nationwide aquifer systems in Egypt (Fig. 1).

The compiled dataset could also be used to inform future field data collection, increase efficiency in field activities, and empower the scientific community to tackle country-scale science questions in a groundbreaking manner.

2 Egypt Major Aquifer Systems

Except for the Mediterranean coast, Egypt experiences an arid climate (Fig. 1). The average annual rainfall in Sinai, the Eastern Desert, and the Western Desert is approximately 40 mm, 20 mm, and 5 mm per year, respectively (Fig. 1). Beneath the Egyptian deserts, seven distinct aquifer systems have been identified, namely: alluvial, Nile Valley and Nile Delta, coastal, Moghra, carbonate, Nubian, and fractured basement aquifer systems (Fig. 1).

2.1 Alluvial Aquifer

The Quaternary alluvial aquifers are distributed along the major wadis in the Eastern Desert and Sinai Peninsula (Fig. 1). These unconfined, shallow, and porous aquifers exhibit varying groundwater productivity [31]. They comprised mainly of sand and gravels with occasional clay lenses. Their thickness ranges from a few meters to several hundred meters, with the maximum thickness aligned with the wadi main axis. Based on their spatial distribution, these aquifers overlay the Nubian, carbonate, or fractured basement aquifer systems [10, 31–36].

2.2 Nile Valley and Nile Delta Aquifers

The Nile Valley aquifer primarily comprises Late Tertiary and Quaternary deposits, running parallel to the Nile River (Fig. 1). In central Egypt, the average aquifer thickness is around 300 m. The thickness of the Nile Valley aquifer decreases toward both the north and south directions in Egypt. This aquifer is separated from the underlying Nubian aquifer by impermeable clay layers [37–39]. Approximately, 40% of Egypt's irrigation, municipal, and domestic water needs are met by the Nile valley aquifer [38, 40]. The Nile Delta aquifer consists of confined Pleistocene sand and gravel deposits, overlain by Holocene clay sediments. In the northern regions of Egypt, the thickness of this aquifer can reach up to 1 km [41–43]. It mainly fulfills irrigation and domestic demands in Egypt [38, 41].

2.3 Coastal Aquifers

The coastal aquifers are distributed along the coastlines of the Mediterranean Sea, Red Sea, Gulf of Suez, and Gulf of Aqaba (Fig. 1). The Mediterranean coastal aquifers are primarily composed of limestone to the west of the Nile Delta and consist of sand, gravel, and limestone in North Sinai [37, 38, 44]). Unfortunately, these aquifers are experiencing a significant contamination in groundwater, especially in the areas east of the Nile Delta due to saltwater intrusion [45, 46]. On the other hand, the Red Sea coastal aquifers are predominantly composed of Quaternary sediments in the deltaic regions of the primary wadis, while Tertiary sand sediments dominate elsewhere.

2.4 Moghra Aquifer

The Moghra aquifer, situated west of the Nile Delta (Fig. 1), is composed of Lower Miocene fluviatile and fluvio-marine coarse sand and gravel sediments. Its thickness can reach up to 930 m, but it gradually decreases toward the north and west [42, 47, 48]. Groundwater resources of this aquifer are used mainly for irrigation. This aquifer primarily discharges through evaporation in local depressions.

2.5 Carbonate Aquifer

The carbonate aquifer, which spans most of Egypt (Fig. 1), is a highly intricate system significantly influenced by geological structures, primarily faults and fractures [37, 49]. This aquifer comprises three main units (lower, middle, and upper) dating from the Upper Cretaceous to Middle Miocene periods. These units are separated by two impermeable shale layers [38, 42, 50].

2.6 Nubian Aquifer

The Nubian aquifer extends across Egypt, Libya, Sudan, and Chad and consists of two major reservoirs separated by low-permeability layers: the Nubian and post-Nubian reservoirs. In Egypt, the Nubian aquifer is found in the Western Desert, the Eastern Desert, and Sinai (Fig. 1). This aquifer predominantly comprises Paleozoic and Mesozoic sandstone with a thickness ranging from 0.5 to 4 km, with intercalated Upper Cretaceous deltaic and shale deposits [51–53]). The Nubian aquifer overlays the Proterozoic basement rocks and is confined in central and northern Egypt beneath thick Upper Cretaceous shale deposits [52, 53].

2.7 *Fractured Basement Aquifer*

The fractured basement aquifer is located at shallow depths within the Pre-Cambrian crystalline basement rocks of the Eastern Desert and southern Sinai (Fig. 1). The potential storage and movement of groundwater in this aquifer are governed by various tectonic elements, such as intersecting faults, intersecting shear zones and faults, or highly fractured mélange settings [8, 38].

3 Data and Methods

A flowchart of the major data and methods used in this study is shown in Fig. 2. The chemical and isotopic compositions of groundwater samples were gathered from 98 articles published between 1974 and 2023. A total of 1,904 groundwater samples were compiled for chemical compositions and 1,618 samples were compiled for isotopic compositions (Fig. 3). For each sample, we have collected the full citation; aquifer name, age, and location; well type, depth, name, elevation, and depth to water table; longitude and latitude; Deuterium (δ^2H) and Oxygen-18 ($\delta^{18}O$) concentrations; and salinity expressed as total dissolved solids (TDS). In cases where latitude and longitude details for water samples were not provided in the articles, a location map depicting the distribution of wells or samples was registered in a Geographic Information System (GIS) environment, and subsequently, the sample locations were manually digitized. Descriptive statistics for chemical and isotopic compositions of the groundwater data from seven aquifers in Egypt are shown in Tables 1 and 2, respectively.

The compiled dataset was employed to examine the chemical (e.g., salinity) and isotopic (δ^2H and $\delta^{18}O$) compositions of groundwater resources within the seven Egyptian aquifer systems. These data were presented based on geographic settings, which include the Eastern Desert, Western Desert, Nile Valley and Delta, and Sinai, for each of the investigated aquifer systems. The isotopic compositions were presented in reference to the Global Meteoric Water Line (GMWL; $\delta^2H = 8\,\delta^{18}O + 10$ [54]) and/or the East Mediterranean Water Line (EMWL; $\delta^2H = 8\delta^{18}O + 22$ [55]).

Modern precipitation's isotopic compositions at six stations in Egypt (Fig. 1) were extracted from the International Atomic Energy Agency (IAEA) WISER database (available at: https://nucleus.iaea.org/wiser/index.aspx).

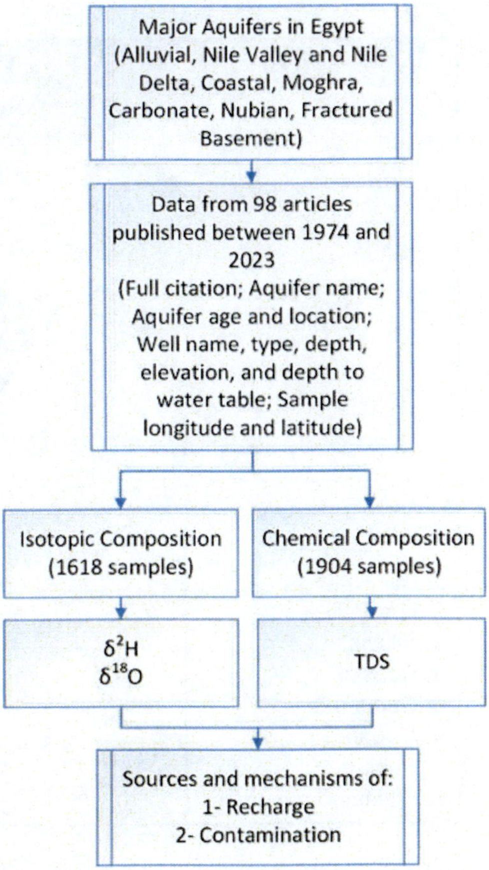

Fig. 2 A flowchart showing methods used to compile historical isotopic (hydrogen and oxygen) and chemical (salinity) compositions of all groundwater samples collected for seven aquifer systems in Egypt (alluvial, Nile Valley and Nile Delta, coastal, Moghra, carbonate, Nubian, fractured basement). The collected isotopic and chemical compositions were then used to investigate the seven Egyptian aquifers' recharge and contamination sources and mechanisms

4 Isotopic and Chemical Compositions of Egypt Groundwater Resources

4.1 *Alluvial Aquifer*

In Sinai, the salinity of the alluvial aquifer varied from 562 ppm (Dahab basin, South Sinai) to 37,343 ppm (Wadi Watir Delta, Nuweiba) with an average of 4,032 ppm (Fig. 4a; Table 1). The high salinity of the alluvial aquifer at Wadi Watir Delta is due to seawater intrusion. At the Eastern Desert, the groundwater salinity ranged from 154.4 ppm (Wadi El Kharit, Idfu-Aswan area) to 18,668 ppm (Safaga-El Quseir District) with an average of 4,765 ppm. The high salinity values, 18,668 ppm, are mainly attributed to the leaching of some evaporates encountered within the alluvial sediments [56].

The isotopic compositions of groundwaters from the alluvial aquifers in Sinai (δ^2H: −23.46‰ ± 5.40, δ^{18}O: −4.27‰ ± 0.89) are plotted between the GMWL and

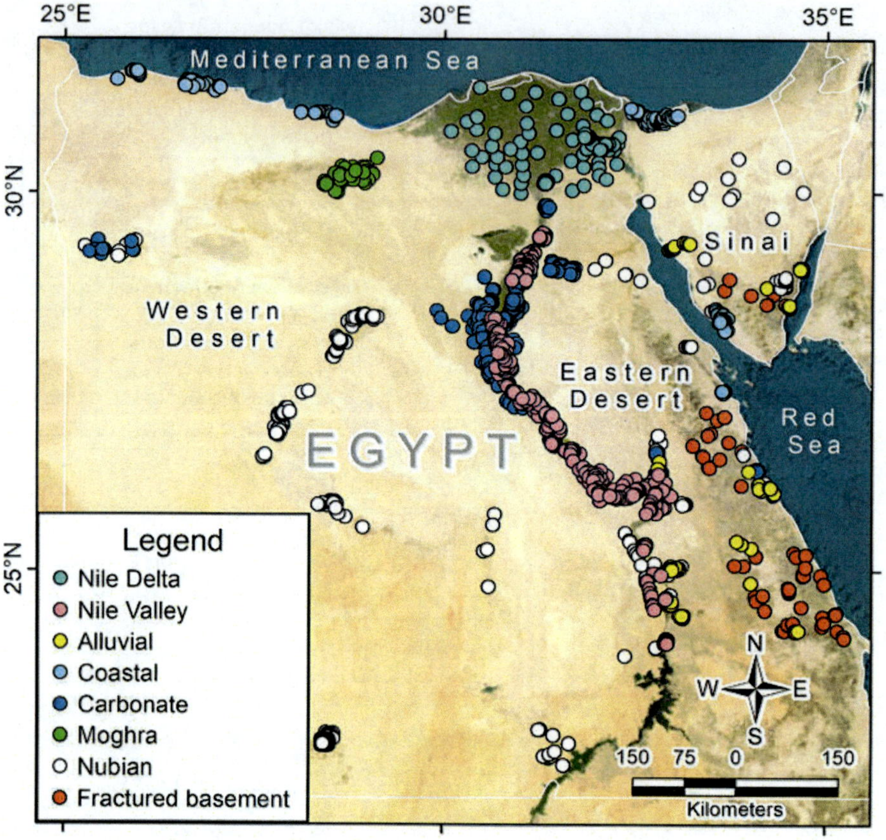

Fig. 3 Spatial distribution of wells used to study the chemical and isotopic compositions of Egypt's groundwater resources, grouped by aquifer names

EMWL, closely aligning with the average isotopic composition of modern precipitation recorded at the Rafah and El Arish stations (δ^2H: −9.86‰ to −8.81‰; δ^{18}O: −3.48‰ to −3.22‰) (Fig. 4b; Table 2). This alignment indicates that these groundwaters primarily recharge from modern precipitation. The lower slope (4.5) and intercept (−3.7) of the line fitted to Sinai groundwater samples may be attributed to evaporation from wadi runoff before infiltration. On the other hand, alluvial groundwaters from the Eastern Desert (δ^2H: −35.68‰ ± 16.55, δ^{18}O: −4.62‰ ± 2.53) are mostly situated below the GMWL with a lower slope (6.2) and intercept (−6.8) (Fig. 4b; Table 2), suggesting the influence of evaporation and/or mixing with evaporated water. The wide variation and depletion in isotopic composition of alluvial aquifers groundwater in the Eastern Desert (−8.9‰ to 5.2‰ for δ^{18}O; −63.2‰ to 32‰ for δ^2H) cannot be solely attributed to modern precipitation and/or draining as wadi runoff from the Red Sea hills. Instead, it is suggested that an

Table 1 Statistical parameters of salinity data for different aquifers across Egypt

Aquifer	Location	No. of observations	Minimum	Maximum	Mean	Standard deviation
Nubian	Sinai	42	307	8,915	1,531.3	1,464.9
	Western Desert	185	79	10,908	516.4	1,017.1
	Eastern Desert	80	1,092	13,827	2,935.2	1,987.3
Coastal	Along the Red Sea	29	5,832	15,680	9,181	3,188
	Along the Gulf of Suez	63	402	9,384	3,417	2,497
	Along the Mediterranean Sea	106	1,210	35,900	6,437.9	5,263
Alluvial	Sinai	94	562	37,343	4,032.7	3,936.3
	Eastern Desert	107	154.4	18,668	4,765.7	4,328.5
Fractured basement	Sinai	24	432	14,156	1,977	2,870
	Eastern Desert	59	405	82,400	7,487.6	12,246.4
Carbonate	Western Desert	139	228.7	8,012	1,494.4	1,333.2
	Eastern Desert	210	193.5	13,900	1,509.6	1,731.7
Nile	Nile Valley aquifer	536	140	8,249	1,108.5	1,036.9
	Nile Delta aquifer	140	225.6	33,517	2,395.9	4,740
Moghra	Western Desert	90	2,236	14,008	5,483.6	2,036.6

additional recharge source for shallow alluvial aquifers in the Eastern Desert and Sinai is upward leakage from deep Nubian aquifers along complex fault systems.

4.2 *Nile Valley and Delta Aquifers*

The salinity distribution along 800 km extension of the Nile Valley aquifer from Cairo, in the north, to Aswan southward is shown in Fig. 5a. The salinity average is 1,108.5 ppm (Table 1). It varies from 140 ppm (at Assiut Governorate) to 8,249 ppm (east Esna city, Aswan Governorate) (Fig. 5a). Generally, low groundwater salinity is owing to recharge from surface water resources (i.e., Nile River, irrigation canals, and drains) that are intensively distributed within the Nile Valley for irrigation and agricultural activities, while high TDS are mostly found in desert margins of the Nile Valley due to limited recharge and prolonged water-rock interactions. On the other hand, within the Nile Delta aquifer, the groundwater salinity varies from 225.6 (at the Delta apex, east Cairo) to 33,517 ppm (middle coastal plain, near Rosetta) with clear northward and eastward increasing trends (Fig. 5b). The average salinity of the Nile Delta aquifer was 2,369 ppm (Table 1). The spatial salinity variation might be related to paleo-hydrogeological conditions, i.e., Holocene marine transgressions, of the Nile Delta for the last 32 ka [57]. This evidence is supported by the

Table 2 Descriptive statistics for the $\delta^{18}O$, $\delta^{2}H$, and d-excess parameters of the groundwater samples collected from seven aquifers in Egypt

Aquifer	Location	Statistics	$\delta^{18}O$ (‰)	$\delta^{2}H$ (‰)	d-excess (‰)
Alluvial	Sinai	No. of observations	37	37	37
		Minimum	−5.9	−32.5	−3.6
		Maximum	−1.3	−9.2	19.5
		1st quartile	−4.8	−28	7.5
		3rd quartile	−4.2	−20	13.8
		Mean	−4.3	−23.5	10.8
		Standard deviation ($n-1$)	0.9	5.4	4.6
	East Nile Valley	No. of observations	85	85	85
		Minimum	−8.9	−63.3	−15.1
		Maximum	3.1	21.1	14.4
		1st quartile	−6.5	−46.8	−2.4
		3rd quartile	−3.4	−29.6	6
		Mean	−4.6	−35.7	1.3
		Standard deviation ($n-1$)	2.5	16.6	6.8
Nile	Nile delta	No. of observations	327	327	327
		Minimum	−3.1	−27.5	−19.9
		Maximum	5.9	42.1	15.8
		1st quartile	−0.4	2.1	−1.3
		3rd quartile	2.5	21.7	6.7
		Mean	1.2	11.3	1.8
		Standard deviation ($n-1$)	1.9	11.9	6
	Nile Valley	No. of observations	103	103	103
		Minimum	−3.8	−30.3	−20
		Maximum	6.5	32	9.2
		1st quartile	−0.9	−6.2	−1.2
		3rd quartile	2.3	20.8	4.7
		Mean	0.8	6.4	0.3
		Standard deviation ($n-1$)	2.1	15.7	6
Coastal	Sinai (Gulf of Aqaba)	No. of observations	27	27	27
		Minimum	−3.8	−20.1	−32.4
		Maximum	6.9	22.5	11.6
		1st quartile	−3.6	−18	−1.4
		3rd quartile	−0.3	−4.7	10.5
		Mean	−1.5	−9.1	3
		Standard deviation ($n-1$)	3	12.7	11.4
	Sinai (Gulf of Suez)	No. of observations	34	34	34

(continued)

Table 2 (continued)

Aquifer	Location	Statistics	$\delta^{18}O$ (‰)	$\delta^{2}H$ (‰)	d-excess (‰)
		Minimum	−6.8	−47	−16
		Maximum	−0.3	−8.1	17.6
		1st quartile	−5.6	−38.4	3.9
		3rd quartile	−4.4	−24.2	12.6
		Mean	−4.7	−30.4	7
		Standard deviation (n−1)	1.5	10.1	7.3
	Sinai (Mediterranean Sea)	No. of observations	30	30	30
		Minimum	−4.1	−25.7	−9.7
		Maximum	4.3	25.5	10.3
		1st quartile	−2.9	−22.4	−2.5
		3rd quartile	−1.3	−11.9	2.6
		Mean	−1.6	−12.9	−0.1
		Standard deviation (n−1)	2.1	15.6	4.6
	Northwestern	No. of observations	72	72	72
		Minimum	−6.6	−33.5	−2.2
		Maximum	3	23.6	22.5
		1st quartile	−5	−23.8	9.5
		3rd quartile	−2.9	−13.9	16.8
		Mean	−3.2	−14	11.9
		Standard deviation (n−1)	2.9	16.7	6.9
	Eastern Desert, Red Sea	No. of observations	9	9	9
		Minimum	−3.6	−27.5	−25.7
		Maximum	5.3	16.7	5.2
		1st quartile	−2.4	−17.9	−5.2
		3rd quartile	1.5	6.2	4.3
		Mean	0	−2.4	−2.7
		Standard deviation (n−1)	2.8	16.7	9.7
Moghra	Western Desert	No. of observations	21	21	21
		Minimum	−2.3	−13.9	−33.2
		Maximum	5.8	13.4	8.1
		1st quartile	−1.2	−5.2	−5.2
		3rd quartile	0.9	4	6.6
		Mean	−0.1	−1.4	−0.3
		Standard deviation (n−1)	1.9	6.6	9.7
Fractured basement	Eastern Desert	No. of observations	27	27	27
		Minimum	−3.5	−32	−20.8
		Maximum	5.4	22.2	14.1

(continued)

Table 2 (continued)

Aquifer	Location	Statistics	$\delta^{18}O$ (‰)	δ^2H (‰)	d-excess (‰)
		1st quartile	−2	−12.3	0.7
		3rd quartile	−0.9	−1.9	6.8
		Mean	−0.9	−4.8	2.6
		Standard deviation (*n*−1)	2.2	11.2	8.5
	Sinai	No. of observations	28	28	28
		Minimum	−5.6	−31.3	−6.8
		Maximum	−0.9	−5.5	17.9
		1st quartile	−4.6	−24.5	8.6
		3rd quartile	−3.4	−17.4	13.8
		Mean	−4	−20.9	10.8
		Standard deviation (*n*−1)	1.2	5.8	5.5
Carbonate	East Nile Valley	No. of observations	8	8	8
		Minimum	−2.1	−10.8	1
		Maximum	0.6	5.4	10.5
		1st quartile	−1.5	−8.9	4.3
		3rd quartile	−0.5	4.9	8.6
		Mean	−0.9	−1.6	5.9
		Standard deviation (*n*−1)	0.9	7	3.3
	West Nile Valley	No. of observations	20	20	20
		Minimum	−8.3	−63.6	−3.8
		Maximum	2.7	22.9	14.7
		1st quartile	−3.1	−15	3.4
		3rd quartile	−2	−5.6	11.2
		Mean	−2.6	−13.5	7.2
		Standard deviation (*n*−1)	2.1	17.4	5.1
	Sinai	No. of observations	8	8	8
		Minimum	−6.7	−34.2	6.4
		Maximum	−4	−24	19.4
		1st quartile	−6	−32.3	10.4
		3rd quartile	−5	−29.5	16.8
		Mean	−5.5	−30.5	13.5
		Standard deviation (*n*−1)	0.8	3.1	4.6
Nubian	Sinai	No. of observations	70	70	70
		Minimum	−9.6	−72.9	−7.9
		Maximum	−3.5	−22	17
		1st quartile	−8	−53.6	6
		3rd quartile	−5.9	−36.4	12.3

(continued)

Table 2 (continued)

Aquifer	Location	Statistics	$\delta^{18}O$ (‰)	$\delta^{2}H$ (‰)	d-excess (‰)
		Mean	−6.8	−45.7	9.1
		Standard deviation (*n*−1)	1.4	12.2	4.2
	Eastern Desert	No. of observations	42	42	42
		Minimum	−9.5	−73.1	−1.1
		Maximum	−4.1	−31.1	15.5
		1st quartile	−7.1	−53.4	2.9
		3rd quartile	−6.1	−39.3	8.8
		Mean	−6.7	−47.7	5.5
		Standard deviation (*n*−1)	1.2	10.1	4.3
	Western Desert	No. of observations	220	220	220
		Minimum	−11.9	−89	−20.4
		Maximum	3.1	20.9	15.2
		1st quartile	−10.8	−81.8	0.3
		3rd quartile	−10	−77.6	5.4
		Mean	−10	−77.1	2.5
		Standard deviation (*n*−1)	2	14.3	5.1

conformity of the brackish groundwater zone extent in the east to the coastline during the maximum transgression at 8 ka [57]. The high northward abnormal salinity, 33,517 ppm, is interpreted as pockets of evapoconcentrated Holocene groundwater.

Groundwaters within the Nile Delta and Valley aquifers exhibit a wide range of variation, with $\delta^{18}O$ ranging from −3.8‰ to 6.5‰ and $\delta^{2}H$ ranging from −30.3‰ to 42.1‰ (Fig. 5c; Table 2). The majority of groundwater samples are plotted on the mixing line connecting the Nile water components (pre- and post-High Dam waters; $\delta^{2}H$: 25.79‰ ± 5.46, $\delta^{18}O$: 3.16‰ ± 0.97; Fig. 5c), indicating that their primary recharge source is from the Nile River, as well as from irrigation canal and drain systems. The deviation of groundwater samples from the GMWL suggests evaporation before recharge and/or mixing with highly evaporated water, such as irrigation return flow and wastewater. This is further emphasized by the lower slopes (6.9 and 5.8) and intercepts (1.1 and 4.5) of the evaporation lines fitted to the Nile Valley and Delta samples in Fig. 5c. In the Nile Valley and Delta region, water undergoes varying degrees of evaporation due to circulation between irrigation canals, irrigated soils, drainage systems, and recycling of drainage water for irrigation. As a result, the water becomes isotopically enriched before infiltrating into the aquifer. For other groundwaters that are more depleted in isotopic composition compared to the Nile waters, it is plausible to interpret them as representing a mixture of pre-High Dam Nile water and fossil water from deeper aquifer systems, such as the carbonate and/or Nubian aquifers. The discharge of these deeper systems, through complex fault

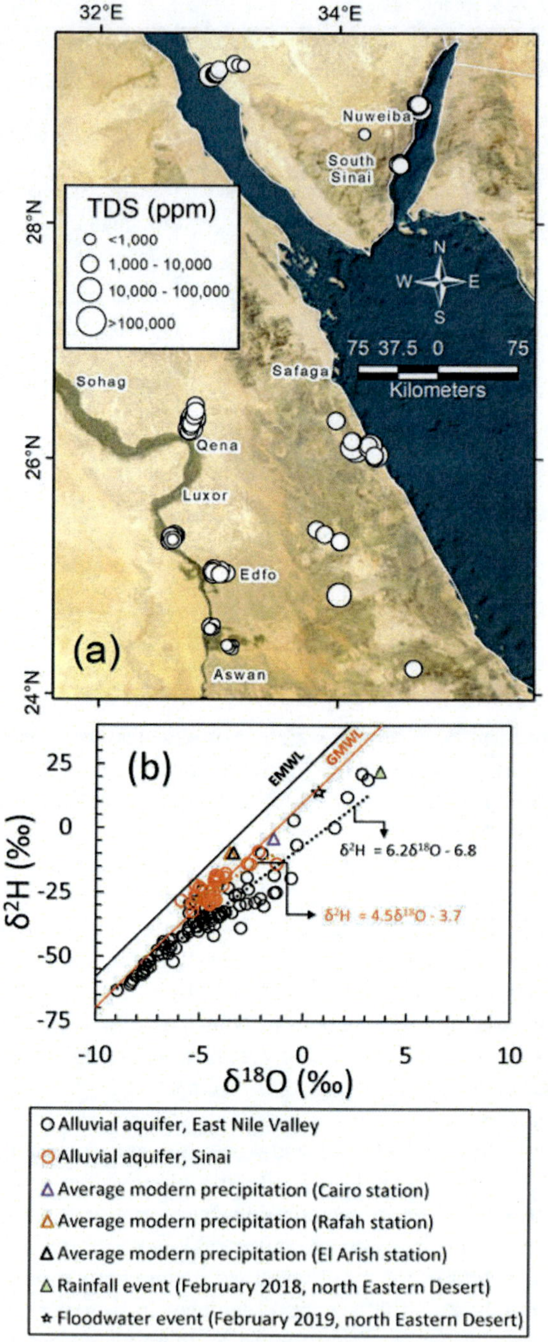

Fig. 4 (**a**) Spatial distribution of salinity values for groundwater samples collected from the alluvial aquifer. (**b**) δ^2H versus $\delta^{18}O$ plot for groundwater samples collected from the alluvial aquifer in the Eastern Desert and Sinai Peninsula

Fig. 5 Spatial distribution of salinity values for groundwater samples collected from the (**a**) Nile Valley and (**b**) Nile Delta aquifers. **c**) δ^2H versus $\delta^{18}O$ plot for groundwater samples collected from the Nile Delta and the Nile Valley aquifers

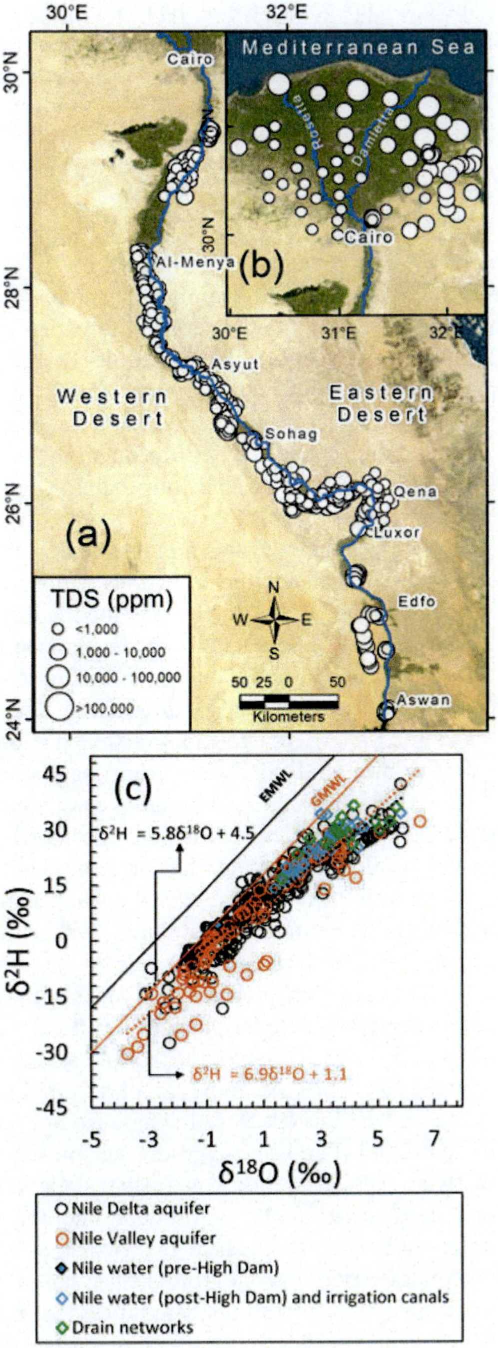

systems, occurs within the alluvial settings on the fringes of the desert surrounding the Nile Valley and Delta. These areas are currently being subjected to reclamation and cultivation.

4.3 Coastal Aquifers

Along the Mediterranean Sea, the salinity of groundwater in coastal aquifer varied from 1,210 ppm to 35,900 ppm with an average of 6,438 ppm (Fig. 6a; Table 1). For the Gulf of Suez, groundwater salinity varied from 402 ppm to 9,384 ppm with an average of 3,416 ppm (Fig. 6a; Table 1). Along the Red Sea, salinity of coastal aquifer ranged from 5,832 ppm to 15,680 ppm with an average of 9,180 ppm (Fig. 6a; Table 1). In addition to seawater intrusion, groundwater abstraction and disposal of high concentrated effluents from desalinization plants (DSP) are the main factors to increase the groundwater salinity, especially along the Mediterranean Sea and Red Sea.

Distinctive characteristics are observed in the isotopic compositions of groundwaters from coastal aquifers in various regions. Groundwaters from Sinai (along the Gulf of Suez, the Gulf of Aqaba, and the Mediterranean Sea; δ^2H: $-18.31‰ \pm 15.9$, $\delta^{18}O$: $-2.72‰ \pm 2.66$; Fig. 6b), the Eastern Desert (along the Red Sea coast; δ^2H: $-2.36‰ \pm 16.71$, $\delta^{18}O$: $0.04‰ \pm 2.79$; Fig. 6b), and the northern portions of the Western Desert (along the Mediterranean Sea; δ^2H: $-14.03‰ \pm 16.69$, $\delta^{18}O$: $-3.23‰ \pm 2.87$; Fig. 6c) generally display isotopic compositions similar to those of modern rainfall and floodwater from local storm events (δ^2H: $-4.52‰ \pm 15.98$, $\delta^{18}O$: $-2.01‰ \pm 3.02$). Additionally, coastal groundwaters from Sinai (Fig. 6b) and the north Western Desert (Fig. 6c) are mostly plotted between the GMWL and EMWL, while those from the Eastern Desert are situated below the GMWL (Fig. 6b). This difference suggests distinct moisture sources of precipitation. The geographic distribution of most samples at the mouths of large ephemeral streams in Sinai and the Eastern Desert supports the concept of modern precipitation recharge. These streams dissect the crystalline rocks of the Sinai massifs and the Red Sea Hills, gathering rainfalls and occasional flash floods, and eventually discharging into the coastal plains of the Mediterranean Sea and the Red Sea. However, the δ^2H-$\delta^{18}O$ diagram also reveals that some samples from Sinai, particularly those along the Gulf of Suez, and the Eastern Desert exhibit isotopic compositions more depleted than the current rainfall (Fig. 6b), suggesting the presence of fossil water, possibly from the underlying Nubian aquifer. It is likely that Nubian aquifer waters are seeping upward along deep-seated faults, infiltrating into the shallow alluvial and/or carbonate coastal aquifers in the coastal plains of the Gulf of Suez and the Red Sea. The deviation of certain coastal groundwater samples from the GMWL may result from various natural and anthropogenic processes, including evaporative isotopic enrichment of wadi flash floods (indicated by dashed lines in Fig. 6b, c), evaporation from the water table, seawater intrusion, mixing with evaporated waters from agricultural

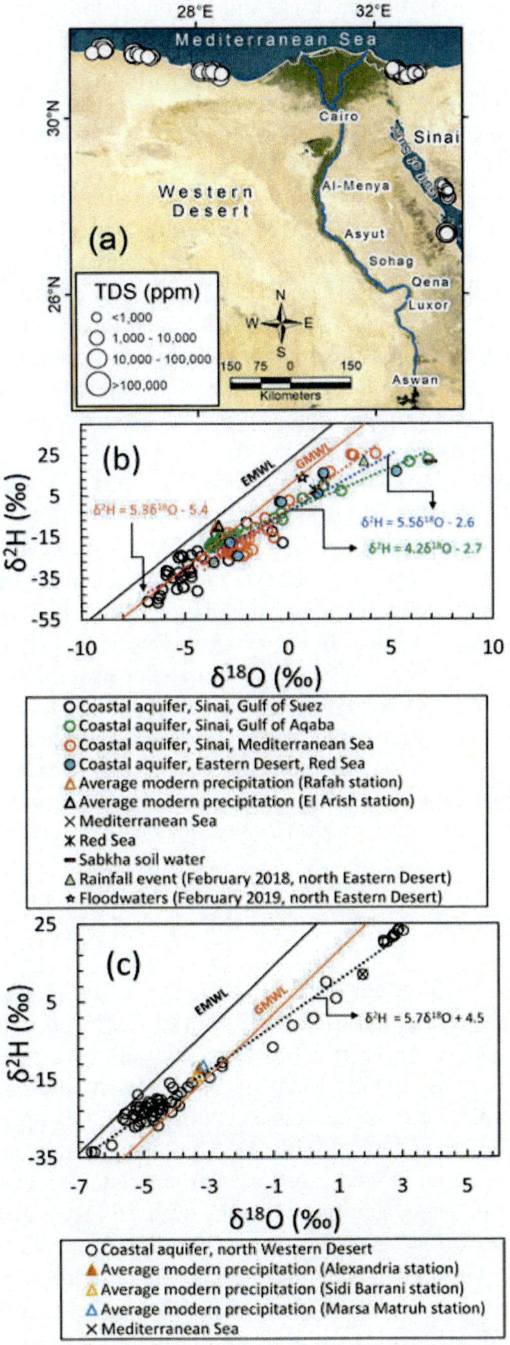

Fig. 6 (**a**) Spatial distribution of salinity values for groundwater samples collected from the coastal aquifers. δ^2H versus $\delta^{18}O$ plot for groundwater samples collected from coastal aquifers in (**b**) the Eastern Desert and Sinai, and (**c**) the Western Desert

or domestic activities, and/or contributions from salt marshes and lakes along the coastal plains.

4.4 Moghra Aquifer

The salinity of the Moghra aquifer varied from 2,236 ppm to 14,008 ppm with an average of 5,483 ppm and an increasing salinity trend toward the Qattara Depression in the west (Fig. 7a; Table 1). The water from the Moghra aquifer is classified as brackish to saline water due to many factors, including seepage of saltwater from the surface saline lakes, low recharge of groundwater and leaching of clay and shale lenses along the groundwater flow direction toward the Qattara Depression.

The isotopic composition of groundwater from the Moghra aquifer (δ^2H: −1.43‰ ± 6.57, δ^{18}O: −0.14‰ ± 1.88) bears considerable resemblance to that of the Nile waters, particularly the pre-High Dam component (−0.60‰ for δ^{18}O; 4.30‰ for δ^2H), indicating a potential contribution from the groundwater of the Nile Delta aquifer (Fig. 7b; Table 2). However, the isotopic depletion observed in certain Moghra groundwater samples cannot be fully explained by the Nile water components (pre- and post-High Dam), hinting at the involvement of a depleted isotopic source. In the Western Desert, the Nubian aquifer groundwaters are under high artesian pressure and experience upward leakage through extensive deep-seated faults into the overlying shallow aquifers. This observation supports the hypothesis that Moghra groundwater represents a mixture of the Nile waters and fossil Nubian groundwaters, as indicated by mixing lines and lower values of slope (3.1) and intercept (−1) of the evaporation line fitted to the samples (Fig. 7b).

4.5 Carbonate Aquifer

At the Western Desert, the salinity of the carbonate aquifer varies from 228.7 ppm (western desert fringe of Assiut) to 8,012 ppm (northeast Siwah Oasis) (Fig. 8a). At the Eastern Desert, however, the salinity ranges from 193.5 ppm (east Al-Minya Governorate) to 13,900 ppm (Helwan area, east Cairo) (Fig. 8a). The average salinity of the carbonate aquifer is 1,500 ppm (same for both the Western and Eastern deserts) (Table 1). The low groundwater salinity is related to flood water recharge, at the Eastern Desert, and lateral recharge from the Nile River and/or the Nile Valley aquifer, while the high salinity is attributed to water–rock interaction in the aquifer as well as the anthropogenic inputs.

Groundwaters from the carbonate aquifer in Sinai display a higher degree of isotopic depletion (δ^2H: −30.50‰ ± 3.14, δ^{18}O: −5.49‰ ± 0.84) compared to those from the carbonate aquifer in the western and eastern desert fringes of the Nile Valley (δ^2H: −10.09‰ ± 16.03, δ^{18}O: −2.11‰ ± 1.95) (Fig. 8b; Table 2). These samples are mostly located between the GMWL and EMWL, exhibiting isotopic

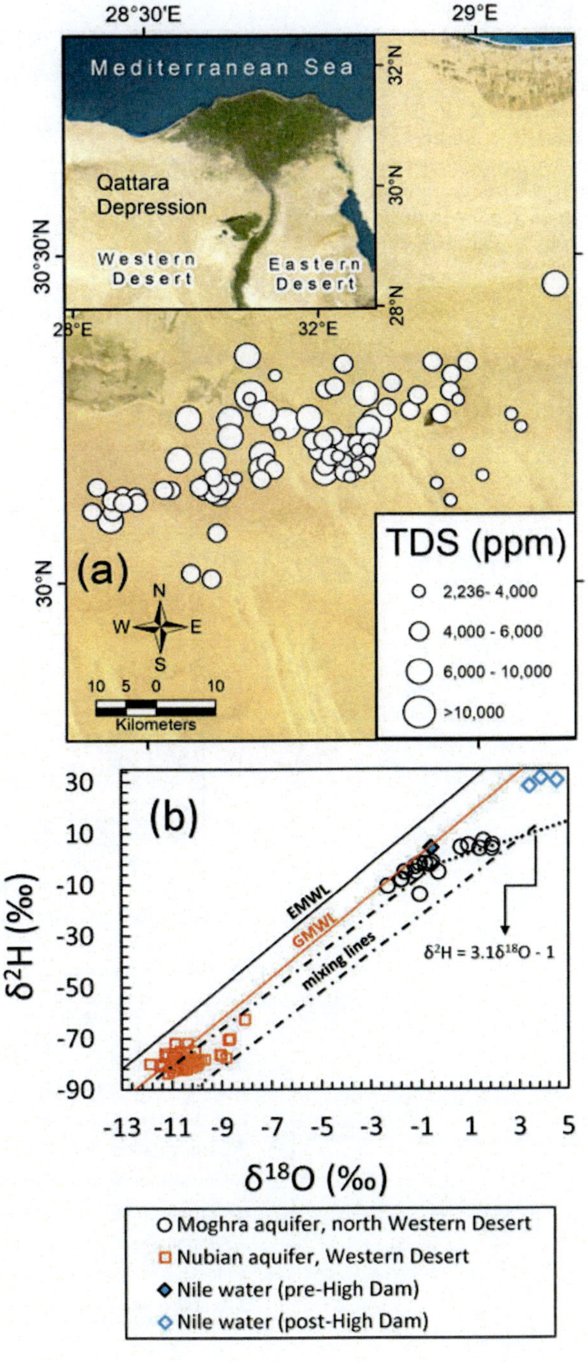

Fig. 7 (**a**) Spatial distribution of salinity values for groundwater samples collected from the Moghra aquifer. (**b**) δ^2H versus $\delta^{18}O$ plot for groundwater samples collected from the Moghra and Nubian aquifers

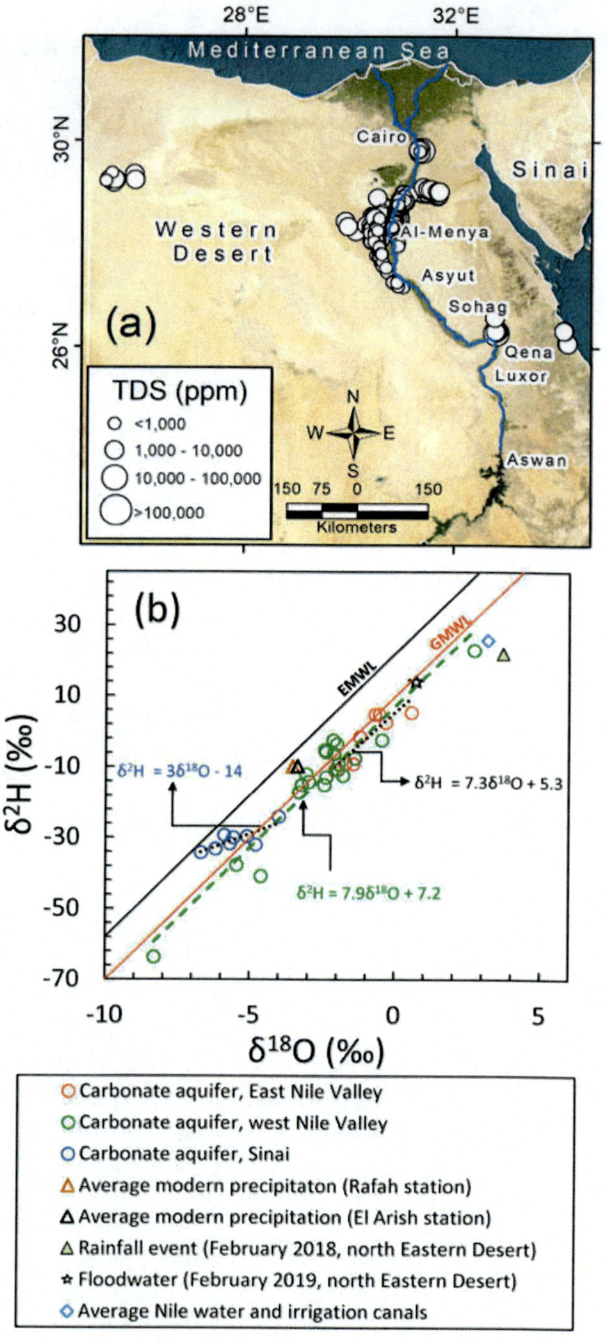

Fig. 8 (**a**) Spatial distribution of salinity values for groundwater samples collected from carbonate aquifer. (**b**) δ^2H versus δ^{18}O plot for groundwater samples collected from carbonate aquifer in the Western Desert, Eastern Desert, and Sinai

compositions that are depleted and comparable to those of modern precipitation (Rafah and El Arish stations; Fig. 8b). The observed pattern suggests that Sinai groundwaters may result from a mixing process between depleted water from the Nubian aquifer and modern meteoric water. Similarly, a mixed origin is indicated for the carbonate aquifer at the eastern and western desert fringes of the Nile Valley (Fig. 8b). As rainfall is scarce over the Western Desert, the most enriched isotopic composition for the west-fringe carbonate groundwaters could be attributed to lateral recharge from the adjacent Nile Valley aquifer. Moreover, Fig. 8b reveals that groundwater from the eastern carbonate aquifer is more enriched in isotopic composition compared to the western one. This variation could be due to significant contributions from modern precipitation that falls over the Red Sea Hills and drains westward through numerous wadi channels to the Nile Valley, where alluvial and carbonate aquifers are dominant.

4.6 *Nubian Aquifer*

At the Eastern Desert, Nubian aquifer salinity varied from 1,092 ppm (Queih basin, near Gebel Duwi) to 13,826 ppm (Wadi Abadi, Idfu area) with an average of 2,935 ppm (Fig. 9a; Table 1). In the Western Desert, it ranges from 79 ppm (Farafra Oasis) to 10,907 ppm (El Bahariya Oasis) with an average of 516 ppm (Fig. 9a; Table 1). At Sinai, the salinity of the Nubian aquifer varied from 307 ppm (Wadi Sidri and Wadi Baba, Abu Zeneima area) to 8,915 ppm (Hammam Musa, South Sinai) with an average of 1,531 ppm (Fig. 9a; Table 1). The Nubian aquifer salinity difference among different settings in Egypt might be related to water-rock interaction with increasing groundwater age and distance from recharge zones, or incomplete flushing of marine water [58].

Groundwater samples from the Nubian aquifer in the Western Desert are situated below the GMWL (δ^2H: 77.07‰ ± 14.27, δ^{18}O: −9.90‰ ± 2.09; Fig. 9b; Table 2). These samples exhibit more depleted isotopic compositions compared to those collected from the Nubian aquifer in the Eastern Desert (δ^2H: −47.41‰ ± 10.11, δ^{18}O: −7.35‰ ± 4.76) and Sinai (δ^2H: −45.71‰ ± 12.17, δ^{18}O: −6.84‰ ± 1.42) (Fig. 9b; Table 2). However, some groundwater samples in the vicinity of East Uweinat and Lake Nasser, in southwest Egypt, display enriched isotopic compositions (δ^2H: 1.85‰ ± 16.75, δ^{18}O: 0.55‰ ± 1.88), suggesting contributions from enriched water from Lake Nasser and/or modern precipitation over northern Sudan through networks of densely fractured and karstified bedrocks. Local precipitation and floodwater events, along with the average modern precipitation from Cairo, and most groundwater samples from the Nubian aquifer in the Eastern Desert are plotted at and below the GMWL (Fig. 9b). On the other hand, most groundwater samples from the Nubian aquifer in Sinai, along with the average modern precipitation along the Mediterranean Sea (e.g., Rafah and El Arish), are plotted between the GMWL and EMWL. This indicates different moisture sources of precipitation contributing to the recharge of the Nubian aquifer in Sinai and the Eastern Desert. The contribution

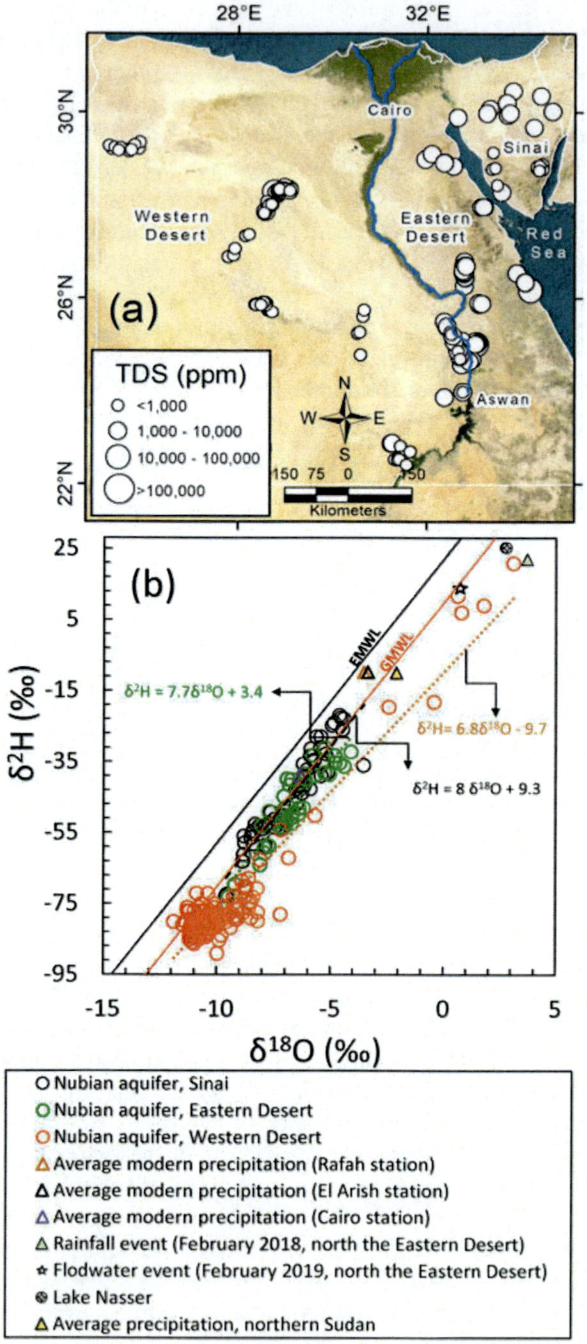

Fig. 9 (**a**) Spatial distribution of salinity values for groundwater samples collected from the Nubian aquifer. (**b**) δ^2H versus $\delta^{18}O$ plot for groundwater samples collected from the Nubian aquifer in the Western Desert, Eastern Desert, and Sinai

from modern precipitation is also supported by regression lines with nearly the same slope as the GMWL (Fig. 9b). The outcrops of the Nubian aquifer at the foothills of the Sinai massif and Red Sea Hills serve as recharge windows, receiving significant rainfall events (e.g., 76 and 100 mm/day recorded for the southern Sinai and the Eastern Desert, respectively) that mix with the depleted fossil groundwater of the Nubian aquifer. The mixing process primarily depends on the structural elements that control the connection between these two waters, as well as the traveling distance from recharge zones.

4.7 *Fractured Basement Aquifer*

At Sinai, the salinity of the fractured basement aquifer varies from 432 ppm to 14,156 ppm with an average of 1,977 ppm (Fig. 10a; Table 1). The high salinity, 14,156 ppm, was recorded at Dahab basin, South Sinai, and might be related to the dissolution of basaltic dike, which act as a barrier for groundwater entrapments [36]. For the Eastern Desert, the salinity ranges from 405 ppm to 82,400 ppm with an average of 7,488 ppm (Fig. 10a; Table 1). The high salinity (82,400 ppm) is recorded within the vicinity of Sukari Gold Mine due to downward leakage from tailing storage facility (TSF) ponds into highly fractured basement rocks beneath [59].

The isotopic compositions of groundwater samples collected from the fractured basement aquifers in the Eastern Desert (δ^2H: −4.97‰ ± 11.24, δ^{18}O: −0.91‰ ± 2.15) and southern Sinai (δ^2H: −20.92‰ ± 5.79, δ^{18}O: −3.96‰ ± 1.15) are plotted close to the average rainfall isotopic compositions of three rainfall stations in Egypt: Cairo, Rafah, and El Arish (Fig. 10b). Samples collected from the Eastern Desert exhibited similar isotopic compositions to recent precipitation events (e.g., February 2018) and floodwaters associated with heavy storms (e.g., February 2019), indicating their modern meteoric origin (Fig. 10b; Table 2). The average modern precipitation from Cairo, recent precipitation events, and floodwater, as well as groundwater from the basement aquifer in the Eastern Desert, are plotted at and below the GMWL, while Sinai groundwaters, together with the average modern precipitation from El Arish and Rafah, were plotted between the GMWL and MMWL (Fig. 10b). This suggests different moisture sources of precipitation over the basement terrain in southern Sinai and the Eastern Desert. The observed deviation from the GMWL for Eastern Desert groundwaters (slope: 4.7, intercept: −0.4) is attributed to varying evaporative isotope enrichment of runoff along wadi channels of the Red Sea Hills. Additionally, more enriched δ^{18}O and δ^2H values of up to 5.35‰ and 22‰, respectively, are related to mixing with the isotopically enriched tailing storage pond waters from Sukari gold mining activities in the Eastern Desert.

Fig. 10 (**a**) Spatial distribution of salinity values for groundwater samples collected from basement aquifers. (**b**) δ^2H versus $\delta^{18}O$ plot for groundwater samples collected from basement aquifers in the Eastern Desert and Sinai

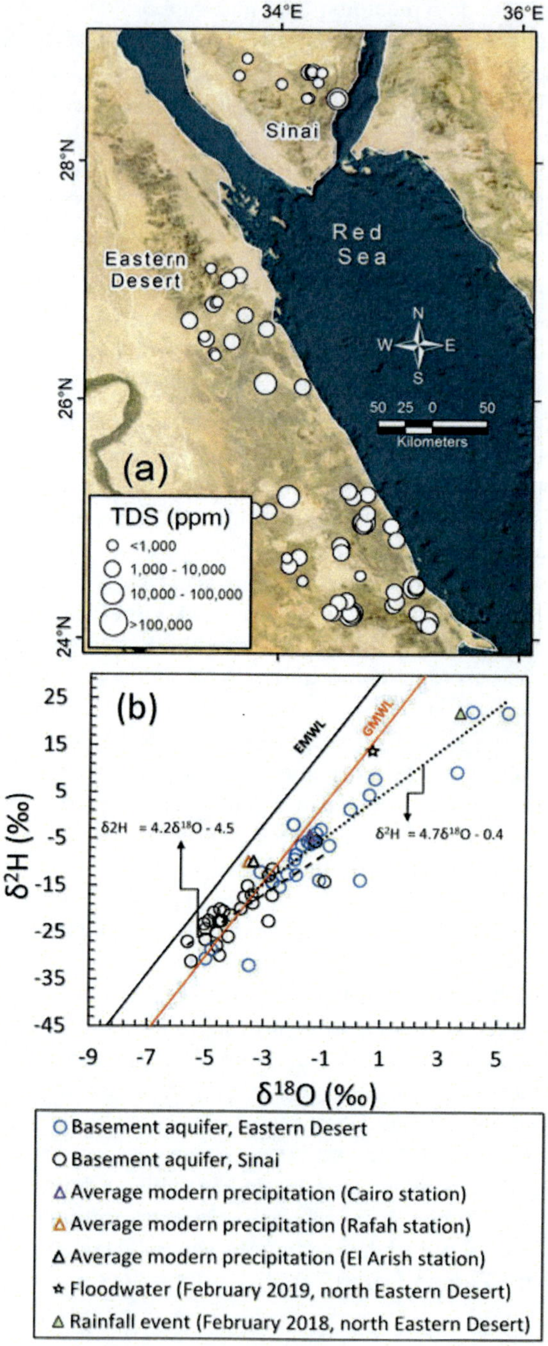

5 Summary and Conclusions

Analysis of historical isotopic and chemical compositions compiled for groundwater samples collected from the seven major aquifer systems (e.g., the Nile Valley and Delta, alluvial, Nubian sandstone, coastal, Moghra, carbonate, and fractured basement) in Egypt reveals several key findings:

- Both alluvial and coastal aquifer systems are recharged from modern rainfall, floodwaters, and upward leakage from deep aquifers.
- The Nile Valley and Delta, Moghra, and carbonate systems receive recharge from the Nile River, modern rainfall, irrigation canals and drains, and upward leakage from deep aquifers.
- The fractured basement aquifers are recharged by modern rainfall and floodwaters.
- The Nubian aquifer may contain mainly fossil water. However, regions that receive present-day recharge are located in areas of higher rainfall rates (e.g., foothills of mountainous areas) or close to surface water bodies (e.g., Lake Nasser and Nile River).
- The Nile Valley and Nubian (at the Western Desert) aquifers have the low average TDS values of 1,108.5 ppm and 516.2 ppm, respectively, while coastal (along the Red Sea) and fractured basement (Eastern Desert) have high average TDS of 9,181 ppm and 7,487.6 ppm, respectively. Alluvial aquifers in Sinai and the Eastern Desert have nearly the same average TDS values (>4,000 ppm). Similarly, carbonate aquifers distributed in the Western and Eastern deserts have average TDS values of 1,500 ppm. The average water salinity of the fractured basement aquifer in the Eastern Desert is about 3.5 times of that in Sinai. The average water salinity of the Nile Delta aquifer is about two times of the Nile valley aquifer.

The compiled datasets could be used by researchers to identify locations for collecting newer samples in areas with limited sample coverage. This approach significantly reduces the time, effort, and resources needed to collect additional field measurements that have already been obtained. Moreover, this database will serve as a valuable guide for field data collection by providing knowledge of the available resources. By compiling all historical isotopic and chemical compositions data for groundwater samples in Egypt, a better understanding of the country-scale isotopic and chemical compositions of water resources will be achieved, along with enhanced insights into recharge mechanisms across the entire country. This practice will enable the scientific community to address unprecedented country- and aquifer-scale scientific questions of significant merit.

6 Recommendations

Key recommendations for sustainable use of groundwater resources in Egypt include:

- Extraction from the alluvial and coastal aquifer systems should be reduced to match the modern recharge rates, and extraction activities should be located near the discharge sources of the deeper aquifers.
- Extraction from the Nile Valley and Delta, Moghra, and carbonate systems should be minimized to match the modern recharge rates from rainfall and Nile waters and should be located close to the discharge locations of the deeper aquifers.
- Extraction from the fractured basement aquifers should be targeted in areas with higher rainfall rates.
- Extraction activities in the Nubian aquifer should target regions that receive present-day recharge in areas of higher rainfall rates (e.g., foothills of mountainous areas) or close to surface water bodies (e.g., Lake Nasser and Nile River).
- Future studies should focus on building upon this compiled database to quantify recharge rates and locations, investigate the aquifer-scale hydrological parameters that control recharge rates and locations, and accurately map locations of deep-seated fault systems along which deeper aquifers discharge to shallower ones.

References

1. Ahmed M, Sultan M, Wahr J, Yan E (2014) The use of GRACE data to monitor natural and anthropogenic induced variations in water availability across Africa. Earth Sci Rev 136:289–300. https://doi.org/10.1016/j.earscirev.2014.05.009
2. Ahmed M (2020) Sustainable management scenarios for northern Africa's fossil aquifer systems. J Hydrol 589:125196. https://doi.org/10.1016/j.jhydrol.2020.125196
3. Khalil MM, Abotalib AZ, Farag MH et al (2021) Poor drainage-induced waterlogging in Saharan groundwater-irrigated lands: integration of geospatial, geophysical, and hydrogeological techniques. Catena 207:105615. https://doi.org/10.1016/J.CATENA.2021.105615
4. Khalil MM, Tokunaga T, Heggy E, Abotalib AZ (2021) Groundwater mixing in shallow aquifers stressed by land cover/land use changes under hyper-arid conditions. J Hydrol:126245. https://doi.org/10.1016/j.jhydrol.2021.126245
5. Ahmed M, Chen Y, Khalil MM (2022) Isotopic composition of groundwater resources in arid environments. J Hydrol 609:127773. https://doi.org/10.1016/J.JHYDROL.2022.127773
6. Ahmed M, Sauck W, Sultan M et al (2013) Geophysical constraints on the hydrogeologic and structural settings of the Gulf of Suez rift-related basins: case study from the El Qaa plain, Sinai, Egypt. Surv Geophys 35:415–430. https://doi.org/10.1007/s10712-013-9259-6
7. Essam D, Ahmed M, Abouelmagd A, Soliman F (2020) Monitoring temporal variations in groundwater levels in urban areas using ground penetrating radar. Sci Total Environ 703: 134986. https://doi.org/10.1016/j.scitotenv.2019.134986

8. Mohamed L, Sultan M, Ahmed M (2015) Structural controls on groundwater flow in basement terrains: geophysical, remote sensing, and field. Surv Geophys 36:717–742. https://doi.org/10.1007/s10712-015-9331-5
9. Scanlon BR, Healy RW, Cook PG (2002) Choosing appropriate techniques for quantifying groundwater recharge. Hydrgeol J 10:18–39. https://doi.org/10.1007/s10040-001-0176-2
10. Sultan M, Metwally S, Milewski A et al (2011) Modern recharge to fossil aquifers: geochemical, geophysical, and modeling constraints. J Hydrol 403:14–24. https://doi.org/10.1016/j.jhydrol.2011.03.036
11. Abdelmohsen K, Sultan M, Ahmed M et al (2019) Response of deep aquifers to climate variability. Sci Total Environ 677:530–544. https://doi.org/10.1016/j.scitotenv.2019.04.316
12. Ahmed M, Abdelmohsen K (2018) Quantifying modern recharge and depletion rates of the Nubian aquifer in Egypt. Surv Geophys 39. https://doi.org/10.1007/s10712-018-9465-3
13. Dailey D, Sauck W, Sultan M et al (2015) Geophysical, remote sensing, GIS, and isotopic applications for a better understanding of the structural controls on groundwater flow in the Mojave Desert, California. J Hydrol Reg Stud 3:211–232. https://doi.org/10.1016/j.ejrh.2014.12.002
14. Doolittle JA, Jenkinson B, Hopkins D et al (2006) Hydropedological investigations with ground-penetrating radar (GPR): estimating water-table depths and local groundwater flow pattern in areas of coarse-textured soils. Geoderma 131:317–329. https://doi.org/10.1016/j.geoderma.2005.03.027
15. Fallatah OA, Ahmed M, Cardace D et al (2019) Assessment of modern recharge to arid region aquifers using an integrated geophysical, geochemical, and remote sensing approach. J Hydrol 569:600–611. https://doi.org/10.1016/j.jhydrol.2018.09.061
16. Fallatah OA, Ahmed M, Save H, Akanda AS (2017) Quantifying temporal variations in water resources of a vulnerable middle eastern transboundary aquifer system. Hydrol Process 31. https://doi.org/10.1002/hyp.11285
17. Manu E, Preko K (2014) Estimation of water table depths and local groundwater flow pattern using the ground Penterating radar. Int J Sci Res Publ 4:1–12
18. Mohamed A, Sultan M, Ahmed M et al (2017) Aquifer recharge, depletion, and connectivity: inferences from GRACE, land surface models, and geochemical and geophysical data. Bull Geol Soc Am 129:534–546. https://doi.org/10.1130/B31460.1
19. Niyazi BA, Ahmed M, Basahi JM et al (2018) Spatiotemporal trends in freshwater availability in the Red Sea Hills, Saudi Arabia. Arab J Geosci 11. https://doi.org/10.1007/s12517-018-4052-y
20. Niyazi BA, Ahmed M, Masoud MZ et al (2019) Sustainable and resilient management scenarios for groundwater resources of the Red Sea coastal aquifers. Sci Total Environ 690:1310–1320. https://doi.org/10.1016/j.scitotenv.2019.07.081
21. Sultan M, Ahmed M, Sturchio N et al (2013) Assessment of the vulnerabilities of the Nubian sandstone fossil aquifer, North Africa. In: Pielke RA (ed) Climate vulnerability: understanding and addressing threats to essential resources. Elsevier, pp 311–333
22. Sultan M, Sturchio NC, Alsefry S et al (2019) Assessment of age, origin, and sustainability of fossil aquifers: a geochemical and remote sensing-based approach. J Hydrol 576. https://doi.org/10.1016/j.jhydrol.2019.06.017
23. Schiavo M, Colombani N, Mastrocicco M (2023) Modeling stochastic saline groundwater occurrence in coastal aquifers. Water Res 235:119885. https://doi.org/10.1016/J.WATRES.2023.119885
24. Yu X, Xin P, Luo Z, Pu L (2023) Thermal effects of freshwater injection on flow and salinity distributions in tidally-affected coastal unconfined aquifers. J Hydrol 622:129739. https://doi.org/10.1016/J.JHYDROL.2023.129739
25. Clark ID, Fritz P (1997) Environmental isotopes in hydrogeology.1st edn. CRC Press/Lewis Publishers

26. Coplen TB (1996) New guidelines for reporting stable hydrogen, carbon, and oxygen isotope-ratio data. Geochim Cosmochim Acta 60:3359–3360. https://doi.org/10.1016/0016-7037(96)00263-3
27. Jasechko S (2019) Global isotope hydrogeology – review. Rev Geophys 57:835–965. https://doi.org/10.1029/2018RG000627
28. Ouarani M, Brahim YA, Mulla D et al (2023) A comprehensive overview of groundwater salinization and recharge processes in a semi-arid coastal aquifer (Essaouira, Morocco). J Hydrol Reg Stud 49:101501. https://doi.org/10.1016/J.EJRH.2023.101501
29. Qiu D, Zhu G, Lin X et al (2023) Dissipation and movement of soil water in artificial forest in arid oasis areas: cognition based on stable isotopes. Catena 228:107178. https://doi.org/10.1016/J.CATENA.2023.107178
30. Smith A, Tetzlaff D, Marx C, Soulsby C (2023) Enhancing urban runoff modelling using water stable isotopes and ages in complex catchments. Hydrol Process 37:e14814. https://doi.org/10.1002/HYP.14814
31. Yousif M, Hussien HM, Abotalib AZ (2020) The respective roles of modern and paleo recharge to alluvium aquifers in continental rift basins: a case study from El Qaa plain, Sinai, Egypt. Sci Total Environ 739:139927. https://doi.org/10.1016/J.SCITOTENV.2020.139927
32. Abdel Moneim AA, Seleem EM, Zeid SA et al (2015) Hydrogeochemical characteristics and age dating of groundwater in the quaternary and Nubian aquifer systems in Wadi Qena, Eastern Desert, Egypt. Sustain Water Resour Manag 13(1):213–232. https://doi.org/10.1007/S40899-015-0018-3
33. Abouelmagd A, Sultan M, Sturchio NC et al (2014) Paleoclimate record in the Nubian Sandstone Aquifer , Sinai. Quatern Res 81:158–167. https://doi.org/10.1016/j.yqres.2013.10.017
34. Sultan M, Yan E, Sturchio N et al (2007) Natural discharge: a key to sustainable utilization of fossil groundwater. J Hydrol 335:25–36. https://doi.org/10.1016/j.jhydrol.2006.10.034
35. Hussien HM, Kehew AE, Aggour T, Korany E, Abotalib AZ, Hassanein A, Morsy S (2017) An integrated approach for identification of potential aquifer zones in structurally controlled terrain: Wadi Qena basin. Egypt. Catena 149:73–85
36. Eissa M, Thomas JM, Pohll GM, Hershey RL, Dahab K, Dawoud M, Gomaa M, El Shiekh A (2013) Groundwater resource sustainability in the Wadi Watir delta. Gulf of Aqaba. Hydrogeol. J. 21(8):1833–1851
37. El Tahlawi MR, Farrag AA, Ahmed SS (2008) Groundwater of Egypt: "An environmental overview". Environ Geol 55:639–652. https://doi.org/10.1007/s00254-007-1014-1
38. Abdel-Shafy HI, Kamel AH (2016) Groundwater in Egypt issue: resources, location, amount, contamination, protection, renewal, future overview. Egypt. J. Chem 59(3):321–362. https://doi.org/10.21608/EJCHEM.2016.1085
39. Dawoud MA, Ismail SS (2013) Saturated and unsaturated River Nile/groundwater aquifer interaction systems in the Nile Valley. Egypt. Arab. J. Geosci. 6:2119–2130. https://doi.org/10.1007/s12517-011-0483-4
40. Omran E-SE (2017) Land and groundwater resources in the Egypt's Nile Valley, Delta, and its fringes. In: Groundwater in the Nile Delta. Springer, Cham, pp 45–103
41. Sherif M (1999) Nile Delta Aquifer in Egypt. Springer, Dordrecht, pp 559–590
42. Negm AM, Sakr S, Abd-Elaty I, Abd-Elhamid HF (2018) An Overview of Groundwater Resources in Nile Delta Aquifer. In: Negm A (ed) Groundwater in the Nile Delta. The Handbook of Environmental Chemistry, vol 73. Springer, Cham. https://doi.org/10.1007/698_2017_193
43. Geirnaert W, Laeven MP (1992) Composition and history of ground water in the western Nile Delta. J. Hydrol. 138(1–2):169–189. https://doi.org/10.1016/0022-1694(92)90163-P
44. Hefny K, Farid MS, Hussein M (1992) Groundwater assessment in Egypt. Int J Water Resour Dev 8:126–134. https://doi.org/10.1080/07900629208722543

45. Elewa HH, Shohaib RE, Qaddah AA, Nousir AM (2013) Determining groundwater protection zones for the quaternary aquifer of northeastern Nile Delta using GIS-based vulnerability mapping. Environ Earth Sci 68:313–331. https://doi.org/10.1007/s12665-012-1740-x
46. Sherif MM, Sefelnasr A, Javad A (2012) Incorporating the concept of equivalent freshwater head in successive horizontal simulations of seawater intrusion in the Nile Delta aquifer. Egypt. J. Hydrol. 464:186–198
47. Abdel Mogith SM, Ibrahim SMM, Hafiez RA (2013) Groundwater potentials and characteristics of EL-Moghra aquifer in the vicinity of Qattara depression. Egypt J Desert Res 62/63:1–20. https://doi.org/10.21608/ejdr.2013.5821
48. Rizk ZS, Davis AD (1991) Impact of the proposed Qattara reservoir on the Moghra aquifer of northwestern Egypt. Ground Water 29:232–238. https://doi.org/10.1111/j.1745-6584.1991.tb00515.x
49. Ibrahim RGM, Lyons WB (2017) Assessment of the hydrogeochemical processes affecting groundwater quality in the Eocene limestone aquifer at the desert fringes of El Minia Governorate. Egypt. Aquat. Geochem. 23:33–52. https://doi.org/10.1007/s10498-016-9298-y
50. Ammar AI, Kamal KA (2018) Resistivity method contribution in determining of fault zone and hydro-geophysical characteristics of carbonate aquifer, eastern desert, Egypt. Appl Water Sci 81(8):1–27. https://doi.org/10.1007/S13201-017-0639-9
51. Hesse K-H, Hissene A, Kheir O et al (1987) Hydrogeological investigations of the Nubian aquifer system, eastern Sahara. Berliner geowissenschaftliche Abhandlungen 75:397–464
52. Thorweihe U (1990) Th Nubian aquifer system. In: Said R (ed) The geology of Egypt. Balkema, Lisse, pp 601–614
53. Bakhbakhi M (2006) Nubian sandstone aquifer system. Non-renewable Groundw. Resour. A Guideb. Soc. Sustain. Manag. Water-Policy Makers, Paris United Nations. Educ. Sci. Cult. Organ. (IHP-VI Ser. Groundw. 10:75–81
54. Craig H (1961) Isotopic variations in meteoric waters. Science 133:1702–1703. https://doi.org/10.1126/SCIENCE.133.3465.1702
55. Gat JR, Carmi I (1970) Evolution of the isotopic composition of atmospheric waters in the Mediterranean Sea area. J Geophys Res 75:3039–3048. https://doi.org/10.1029/JC075I015P03039
56. Ibrahim RGM (2019) Factors affecting groundwater quality in the area between Safaga and El Quseir, Eastern Desert, Egypt. Arab J Geosci 12:1–16. https://doi.org/10.1007/S12517-019-4411-3/TABLES/7
57. Van Engelen J, Verkaik J, King J et al (2019) A three-dimensional palaeohydrogeological reconstruction of the groundwater salinity distribution in the Nile Delta aquifer. Hydrol Earth Syst Sci 23:5175–5198. https://doi.org/10.5194/HESS-23-5175-2019
58. Sherif MI, Sturchio NC (2018) Radionuclide geochemistry of groundwater in the Eastern Desert, Egypt. Appl Geochem 93:69–80. https://doi.org/10.1016/J.APGEOCHEM.2018.04.004
59. Abdelaal A, Sultan M, Elhebiry M et al (2021) Integrated studies to identify site-specific parameters for environmentally benign mining operations: a case study from the Sukari gold mine, Egypt. Sci Total Environ 750:141654. https://doi.org/10.1016/J.SCITOTENV.2020.141654

Understanding Seawater Intrusion by Hydrochemical Parameters and Stable Water Isotopes Along the Coastal Alluvial Aquifers of the Essaouira Basin, Morocco

Mohammed Bahir, Otman el Mountassir, and Shakir Ali

Contents

1 Introduction 268
2 Material and Methods 269
2.1 Study Area 269
2.2 Sampling, Data Collection, and Analysis 270
3 Results and Discussion 271
3.1 Hydrochemical Facies 271
3.2 Ionics Ratio 271
3.3 Origin of Saline Water 282
4 Conclusion 287
5 Recommendation 287
References 288

Abstract The agricultural and domestic water supply of Essaouira's coastal region depends on groundwater resources drawn from the Turonian aquifer. However, due to seawater intrusion and anthropogenic activities, the water supply of coastal aquifers is frequently threatened by salinization. To investigate this matter comprehensively, the present study involved the collection of a total of 34 water samples during the year 2020, with the inclusion of results from 2019 samples. This study used hydrochemical parameters along with stable water isotopes ($\delta^{18}O$ and $\delta^{2}H$) to infer salinization processes. The major ions chemistry showed that the groundwater

M. Bahir (✉) and O. el Mountassir
High Energy and Astrophysics Laboratory, Faculty of Sciences Semlalia, Cadi Ayyad University, Marrakech, Morocco
e-mail: bahir@uca.ac.ma; Otman.elmountassir@ced.uca.ma

S. Ali
Department of Geology, University of Delhi, Delhi, India

Shakir Ali and Abdelazim Negm (eds.), *Groundwater Quality and Geochemistry in Arid and Semi-Arid Regions*, Hdb Env Chem (2024) 126: 267–294, DOI 10.1007/698_2023_1059,

Published online: 29 December 2023

of the Plio-Quaternary aquifer belongs to Cl-Ca-Mg and Cl-Na water types. The ionic ratios, i.e., Br/Cl ≈ 1.5 to 1.7‰, Na/Cl close to 0.86, Mg/Ca and SO_4/Cl showed that the seawater begins to invade the freshwater of the Plio-quaternary aquifer of Essaouira basin. The presence of evaporation and maritime intrusion and the contribution of recent precipitations to aquifer recharging were demonstrated by stable water isotopes ($\delta^{18}O$ and δ^2H). The results show that most of the groundwater samples are saline due to the intrusion of Atlantic Ocean seawater in the coastal aquifer. Hence, an immediate imperative arises for the implementation of an integrated approach to water resource management within this region, employing contemporary methodologies and technologies, in order to secure their enduring viability.

Keywords Coastal aquifers, Environmental isotopes, Essaouira basin, Ionic ratios, Seawater intrusion

1 Introduction

Seawater intrusion in coastal aquifers is a major problem affecting groundwater quality and a threat to water security [1–6]. However, the scenario becomes notably intricate within arid and semi-arid regions, characterized by declining aquifer water levels, deteriorating groundwater quality, diminished crop yields, and ecosystem degradation [7–18]. The infiltration of seawater into the aquifer, primarily driven by excessive pumping from high-capacity wells, has led to the abandonment of wells in areas experiencing heightened salinity [19, 20]. Consequently, the intrusion of saline water into freshwater aquifers has been the focal point of extensive research spanning over a century in Morocco. Broadly, seawater intrusion in coastal aquifers can be attributed to two principal factors: natural and anthropogenic influences [21–25]. The natural component relates to the elevation of sea levels since the Holocene Epoch [1]. Those aquifers that formed prior to the Holocene sea-level rise, during the Quaternary glaciation period, witnessed inundation by seawater, thereby facilitating its ingress into the aquifer systems [26–28]. Subsequently, after the stabilization of sea levels, aquifers characterized by robust hydraulic gradients were capable of resisting marine intrusion, whereas aquifers exhibiting weak hydraulic gradients were unable to repel the ingress of marine waters [1, 29, 30].

The anthropogenic component stems from intensive aquifer pumping, leading to a decline in the piezometric level due to groundwater over-exploitation. As Custodio [29] elucidates, the extent of marine intrusion is contingent upon various hydrogeological parameters inherent to the aquifer, including its geometry, permeability, hydraulic gradient, and others. Nevertheless, the significance of marine intrusion exhibits considerable spatial variation. A comprehensive understanding of aquifer hydrogeology fosters informed management, allowing for judicious groundwater utilization and the prevention or substantial mitigation of saline intrusion.

With these considerations in mind, the current investigation has been initiated with the overarching objective of scrutinizing groundwater conditions and assessing the status of seawater intrusion within the coastal aquifer systems of the Essaouira basin.

2 Material and Methods

2.1 *Study Area*

The research region, recognized as the Essaouira Basin, is situated in the lower reaches of the Essaouira basin [31–36]. It is demarcated by the Ksob Wadi to the north, the Tidzi Wadi to the south, the Tidzi diapir to the east, and the Atlantic Ocean to the west, encompassing an expanse of approximately 300 km^2 (Fig. 1). The climatic conditions prevailing in the Essaouira basin range from semi-arid to arid,

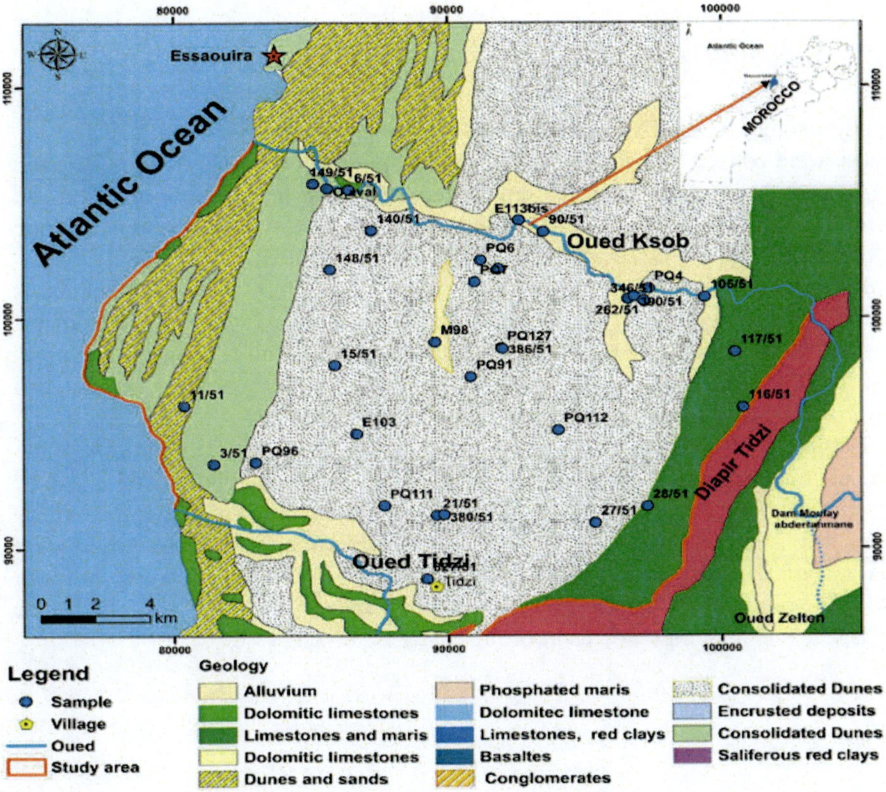

Fig. 1 Geological map of the study area showing sampling locations

characterized by an average temperature of 20°C and an annual mean precipitation of 300 mm.

The mean annual temperature of the study area is about 20°C, with a variation of 17°C between January and August [15–17, 32, 37, 38]. In this region, the groundwater resources are mainly represented by two main reservoirs: the Plio-Quaternary and Turonian aquifers [39–43] (Fig. 1). The Plio-Quaternary aquifer, hosted by gray limestone marl rocks, has a permeability of 3.2 × 10 m/s and contains an important phreatic aquifer (unconfined) with a wall formed in the syncline structure by Senonian marls and a thickness of 60 m [26, 27, 44]. This aquifer is exploited for food crops and potable water supply. The transmissivity value found from pumping tests performed within the aquifer was approximately 4.5×10^{-2} m^2/s [41, 45, 46]. High transmissivity wells are located to the north near the Ksob Wadi, this reflects the effect of recharging the aquifer from this Wadi, while the low transmissivity was recorded in the south part of the aquifer. According to measurements of the piezometric level of the 2020 campaign, the regional groundwater flow is from E–SE to W–NW with a hydraulic gradient of about 2.5% (upstream part) and 1.25% (downstream part).

2.2 Sampling, Data Collection, and Analysis

For this study, a total of 34 water samples (3 spring, 1 surface water, and 30 groundwater) were collected from the alluvial aquifer of Essaouira basin during 2019 and 2020 and have been used to investigate the seawater intrusion based on hydrochemical parameters and stable water isotopes. The physical parameters (like temperature, pH, and electrical conductivity) were measured in situ, and the chemical parameters (Ca^{2+}, Mg^{2+}, Na^{+}, K^{+}, Cl^{-}, $SO_4{}^{2-}$, and $NO_3{}^{-}$) were determined at the Faculty of Science, Semlalia of Marrakech using atomic absorption spectrometry (AAS) for $Ca^{2+}$and Mg^{2+}; emission spectrometry for Na^{+} and K^{+}; ion chromatography for $SO_4{}^{2-}$, $NO_3{}^{-}$, Cl^{-}, and Br. $HCO_3{}^{-}$ and $CO_3{}^{-}$ concentrations were analyzed by titration using 0.1 N HCl acid [47].

Isotopic assessments conducted during the 2020 campaign were performed at the Nuclear Science and Technology Center in Lisbon, Portugal. The stable isotopes, specifically δ^2H and $\delta^{18}O$ (‰ vs VSMOW), were determined utilizing instrumentation consisting of a Finnigan Mat 250 and a VG Micromass instrument, following the analytical protocols as outlined by Friedman [48] and Epstein and Mayeda [49]. The precision of the measurements was within ±1‰ for δ^2H and ±0.1‰ for $\delta^{18}O$. The methodology flowchart of this study is shown in Fig. 2.

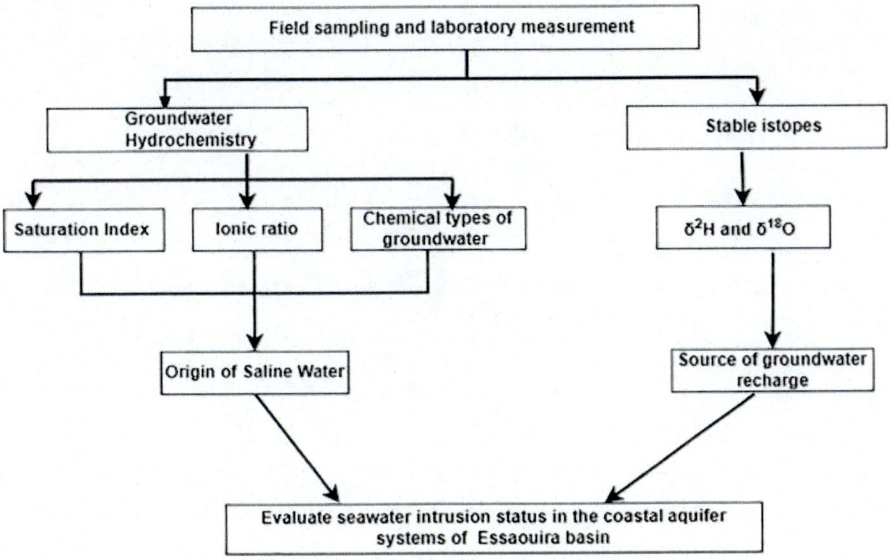

Fig. 2 The methodology used for this study is shown as flowchart

3 Results and Discussion

3.1 Hydrochemical Facies

Various mechanisms, including groundwater flow dynamics, recharge-discharge dynamics, and water-rock interactions, govern the hydrochemical composition of groundwater [50, 51]. Additionally, the mineral weathering processes occurring in the longitudinal trajectory of groundwater flow exert a discernible influence on hydrochemistry [52]. The Piper trilinear diagram [53] applied to the 2020 sample dataset in the study area elucidates that the groundwater within the shallow aquifer of the Essaouira basin predominantly falls into two distinct categories: the Cl-Ca-Mg type (comprising 64%) and the Cl-Na type (representing 36%) (Fig. 3). This transition between these facies underscores the intricate nature of the hydrogeochemical processes that underlie the mineralization patterns observed in the aquifer's groundwater.

3.2 Ionics Ratio

To underscore the significance of marine intrusion in the salinization process of groundwater within the Plio-Quaternary strata of the Essaouira basin, we integrated examinations of ionic relationships alongside the trace element "bromide" within the analyzed samples.

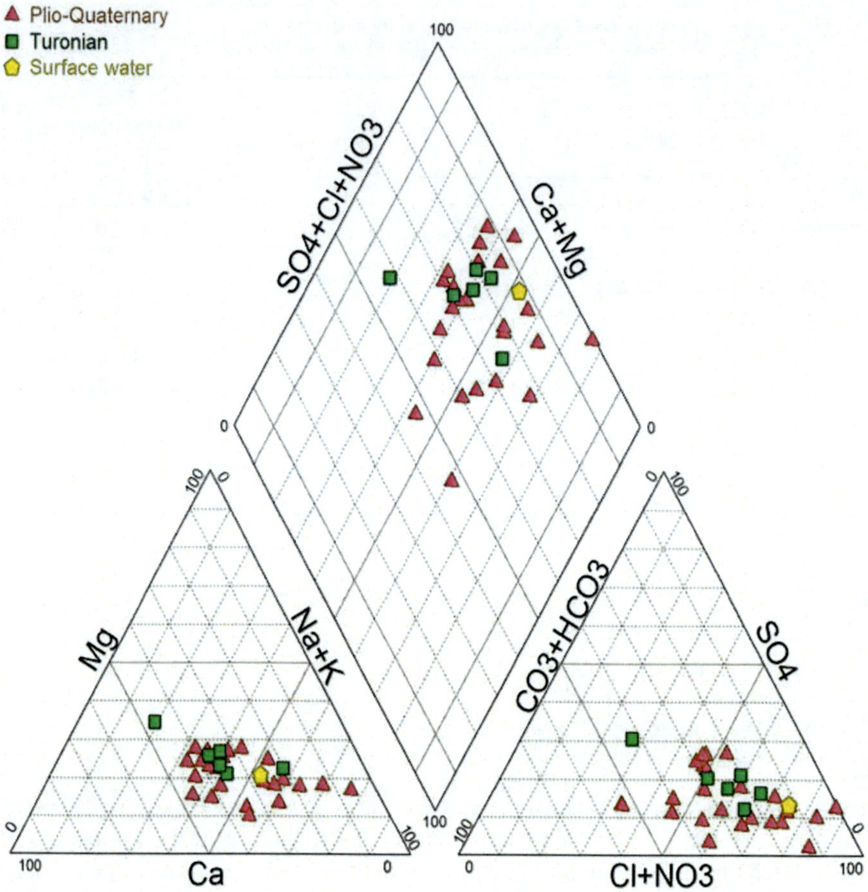

Fig. 3 Piper trilinear diagram of the study area of the campaign 2020

3.2.1 Br/Cl Couple

Bromide functions as a reliable indicator for evaluating the presence of marine intrusion phenomena [54, 55]. Similar to chloride, bromide is classified as a conservative element, demonstrating limited reactivity with the aquifer matrix unless organic matter is notably abundant [56]. These two conservative elements facilitate investigations into solution origins and the potential influence of marine water contributions, given their resistance to alteration by redox processes and their independence from minerals characterized by low solubility [57]. Given the extensive duration of bromide and chloride residence within oceanic masses, the Br/Cl ratio in contemporary seawater remains relatively consistent, typically falling within the range of 1.5 to 1.7×10^{-3} [54, 55]. This ratio remains stable when these two elements share a common origin. However, distinctions emerge when comparing seawater to remnants of evaporated seawater or hypersaline waters [58], arising from the dissolution of evaporite formations, or anthropogenic sources like wastewater effluents [58] and the return of irrigation water. During the process of seawater evaporation, the Br/Cl ratio remains constant until the onset of halite precipitation. As halite precipitates, the solution becomes enriched in bromide, leading to an elevation in the Br/Cl ratio [59]. However, the Br/Cl ratio continues to increase with the progression of halite precipitation within residual brine. Consequently, a solution resulting from the straightforward concentration of seawater prior to halite saturation will exhibit a Br/Cl ratio identical to that of seawater. Conversely, brine that has undergone more extensive concentration beyond the halite precipitation phase will manifest a higher Br/Cl ratio than seawater. Hence, freshwater engaged in dissolving halite until reaching saturation will exhibit a Br/Cl ratio lower than that of seawater. It is important to note that primary halite is the sole chlorinated salt characterized by a Br/Cl ratio lower than that of seawater. Conversely, a blend of freshwater and brine that has passed the halite precipitation phase will present a Br/Cl ratio exceeding that of the marine ratio. The Br versus Cl correlation diagram (Fig. 4a) exhibits a robust positive correlation ($r^2 = 0.99$) between these two ionic species, implying a common origin for bromides and chlorides. On the Br/Cl diagram as a function of Cl (Fig. 4b), samples 11/51, 45/51, 149/51, O94, and O114 (Fig. 1) fall within the dilution field of seawater, characterized by a Br/Cl ratio ranging from 1.5 to 1.7. This range signifies a salinity of marine origin. The presence of brine formation in certain wells can be attributed to the absence of aquifer recharge from precipitation, as evidenced by higher Br/Cl ratios in those specific wells. Conversely, the remaining data points exhibit molar ratios lower than that of seawater, indicating their divergence from the marine intrusion phenomenon and thereby implying an alternative source of salinization, such as salt dissolution.

3.2.2 Ca/Mg Couple

One prominent indicator of seawater intrusion is the notable increase in calcium concentration compared to that of seawater. The Mg/Ca ratio emerges as a valuable natural tracer for discerning the marine intrusion phenomenon within coastal aquifers [60, 61]. This ratio exhibits a direct correlation with the proportion of marine water present in the mixture, with chloride concentrations serving as a marker for salinity [61]. This association is rooted in the fact that seawater possesses a Ca/Mg ratio of 0.2, whereas freshwater exhibits a ratio exceeding 1. Importantly, saline water characterized by high calcium concentrations can result from mechanisms unrelated to cation exchange phenomena [62]. As the proportion of seawater introduced into the mixture increases, the Ca/Mg ratio diminishes. In the context of this case study, a reduction in the Ca/Mg ratio coincides with an elevation in chloride content (Fig. 5a). This phenomenon is evident in data points 11/51 and 27/51 (Fig. 1).

This observation potentially underscores the marine origin of mineralization, notably in the case of wells 11/51, 149/51, and O94. Points representing areas

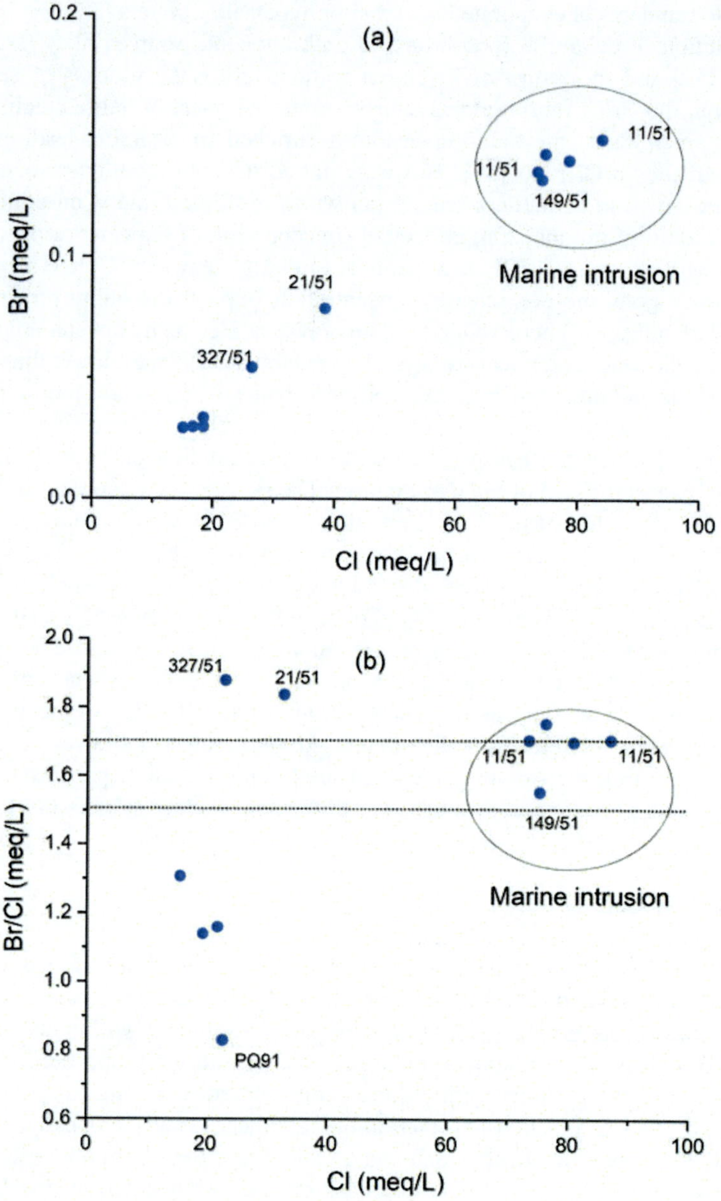

Fig. 4 Correlation diagram (**a**) Br vs Cl and (**b**) Br/Cl vs Cl of analyzed samples of the campaign 2019

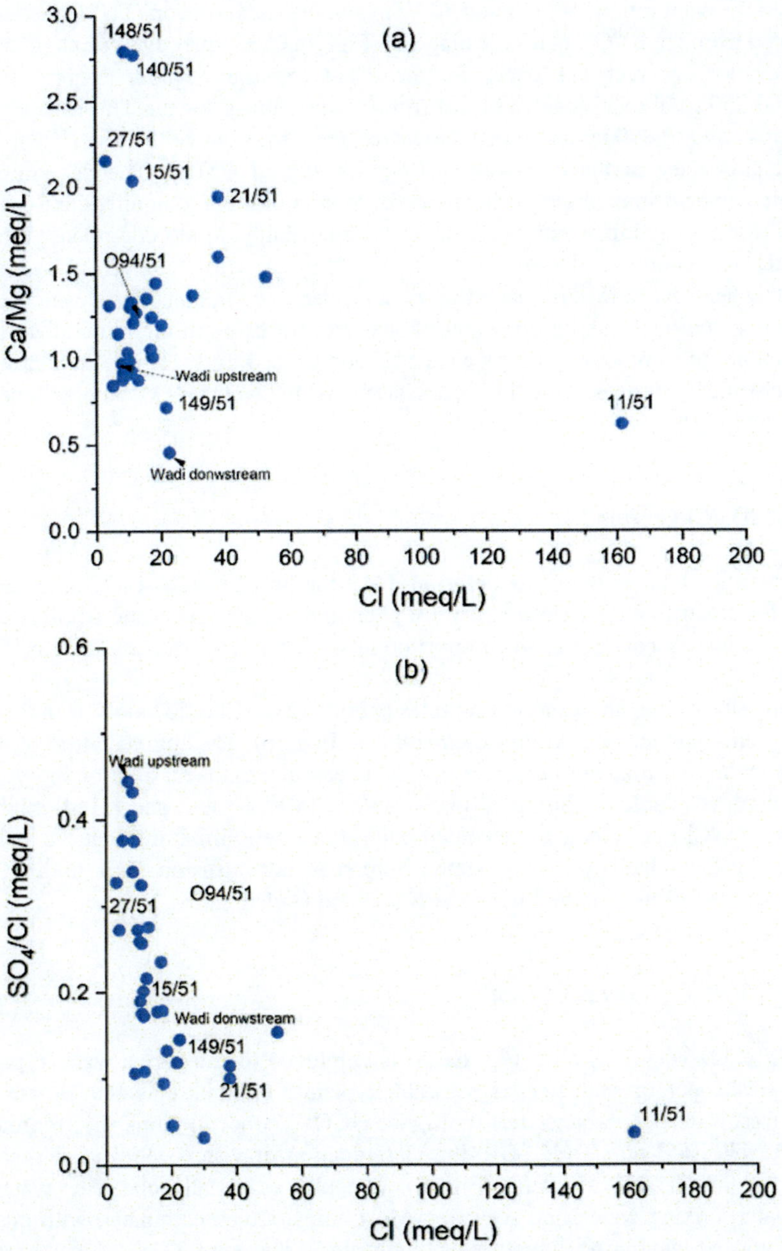

Fig. 5 Correlation diagram (**a**) Ca/Mg vs Cl and (**b**) SO_4/Cl vs Cl of analyzed samples of the campaign 2019

affected by evaporation (105/51 and 346/51) and marine intrusion (11/51) have been excluded from the $\delta^{18}O$ vs altitude diagram (Fig. 9). Consequently, we can infer that the elevations of recharge zones for the Plio-Quaternary aquifer range between 240 and 950 m above mean sea level (masl), while those for the Turonian aquifer span from 590 to 890 m above msl. However, the elevations of the Tidzi diapir (the eastern boundary of the two aquifers) do not exceed 400 m. The disparities in isotopic compositions observed between the waters of the two aquifers and precipitation (used for gradient estimation) cannot be attributed to altitude-related effects. Instead, this isotopic depletion is more likely attributable to recharge during winter periods where the ratios approach that of seawater, accompanied by elevated chloride concentrations. The involvement of seawater intrusion in the mineralization is improbable for samples exhibiting a Ca/Mg ratio exceeding 1. Hence, an alternative hydrochemical process likely associated with water-rock interactions is implicated [63].

3.2.3 SO_4/Cl Couple

In accordance with Pulido-Lebouef et al. [61], the SO_4/Cl ratio serves as a natural tracer for discerning the marine intrusion phenomenon within coastal aquifers. This ratio experiences a decline as the proportion of seawater in the mixture increases [60, 61, 64].

In the study area, all examined samples exhibit a SO_4/Cl ratio below 1, indicative of the predominance of chlorides over sulfates (Fig. 5b). The augmentation of these samples with SO_4 ions implies alternative sources, such as the dissolution of gypsum and anhydrite (Table 1). Samples 11/51, 45/51, 149/51, O94, and O114 exhibit a modest SO_4/Cl ratio alongside elevated chloride content, implying that the salinity increase in these wells primarily stems from seawater intrusion. This finding corroborates the results obtained from the Na/Cl and Ca/Mg ratios.

3.2.4 Cl/HCO_3 Versus Cl Plot

The Cl/HCO_3 versus Cl plot (Fig. 6a, b) is employed to categorize water types as either freshwater or seawater, as seawater typically exhibits elevated Cl^- levels, while freshwater is characterized by higher HCO_3^- concentrations. In accordance with Revelle [65] and Todd [66], the classification of water salinization types is based on the Cl^-/HCO^-_3 ratio (Fig. 6). The plot essentially classifies water as freshwater, mixing water, and seawater. Most points are contaminated with marine intrusions and leaving no signature of freshwater. Like point 27/51, it is characterized by the recharge of the aquifer, which is also contaminated.

Table 1 Physicochemical and isotopic results of analyzed samples of the two campaigns, 2019 and 2020

	X	Y	Z	H	T		EC	Ca	Mg	Na	K	HCO_3	Cl	SO_4	NO_3	Br
Sample	m				°C	pH	µS/cm	mg/L								
Campaign 2019																
3/51	81,472	93,666	7	−2.94	21.3	9.2	2,011	108.5	50.51	265.3	5.75	402.04	369.87	95.3	145.33	0.27
11/51	80,383	96,193	3	−1.4	20.6	9.5	16,300	544.77	533.17	2,546.54	118.94	204.4	5,728.15	281	114.88	1.05
15/51	85,890	97,935	65	56.47	20.4	7.3	1,780	132.71	39.46	217.88	18.53	314.2	405.01	110.7	54.7	0.22
21/51	89,582	91,439	128	99.9	23	8.8	4,460	355.73	110.84	500.51	6.54	204.4	1,346.64	209.6	8.89	0.32
27/51	95,370	91,109	201	170.7	22.5	9.2	863	60.02	16.89	88.64	1.7	167.8	113	50.2	87.38	0.18
28/51	97,284	91,810	234	184.4	23.2	8.9	941	57.54	26.61	90.4	2.79	277.6	143.36	52.9	23.68	0.17
90/51	93,482	103,717	88	69.25	22.4	7.64	1,396	89.79	47.43	130.77	3.47	253.2	240.66	145	17.62	0.1
105/51	99,400	100,885	115	104.3	23	7.5	1,898	92.55	61.37	238.47	5.45	323.96	387.84	170.7	0.8	0.15
137/51	86,660	94,478	85	58.8	22.65	9.4	4,888	276.96	105.4	661.64	5.34	245.88	1,344.83	182.6	200.17	0.63
140/51	87,205	103,776	60	43.7	20.8	8	1,942	122.73	26.86	254.84	56.28	504.52	415.3	60.9	4.1	0.39
148/51	85,703	102,084	58	43.4	21.43	7.55	1,310	132.8	28.88	132.73	13.27	353.24	312.16	44.9	0.65	0.13
149/51	85,100	105,800	36	−2.68	22.83	9.27	3,169	90.13	76.08	571.22	8.35	424	767.07	123	91.78	0.55
261/51	96,540	99,238	118	72.4	24	7.4	1,962	137.38	68.81	194.46	5.69	260.52	409.22	95.7	199.75	0.15
272/51	97,195	100,703	105.5	80	22	7.47	1,620	84.61	57.38	187.96	4.56	304.44	306.79	156	5.15	0.12
327/51	89,220	88,690	124	95.4	22.3	9.5	3,500	194.15	85.89	503.7	9.64	326.4	1,062.23	45.2	0.5	0.43
O4	97,352	101,259	102	80.3	21.5	7.6	1,600	83.53	57.59	179.43	4.14	258.08	291.38	170.7	1.54	0.11
PQ5	91,853	102,137	83	55.6	25	7.64	1,610	98.26	64.98	155.28	2.68	326.4	293.33	135.3	17.63	0.1
PQ6	91,193	102,500	70	54	22.55	7.36	1,910	106.23	69.28	213.83	9.54	363	389.58	135.6	28.55	0.16
PQ7	90,976	101,554	75	53.6	22.6	7.85	1,236	47.17	33.87	158.02	6.44	314.2	184.61	94.1	36.81	0.23
PQ8	91,990	97,737	106	–	23.2	9.6	2,480	199.69	83.92	239.86	3.61	184.88	659.86	118.4	77.38	0.14
PQ91	90,829	97,430	102	61.8	23.2	9.7	2,621	155.37	78.77	295.52	6.38	260.52	719.38	44.1	58.03	0.44
O94	81,369	93,593	4	−0.43	22	9.8	2,116	87.38	41.77	299.84	22.28	289.8	442.45	129.7	120	0.29
PQ96	83,000	93,735	21	14.1	21	9.5	1,820	83.03	37.77	256.91	5.72	226.36	391.3	93	117.53	0.23

(continued)

Table 1 (continued)

	X	Y	Z	H	T		EC	Ca	Mg	Na	K	HCO_3	Cl	SO_4	NO_3	Br
Sample	m				°C	pH	μS/cm	mg/L								
PQ111	87,675	91,861	112	79.62	22.8	8.58	2,470	161.84	96.83	257.57	4.89	397.16	623.78	79.7	22.44	0.14
PQ113	87,063	95,467	83	65.45	22.4	9.5	5,930	237.6	97.49	975.88	23.11	216.6	1,862.48	388	43.07	1.03
O127	91,965	98,689	102	–	24.2	9	2,509	166.96	95.72	262.42	3.06	289.8	600.56	191.4	53.36	0.16
390/51	96,796	100,921	106	–	28	7.19	1,857	112.3	68.81	190.61	3.57	353.24	368.68	129.4	17.01	0.12
346/51	97,149	100,742	106	–	29	7.18	1,768	109.63	63.98	172.51	3.73	363	344.18	127	9.64	0.11
380/51	89,840	91,466	131	–	25	8.8	2,190	92.17	63.74	304.75	12.23	331.28	463.01	173.1	0.5	0.25
6/51	86,390	105,543	20	12.6	22.45	8.85	2,457	174.57	78.27	265.29	3.13	338.6	558.67	134.5	91.62	0.2
M98	89,527	98,955	94	–	24.5	9.1	2,510	164.4	80.24	291.9	2.79	262.96	612.61	148.3	73.57	0.19
O98 (Wadi upstream)	97,273	101,093	100	–	26.7	8.04	1,500	64.8	40.35	186.07	3.89	192.2	281.27	154.2	0.5	0.11
O36 (Wadi downstream)	86,400	105,580	12	–	27.11	9.9	3,040	57.41	76.41	503.02	5.97	192.2	799.67	157.5	0.83	0.32
			Min	−2.94	20.4	7.18	863	47.17	16.89	88.64	1.7	167.8	113	44.1	0.5	0.1
			Max	184.4	29	9.9	16,300	544.77	533.17	2,546.54	118.94	504.52	5,728.15	388	200.17	1.05
			Average	61.93	23.20	8.54	2,717.09	141.90	77.73	357.38	11.92	294.75	704.21	134.17	52.28	0.27
Campaign 2020																
11/51	80,383	96,193	3	−1.35	21.92	8.1	23,850	454.71	715.93	6,079.29	214.3	143.96	10,741.23	2,098	50.67	
3/51	81,472	93,666	7	−3.1	22	6.5	2,033	87.65	47.43	267.95	5.71	407.48	391.53	125.9	129.63	
O96	83,000	93,735	21	13.8	21.7	7.41	2,504	90.52	52.69	355.69	5.83	226.92	583.22	187.3	98.37	
15/51	85,890	97,935	65	56.5	20.5	7.5	995	51.57	16.32	125.28	23.82	356.24	125	69.4	7.58	
327/51	89,220	88,690	124	94.7	22.8	7.6	3,876	187.33	82.88	490.94	10.33	301.5	1,192.46	42.5	10.59	
27/51	95,370	91,109	201	171.8	23	7.7	775	60.86	12.92	70.89	1.75	265.96	80.73	50.7	27.7	
28/51	97,284	91,810	234	184.55	24	7.6	945	64.6	23.66	80.8	2.07	256.2	121.11	52.9	78.01	
21/51	89,582	91,439	128	99.9	23.5	7.4	4,600	356.85	100.16	466.36	6.98	195.2	1,417.14	233.4	11.56	
O111	87,675	91,861	112	79.48	24.1	7.2	2,476	161.15	81.82	233.5	6.57	395.28	557.48	92.1	28.24	

O112	94,027	95,124	161	–	25	7.5	2,270	187.48	52.39	177.55	4.66	212.28	546.42	88.6	58.38	
PQ91	90,829	97,430	102	61.65	–	–	–	–	–	–	–	–	–	–	–	
PQ127	91,965	98,689	102	102	25.6	7.7	2,606	198.41	72.51	203.44	5.33	185.44	639.32	159.2	60.45	
E103	86,662	94,976	112	79.5	23.6	7.5	2,372	131.17	55.85	296.01	13.83	317.2	518.54	206.1	2.3	
117/51	100,504	98,494	180	151.2	22.3	7.9	1,512	108.3	22.93	187.21	8.12	207.4	218.55	208.2	148.45	
116/51	100,807	96,086	200	177.1	23	7.5	2,442	198.39	71.38	227.03	4.67	280.6	550.99	100.8	149.67	
105/51	99,400	100,885	115	103.7	23	7.3	4,317	255.13	150.34	452.56	9.91	366	1,199.79	250	17.21	
O4	97,352	101,259	102	79.3	21.5	8	1,653	118.26	52.69	158.37	6.33	280.28	289.81	217.8	3.71	
272/51	97,195	100,703	105.5	80.7	22.1	7.5	1,689	140.71	57.82	170.55	7.53	314.76	311.51	205.4	5.18	
262/51	96,583	100,817	112	78.8	22.7	7.38	1,714	147.25	64.53	160.25	6.65	324.52	331.37	205.5	5.75	
PQ5	91,853	102,137	83	55.35	30.8	7.5	1,693	144.4	61.25	134.02	6.02	326.96	284.85	212.4	20.04	
E113bis	92,600	104,215	77	59.6	–	–	–	–	–	–	–	–	–	–	–	
90/51	93,485	103,716	88	68	22.7	7.6	1,500	147.59	49.52	123.12	7.18	268.4	263.67	201.7	8.09	
PQ6	91,193	102,500	70	53.9	23.2	7.5	1,947	162.47	68.87	194.21	6.01	397.72	370.77	173.5	30.18	
PQ7	90,976	101,554	75	53.6	23	7.6	1,266	55.34	36	144.58	1.85	317.2	176.21	91.5	59.33	
148/51	85,703	102,084	58	42.3	22.3	7.5	1,425	127.34	25.46	114.74	9.32	317.2	297.23	23.5	6.11	
140/51	87,205	103,776	60	43.5	21.4	7.6	2,100	148.69	25.39	237.75	51.42	470.92	424.54	94.1	1.93	
149/51	85,100	105,800	36	−1.7	23.2	7.6	3,219	80.27	69.77	497.61	8.36	468.48	760.52	150.8	51.66	
O.Aval (PQ36)	85,610	105,600	14		32	8.4	2,804	155.72	72.6	344.57	7.53	212.28	745.23	180.5	21.2	
M98	89,527	98,955	94		27	7.6	2,570	181.21	66.19	258.18	5.11	280.6	575.77	206.2	74.27	
380/51	89,840	91,466	131		26	7.6	2,166	85.78	57.49	270.27	12.44	346.48	459.18	189.7	0.17	
386/51	92,000	98,650	102	–	28.1	7.6	2,397	167.07	80.12	218.73	4.59	309.88	501.95	256.4	49.32	
390/51	96,796	100,921	106	–	28.3	7.06	1,794	148.78	62.42	165.35	6.59	366	368.6	201.7	14.76	
346/51	97,149	100,742	106	–	25	7.7	817	72.85	32.6	31.92	4.34	212.28	74.44	120	8.88	
6/51	86,387	105,547	19	11.95	26.9	7.2	2,472	180.96	71.4	236.96	3.24	356.24	539.48	145	52.87	
			Min	−3.1	20.5	6.5	775	51.57	12.92	31.92	1.75	143.96	74.44	23.5	0.17	
			Max	184.55	32	8.4	23,850	454.71	715.93	6,079.29	214.3	470.92	10,741.23	2,098	149.67	
			Average	73.95	24.13	7.54	2,837.46	151.83	78.54	411.74	14.94	302.74	801.83	213.77	40.38	

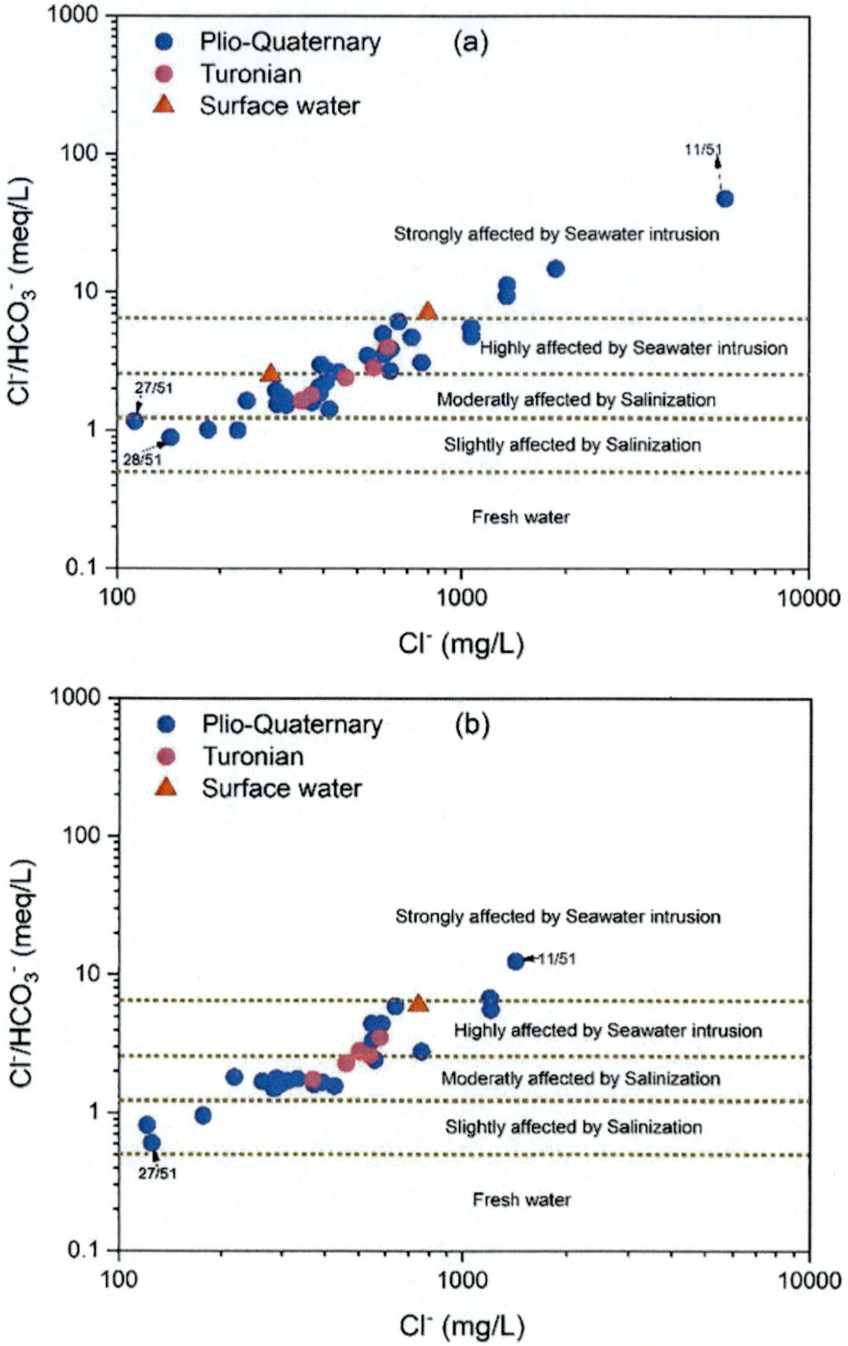

Fig. 6 Cl^-/HCO_3^- vs Cl^- plot for the campaign 2019 (**a**), and 2020 (**b**)

3.2.5 Saturation Index (SI)

The Eq. (1) can be used to calculate a mineral's saturation index (SI):

$$SI = \log\left(\frac{K_{IAP}}{K_{SP}}\right) \quad (1)$$

where K_{SP} denotes the mineral's solubility product and K_{IAP} denotes the ion's activity product in a mineral equilibrium process.

The code PHREEQC [67] was used to calculate the saturation index for carbonate mineral elements (calcite and dolomite) and gypsum (Fig. 7), and the groundwater samples were classified into two groups according to the Mg/Ca ratio and saturation indices. Freshwater is generally dominated by calcium, while seawater is dominated by magnesium. The Mg/Ca ratio can provide an indicator of seawater intrusion. For the first group, the calcite or dolomite saturation index of groundwater samples grouped in a brown dotted circle (Fig. 7) near or above 0 or the equilibrium line is interpreted as saline groundwater with more Mg^{2+} content. This is due to the mixing of underground freshwater and saltwater. For the 2020 companion, there is water sample 11/51 that shows high contamination by the marine intrusion, and regarding the 2019 companion, there are 10 water samples (11/51, 105/51, 149/51, 272/51, O4, O5, O6, O7, 390/51, and 380/51) that were contaminated by high salinity. It is noted that the marine intrusion also contaminated the entire route of Oued Ksob [O98

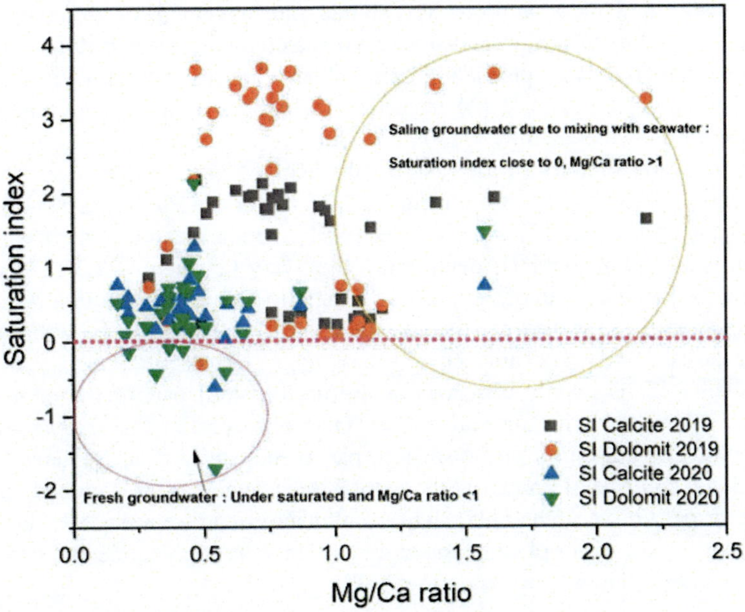

Fig. 7 Saturation index vs Mg/Ca ratio plot for the campaign 2019 (**a**), and 2020 (**b**)

(downstream), O38 (upstream)]. Groundwater samples with Mg/Ca ratio > 1 and SI > 1 suggest that Mg^{2+} was derived from seawater and adsorbed by fresh groundwater with the release of Na ion due to the cation exchange process.

The Mg/Ca ratio for the second group is less than 1 and SI < 0. This indicates the presence of fresh groundwater in the aquifer system, which is also unaffected by seawater. Ca-rich fresh groundwater in the upper part of the study area becomes saline groundwater rich in Mg in the lower region along the groundwater flow direction.

3.3 *Origin of Saline Water*

When paired with the ionic content of water, the stable isotope of water such as $\delta^{18}O$ and $\delta^{2}H$ provides a valuable tool for studying the mixing between water masses of different salinities and therefore tracing the origin of salinity [6]. The stable isotopes $\delta^{18}O$ and $\delta^{2}H$ determine how groundwater forms and evolves [68].

3.3.1 Isotopic Signature of Groundwater in the Area

Isotopic analysis is a useful tool for deciphering the mechanisms that drive an aquifer system's hydrogeological and hydrochemical evolution [11, 69, 70]. According to Kim et al. [55], stable isotopes of oxygen and hydrogen are thought to be transported conservatively in shallow aquifers. The source and movement of groundwater, as well as the technique of aquifer recharge, are revealed by these two isotopes [55, 71–74]. They can also provide physical processes that alter the water, such as evaporation and mixing [75]. The stable isotopes are highly useful in determining the sources of groundwater salinity [71, 72, 74, 76].

In groundwater samples of the KSOB sub-basin, $\delta^{18}O$ varies between −5.36‰ (28/51) and 0.08‰ (11/51) with a mean of −3.7‰ for the Plio-quaternary aquifer and between −5.9‰ (O121) and −1.79‰ (6/51), with a mean of −4.59‰ for the Turonian aquifer. For the $\delta^{2}H$ (deuterium), they vary between −29.3 (27/51) and 3.13‰ (11/51) with a mean of −18.64‰ for the Plio-Quaternary aquifer. As for the Turonian aquifer, the values oscillate between −29.9 (380/51) and −8.8‰ (6/51) with a mean of −24.13‰ (Table 2).

Depleted $\delta^{18}O$ is mainly observed in the northeastern part of the area in the sample of rainwater, whereas enriched $\delta^{18}O$ is observed in the sample near of Atlantic Sea 11/51 location of the western part of the study area. Depleted $\delta^{2}H$ is observed in the northeastern part at the point 380/51, whereas the enriched value in the western part at the point 11/51. The relationship trend between $\delta^{2}H$ and $\delta^{18}O$ (Fig. 8a) allows the major phenomena involved in the hydrodynamic and geochemical functioning of aquifers to be defined [19].

Table 2 Isotope signatures of δ^2H and $\delta^{18}O$ (VSMOW ‰) of campaign 2020

Sample	Nature aquifer	X (m)	Y (m)	Z (m)	H (m)	pH	EC (µS/cm)	T (°C)	Cl^- (meq/L)	$\delta^{18}O$ (‰)	δ^2H (‰)	d-excess (‰)
11/51	PQ	80,383	96,193	3	−1.35	8.1	23,850	21.92	303	0.08	3.13	2.49
3/51	PQ	81,472	93,666	7	−3.1	6.5	2,033	22	11.04	−3.88	−20	11.04
15/51	PQ	85,890	97,935	65	56.5	7.5	995	20.5	3.53	−3.76	−13	17.08
327/51	PQ	89,220	88,690	124	94.7	7.6	3,876	22.8	33.64	−3.61	−16.9	11.98
27/51	PQ	95,370	91,109	201	171.8	7.7	775	23	2.28	−5.26	−29.3	12.78
28/51	PQ	97,284	91,810	234	184.55	7.6	945	24	3.42	−5.36	−25.6	17.28
E103	PQ	86,662	94,976	112	79.5	7.5	2,372	23.6	14.63	−4.91	−24.4	14.88
116/51	PQ	100,807	96,086	200	177.1	7.5	2,442	23	15.54	−5.16	−25	16.28
105/51	PQ	99,400	100,885	115	103.7	7.3	4,317	23	33.84	−1.92	−13.8	1.56
262/51	PQ	96,583	100,817	112	78.8	7.38	1,714	22.7	9.35	−3.24	−18.4	7.52
O5	PQ	91,853	102,137	83	55.35	7.5	1,693	30.8	8.04	−4.9	−27.9	11.3
148/51	PQ	85,703	102,084	58	42.3	7.5	1,425	22.3	8.38	−3.87	−17.8	13.16
149/51	PQ	85,100	105,800	36	−1.7	7.6	3,219	23.2	21.45	−1.95	−13.4	2.2
380/51	T	89,840	91,466	131	–	7.6	2,166	26	12.95	−5.29	−29.9	12.42
386/51	T	92,000	98,650	102	–	7.6	2,397	28.1	14.16	−4.29	−20.6	13.72
390/51	T	96,796	100,921	106	–	7.06	1,794	28.3	10.4	−4.62	−26	10.96
346/51	T	97,149	100,742	106	–	7.7	817	25	2.1	−4.74	−26.3	11.62
6/51	T	86,387	105,547	19	11.95	7.2	2,472	26.9	15.22	−1.79	−8.8	5.52
E3	T	119,340	112,500	330	–	7.7	1,752	21	11	−5.32	−27.7	14.86
O99	Downstream Wadi	85,610	105,600	14	–	8.4	2,804	32	21.02	−4.81	−26.6	11.88
O121	Rainwater	127,760	103,090	450	–	6	40	23.8	–	−5.9	−27.2	20

PQ Plio-Quaternary
T Turo

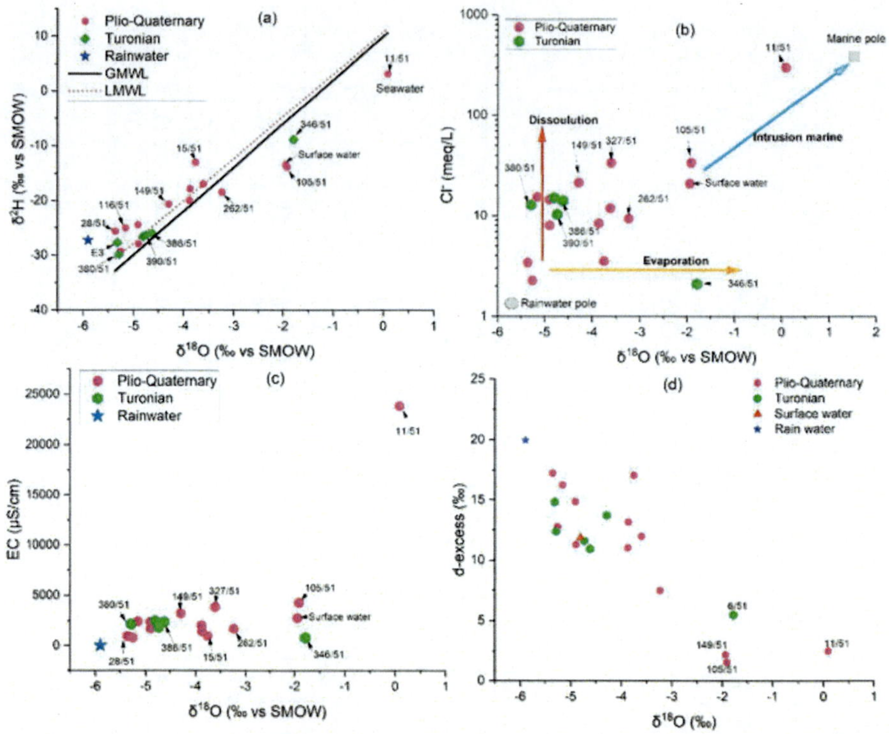

Fig. 8 (**a**) Deuterium vs oxygen-18 diagram of the waters of the Plio-Quaternary and Turonian aquifers of the Essaouira basin (July 2020) (**b**); oxygen-18 vs chloride diagram (**c**); oxygen-18 vs electrical conductivity; (**d**) δ^{18}O vs d-excess

The equation of a local meteoric water line (LMWL) determined from the equation for Essaouira basin (Eq. 2) [41] and the equation of a global meteoric water line (GMWL) are calculated from the following relation (Eq. 3) [77]:

$$\delta 2H = 7.96 \times \delta^{18}O + 11.30 \quad (2)$$

$$\delta 2H = 8 \times \delta^{18}O + 10 \quad (3)$$

Most of the 2020 campaign samples in the two aquifers were found above the GMWL and around the LMWL (Fig. 8a). This indicates that the penetration of precipitation of Atlantic origin with minimal evaporation ensures the aquifer's recharge. This recharge minimizes salinity in some wells like 27/51 and 28/51, which have a low electrical conductivity of 775 and 945 μS/cm, respectively. The closest well to the freshwater pole is rainwater (O121). Other points are close to rainwater in their chemical composition, such as point 15/51 with electrical conductivity which equals to 975, located between Ahmed Ouhmad and Sidi Kawki near the road. The lithological structure of the unsaturated zone and its limited thickness

(5–60 m) probably favor this recharge. The vast majority of samples from the shallow Plio-Quaternary aquifer and all water points from the deep Turonian aquifer are included in this category. However, some other points are located below the LMWL with more enriched isotopic compositions, notably 346/51, 262/51, 11/51, and surface water (105/51) reflecting the intervention of another phenomenon in the groundwater mineralization, such as evaporation and marine effect. This last phase mainly affects surface waters (O99) and wells 105/51 and 346/51. Evaporation is more likely to occur before water infiltration, in the unsaturated zone, or during recharge of the aquifer [68]. The isotopic signature of rainwater and saltwater from the Atlantic Ocean are used to draw the freshwater-seawater border [78–80]. Sample 11/51 is aligned with the mixed freshwater-saltwater line in the same pattern. High salinities are associated with a drop in the piezometric level on this site near Cap Sim, located on the coastal fringe. This demonstrates that the phenomenon of saltwater intrusion is the main driver of the increase in mineralization in these wells and confirms what is being discussed earlier in the ionic ratios sections.

The samples have high salinization and chloride concentrations and are located between the "rainfall" and "marine" poles, implying that seawater has advanced in the studied aquifer. Previously obtained results can be supported by using chloride and oxygen-18 content together. Indeed, the Cl^- versus $\delta^{18}O$ graph (Fig. 8b) reveals that the groundwater in the study area is dominated by maritime intrusion processes and the dissolution of evaporitic formations. The EC vs $\delta^{18}O$ plot (Fig. 8c) depicts the recharging region of the aquifer based on the EC value. The lowest value is 40 μS/cm O121 (rainwater) with –5.9 $\delta^{18}O$ and the highest value is 23,850 μS/cm (11/51) (Cap sim) with 0.08 for $\delta^{18}O$.

The d-excess is a suitable method to trace the influence of evaporation on changing the isotopic composition of rainwater before groundwater recharge [81]. The following Eq. (4) represents the d-excess [82]:

$$d - excess = \delta D - 8 \times \delta^{18}O \tag{4}$$

The d-excess value indicates the conditions that contribute to kinetic isotope fractionation between water and vapor during the primary evaporation process in the ocean [83]. The range of d-excess value of precipitation for the entire study area is provided in Fig. 7d. In groundwater samples of the KSOB sub-basin, d-excess varies between 1.56‰ (105/51) and 17.28‰ (28/51) with a mean of 10.73‰ for the Plio-Quaternary aquifer and between 5.52‰ (6/51) and 20‰ (O121), with a mean of 12.62‰ for the Turonian aquifer.

Several values of d-excess were consistently >10‰ (Fig. 8d). The influence of recycled vapor from surface water masses in the condensation vapor can be inferred from the high d-excess value. The high excess values indicate lower evaporative enrichment of the water samples and show its connection with the water bodies. Although the study site is close to the sea, evaporated vapor from other water sources is evident in the precipitation. The isotope ratios of the water were also enriched, probably because the amount of rainfall was less in this area.

3.3.2 Recharge Area

In order to know the possible recharge areas for the Plio-Quaternary and Turonian aquifers of the Essaouira basin, we considered the regional altitudinal gradient in oxygen-18 of −0.26‰ [84] and validated by Bahir [76]. All the samples contaminated by marine intrusion with a value greater than −3‰ are neglected in the graph identifying the recharge areas. Points representing areas affected by evaporation (105/51 and 346/51) and marine intrusion (11/51) have been excluded from the $\delta^{18}O$ vs altitude diagram (Fig. 9). Consequently, we can infer that the elevations of recharge zones for the Plio-Quaternary aquifer range between 240 and 950 m above mean sea level (msl), while those for the Turonian aquifer span from 590 to 890 m above msl. However, the elevations of the Tidzi diapir (the eastern boundary of the two aquifers) do not exceed 400 m. The disparities in isotopic compositions observed between the waters of the two aquifers and precipitation (used for gradient estimation) cannot be attributed to altitude-related effects. Instead, this isotopic depletion is more likely attributable to recharge during winter periods.

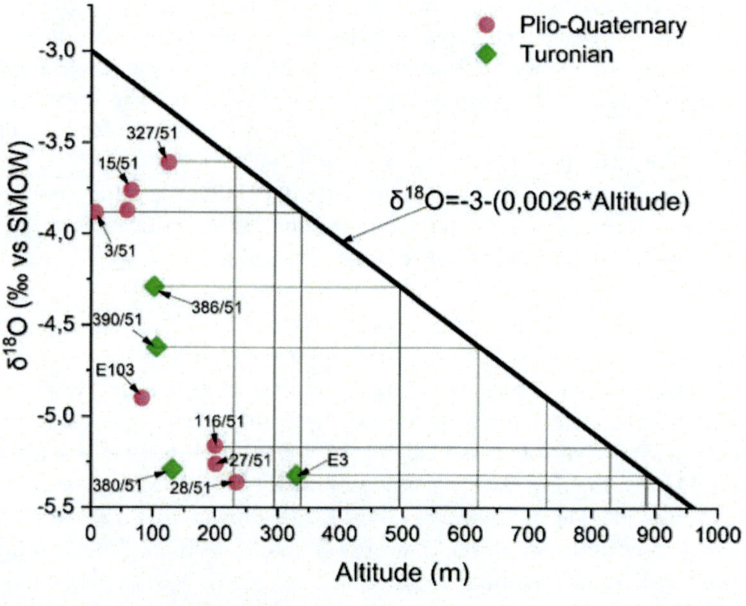

Fig. 9 Relationship of oxygen-18 content and altitude of water points capturing the Plio-Quaternary and Turonian aquifers during 2020

4 Conclusion

In this study, hydrogeochemical and isotopic methodologies involving the analysis of ion pairs (Na, Cl), (Ca, Mg), and (Br, Cl) were employed to investigate the salinization of groundwater. The ionic ratios revealed that seawater intrusion into the freshwater portion of the Plio-Quaternary aquifer within the Essaouira basin has commenced, whereas the Turonian aquifers exhibit comparatively lower levels of seawater contamination. The notable instances of marine water intrusion are dispersed along the coastline, with the saline front advancing approximately 2 km inland. This intrusion is likely attributed to the decline in the piezometric level due to reduced precipitation over recent decades in the study area, compounded by rising sea levels attributed to climate change. Consequently, there is a pressing need to monitor and regulate groundwater over-extraction along the coastal region to ensure the sustainable development of groundwater resources.

5 Recommendation

- The study provides valuable insights into seawater intrusion and salinization in the Essaouira basin's coastal aquifers. Implementing the recommendations can safeguard groundwater resources, ensuring their availability for present and future generations. Collaboration among government agencies, local communities, and researchers is crucial for sustainable water management in the region.
- Continued Monitoring: To ensure the long-term sustainability of water resources, it is imperative to maintain regular monitoring of hydrochemical and isotopic parameters. This ongoing monitoring will facilitate tracking changes over time and evaluating the effectiveness of any implemented water resource management strategies.
- Recharge Enhancement: Considering the significant role of precipitation in recharging the aquifers, exploring methods for enhancing recharge is of utmost importance. Implementing rainwater harvesting techniques and artificial recharge measures can alleviate the adverse effects of seawater intrusion and augment freshwater resources.
- Research and Innovation: Promoting further research and innovation in hydrogeology and water resource management will foster a deeper understanding of the intricate processes involved in seawater intrusion. Novel technologies and approaches may emerge as promising solutions to address the challenges of water scarcity in coastal regions.

Acknowledgment The authors would like to sincerely thank the reviewers and editor for their invaluable efforts in thoroughly reviewing and editing our work. We are also very grateful to Professor Carreira Paula for analyzing our stable isotopes in their laboratory.

Declaration of Non-infringement of Third-Party Copyrights All the materials presented in the article are free of any third-party material and therefore, no copyright clearance is needed.

References

1. Edmunds WM, Milne CJE (2001) Palaeowaters in coastal Europe: evolution of groundwater since the late Pleistocene. Geol Soc Spec Publ 189:289–311
2. Berhail S (2019) The impact of climate change on groundwater resources in northwestern Algeria. Arab J Geosci 12:770. https://doi.org/10.1007/s12517-019-4776-3
3. Custodio E (1997) Seawater intrusion in coastal aquifers. Guidelines for study. Monitoring and control. Water report N°11. Food and agriculture Organization of the United Nations, Rome, Italy, 152 p
4. Fernandes PG, Carreira PM, Bahir M (2010) Mass balance simulation and principal components analysis applied to groundwater resources: Essaouira basin (Morocco). Environ Earth Sci 59:1475–1484. https://doi.org/10.1007/s12665-009-0133-2
5. Fernandes PG, Bahir M, Mendonça J, Carreira PM, Fakir Y, Silva M (2005) Anthropogenic features in the Sines (Portugal) and Essaouira (Morocco) coastal aquifers: a comparative study of their hydrochemical evolution by a principal component analysis. Estud Geol 61:207–219
6. Freeze RA, Cherry JA (1979) Groundwater. Prentice-Hall, Englewood Cliffs. 604 p
7. Bahir M, Ouhamdouch S, Carreira PM, Chkir N, Zouari K (2018) Geochemical and isotopic investigation of the aquifer system under semiarid climate: case of Essaouira basin (southwestern Morocco). Carbonates Evaporites 33:65–77. https://doi.org/10.1007/s13146-016-0323-4
8. Bahir M, Ouazar D, Ouhamdouch S (2019) Hydrogeochemical mechanisms and recharge mode of the aquifers under semiarid climate from Morocco. Appl Water Sci 9:103. https://doi.org/10.1007/s13201-019-0988-7
9. Bahir M, Ouazar D, Ouhamdouch S (2019) Hydrogeochemical investigation and groundwater quality in Essaouira region. Morocco Mar Freshwater Res 70:1317–1332
10. El Mountassir O, Bahir M, Ouazar D, Ouhamdouch S, Chehbouni A, Ouarani M (2020) The use of GIS and water quality index to assess groundwater quality of Krimat aquifer (Essaouira, Morocco). SN Appl Sci 2:871. https://doi.org/10.1007/s42452-020-2653-z
11. El Mountassir O, Bahir M, Ouazar D, Carreira PM (2021) Nitrate pollution in groundwater of the Ouazi Basin: Case of Essaouira (Southwestern Morocco). In: Abrunhosa M, Chambel A, Peppoloni S, Chaminé HI (eds) Advances in geoethics and groundwater management: theory and practice for a sustainable development. Advances in science, technology & innovation. Springer, Berlin, pp 239–245. https://doi.org/10.1007/978-3-030-59320-9_49
12. El Mountassir O, Bahir M, Ouazar D, Carreira PM (2021) For a better understanding of recharge and salinization mechanism of a Cenomanian–Turonian aquifer. In: Abrunhosa M, Chambel A, Peppoloni S, Chaminé HI (eds) Advances in geoethics and groundwater management: theory and practice for a sustainable development. Advances in science, technology & innovation. Springer, Berlin, pp 201–206. https://doi.org/10.1007/978-3-030-59320-9_42
13. El Mountassir O, Bahir M, Ouazar D, Chehbouni A, Carreira PM (2021) GIS-based assessment of aquifer vulnerability using DRASTIC model and stable isotope: a case study on Essaouira basin. Arab J Geosci 14:321. https://doi.org/10.1007/s12517-021-06540-6
14. El Mountassir O, Bahir M, Ouazar D, Chehbouni A, Carreira PM (2021) Geochemical and isotopic evidence of groundwater salinization processes in the Essaouira region, north-west coast, Morocco. SN Appl Sci 3:698. https://doi.org/10.1007/s42452-021-04623-3

15. El Mountassir O, Bahir M, Chehbouni A, Dhiba D, El Jiar H (2022) Assessment of groundwater quality and the Main controls on its hydrochemistry in a changing climate in Morocco (Essaouira Basin). Sustainability 14:8012. https://doi.org/10.3390/su14138012
16. El Mountassir O, Bahir M, Ouazar D, Chehbouni A, Carreira PM (2022) Evaluation of nitrate source and its distribution in the groundwater of Essaouira basin. Sustain Water Resour Manag 8:28. https://doi.org/10.1007/s40899-022-00609-0
17. El Mountassir O, Bahir M, Ouazar D, Chehbouni A, Carreira PM (2022) Temporal and spatial assessment of groundwater contamination with nitrate using nitrate pollution index (NPI), groundwater pollution index (GPI), and GIS (case study: Essaouira basin, Morocco). Environ Sci Pollut Res 29:17132–17149. https://doi.org/10.1007/s11356-021-16922-8
18. El Mountassir O, Bahir M (2023) The assessment of the groundwater quality in the coastal aquifers of the Essaouira Basin, southwestern Morocco, using hydrogeochemistry and isotopic signatures. Water 15(9):1769
19. Bahir M, El Mountassir O, Chehbouni A, Dhiba D, El Jiar H, Carreira PM (2022) Hydrogeochemical and isotopic assessment for characterizing groundwater quality and recharge processes in the Essaouira Basin, Northwestern Morocco. Arab J Geosci 15:603. https://doi.org/10.1007/s12517-022-09817-6
20. Bahir M, Mountassir OEL, Ouazar D, Chehbouni A, Carreira PM (2021) Stable isotope and quality of groundwater around Ksob sub-basin. Essaouira Morocco Sustain Water Resour Manag 7:73. https://doi.org/10.1007/s40899-021-00553-5
21. Bahir M, El Moukhayar R, Youbi N, Chamchati H, Chkir Ben Jemaa N (2013) Management and protection of groundwater resources in semiarid zones: application of hydrochemical methodologies to Essaouira Synclinal Basin, Morocco. Int J Hydrol Sci Technol 3:160–175
22. Bahir M, El Moukhayar R, Chamchati H, Youbi N, Carreira P, Chkir N (2014) Recharge and hydrogeochemical evolution groundwater in semiarid zone (Essaouira Basin. Morocco). Comun Geol 3:651–653. http://www.lneg.pt/iedt/unidades/16/paginas/26/30/185
23. Bahir M, El Moukhayar R, Chkir N, Chamchati H, Fernandes PG, Carreira P (2013) Groundwater chemical evolution in the Essaouira Aquifer Basin – NW Morocco. Open J Mod Hydrol 3:8. https://doi.org/10.4236/ojmh.2013.33017
24. Jalal M, Blavoux B, Bahir M, Bellion Y, Laftouhi N, Puig J, Mennani A, Daniel M (2001) Study of the functioning of the Cenomanian-Turonian karstic aquifer system of Oued Igrounzar, Essaouira Basin, Morocco (Etude du fonctionnement du système aquifère karstique Cenomanien-Turonien de l'oued Igrounzar, bassin d'Essaouira, Maroc). J African Earth Sci 32:803–817. https://doi.org/10.1016/S0899-5362(02)00056-8
25. Meddi M, Talia M, Martin C (2009) Recent evolution of weather conditions and flows on the basin of Macta (northwest of Algeria). PhysioGeo 23:61–84
26. Bahir M, Mennani A, Jalal M, Youbi N (2000) Water resources of the Essaouira synclinal basin (Morocco) (Ressources hydriques du bassin synclinal d'Essaouira (Maroc)). Estud Geol 56:185–195
27. Bahir M, Mennani A, Jalal M, Youbi N (2000) Contribution to the study of water resources in the Essaouira synclinal basin (Morocco) (contribution à l'étude des ressources hydriques du bassin synclinal d'Essaouira (Maroc)). EstudGeol 56:185–195
28. Bahir M, Ouhamdouch S, Ouazar D, El Moçayd N (2020) Climate change effect on groundwater characteristics within semiarid zones from western Morocco. Groundw Sustain Dev 11:1–14. https://doi.org/10.1016/j.gsd.2020.100380
29. Custodio E (2002) Coastal aquifers as important natural hydrogeological structures. In: Bocanegra M, Massone H (eds) Groundwater and human development, pp 1905–1918
30. Mendonça JL, da Silva MO, Bahir M (2004) Considerations concerning the origin of the Estoril (Portugal) thermal water. Estud Geológicos 60:153–159
31. Bahir M, Mennani A (2002) Issues in groundwater management in Morocco (Problématique de la gestion des eaux souterraines au Maroc). Estud Geológicos 58:103–108

32. Bahir M, El Mountassir O, Ouazar D, Carreira PM (2021) Hydrochemical analysis and evaluation of groundwater quality in Ouazi Basin (Essaouira, Morocco). In: Abrunhosa M, Chambel A, Peppoloni S, Chaminé HI (eds) Advances in geoethics and groundwater management : theory and practice for a sustainable development. Advances in science technology & innovation. Springer, Berlin, pp 247–250. https://doi.org/10.1007/978-3-030-59320-9_50
33. Bahir M, Ouhamdouch S, Carreira PM (2016) Water resources in Morocco facing climate change; case study of the Plio-quaternary aquifer in the Essaouira synclinal basin (La ressource en eau au Maroc face aux changements climatiques; cas de la nappe Plio-Quaternaire du bassin synclinale d'Essaouira). Comun Geol 103(1):35–44
34. Bahir M, Ouhamdouch S, Carreira PM (2018) Geochemical and isotopic approach to decrypt the groundwater salinization origin of coastal aquifers from semiarid areas (Essaouira basin, Western Morocco). Environ Earth Sci 77:485. https://doi.org/10.1007/s12665-018-7663-4
35. Bahir M, Ouhamdouch S, Carreira PM (2018) Isotopic and geochemical methods for studying water–rock interaction and recharge mode: application to the Cenomanian-Turonian and Plio-quaternary aquifers of Essaouira Basin. Morocco Mar Freshwater Res 69:1290–1300
36. Bahir M, Ouazar D, Ouhamdouch S (2018) Characterization of mechanisms and processes controlling groundwater salinization in coastal semiarid area using hydrochemical and isotopic investigations (Essaouira basin. Morocco). Environ Sci Pollut Res 25:24992–25004
37. Ouhamdouch S, Bahir M, Carreira P (2018) Impact of climate change on water resources in semi-arid environments: example of the Essaouira Basin (Morocco) (Impact du changement climatique sur la ressource en eau en milieu semi-aride: exemple du bassin d'Essaouira (Maroc)). Rev Des Sci De L'Eau 31:13–27
38. Bahir M, Ouhamdouch S (2020) Groundwater quality in semiarid environments (Essaouira Basin. Morocco). Carbonates Evaporites 35:41
39. Bahir M, Jalal M, Mennani A, Laftouhi N (2001) Hydrogeological potential of the Kourimat syncline (Essaouira Basin, Morocco) (Potentialités hydrogéologiques du synclinal de Kourimat (bassin d'Essaouira. Maroc)). Estud Geol 57:47–52
40. Bahir M, Jalal M, Blavoux B (2001) Contribution of isotopic hydrochemistry to the understanding of recharge areas in arid and semi-arid zones: case study of the Meskala-Kourimat basin (Apport de l'hydrochimie isotopique à la connaissance des zones de recharge en zones aride et semi-arides : Cas du bassin de Meskala-Kourimat). J Environ Hydrol
41. Mennani A, Blavoux B, Bahir M, Bellion Y, Jalal M, Daniel M (2001) Contribution of chemical and isotopic analysis to the understanding of the functioning of Plio-quaternary and Turonian aquifers in the synclinal zone of Essaouira (Western Morocco) (Apports des analyses chimiques et isotopiques à la connaissance du fonctionnement des aquifères plio-quaternaire et turonien de la zone synclinale d'Essaouira (Maroc occidental)). J African Earth Sci 32:819–835
42. Bahir M, Ouhamdouch S (2020) Groundwater quality in semiarid environments (Essaouira Basin. Morocco). Carbonates Evaporites 35:1–16. https://doi.org/10.1007/s13146-020-00576-7
43. Bahir M, Ouhamdouch S, Ouazar D, Chehbouni A, Ouarani M, El Mountassir O (2021) Groundwater quality of the alluvial and carbonate aquifers of Essaouira basin (Morocco). Carbonates Evaporites 36:23. https://doi.org/10.1007/s13146-021-00697-7
44. Fekri A (1993) Contribution à l'étude hydrogéologique et hydrogéochimique de la zone synclinale d'Essaouira (Bassin synclinal d'Essaouira). Thèse de doctorat, Université Cadi Ayyad de Marrakech, Maroc. 172 p
45. Bahir M, Chkir N, Trabelsi R, Ammar Friha H, Zouari K, Chamchati H (2012) Hydrogeochemical behaviour of two coastal aquifers under severe climatic and human constraints: comparative study between Essaouira basin in Morocco and Jeffara basin in Tunisia. Int J Hydrol SciTechnol 2:75–100. https://doi.org/10.1504/IJHST.2012.045940
46. Mennani A, Blavoux B, Bahir M, Bellion Y, Jalal M, Daniel M (2001) Contribution of chemical and isotopic analysis to the understanding of the functioning of Plio-quaternary and Turonian aquifers in the synclinal zone of Essaouira (Western Morocco) (Apports des analyses chimiques et isotopiques à la connaissance du fonctionnement des aquifères plio-quaternaire et turonien de la zone synclinale d'Essaouira (Maroc occidental)). J Afr Earth Sci 32:819–835

47. Ali S, Shekhar S, Bhattacharya P, Verma G, Chandrasekhar T, Chandrashekhar AK (2018) Elevated fluoride in groundwater of Siwani Block. Western Haryana. India: a potential concern for sustainable water supplies for drinking and irrigation. Groundw Sustain Dev 7:410–420. https://doi.org/10.1016/j.gsd.2018.05.008
48. Friedman I (1953) Deuterium content of natural waters and other substances. Geochim Cosmochim Acta 4:89–103
49. Epstein S, Mayeda T (1953) Variation of $\delta^{18}O$ content of waters from natural sources. Geochim Cosmochim Acta 4:213–224
50. Rabeiy RE (2018) Assessment and modeling of groundwater quality using WQI and GIS in upper Egypt area. Environ Sci Pollut Res 25(31):30808–30817. https://doi.org/10.1007/s11356-017-8617-1
51. Teshome F (2020) Seasonal water quality index and suitability of the water body to designated uses at the eastern catchment of lake Hawassa. Environ Sci Pollut Res 27(1):279–290. https://doi.org/10.1007/s11356-019-06794-4
52. Selvakumar S, Chandrasekar N, Kumar G (2017) Hydrogeochemical characteristics and groundwater contamination in the rapid urban development areas of Coimbatore, India. Water Resour Ind 17:26–33. https://doi.org/10.1016/j.wri.2017.02.002
53. Piper AM (1944) A graphic procedure in the geochemical interpretation of water analyses. Trans Am Geophys Union 25:914–923
54. De Montety V, Radakovitch O, Vallet-Coulomb C, Blavoux B, Hermitte D, Vallès V (2008) Origin of groundwater salinity and hydrogeochemical processes in a confined coastal aquifer: case of the Rhône delta (southern France). Appl Geochem 23:2337–2349
55. Kim Y, Lee K, Koh D, Lee D, Lee S, Park W, Koh G, Woo N (2003) Hydrogeochemical and isotopic evidence of groundwater salinization in a coastal aquifer: a case study in Jeju volcanic Island, Korea. J Hydrol 270:282–294
56. Davis SN, Whittemore DO, Fabryka-Martin J (1998) Uses of chloride/bromide ratios in studies of potable water. Groundwater 36:338–350
57. Fedrigoni L, Krimissa M, Zouari K, Maliki A, Zuppi GM (2001) Origin of mineralization and hydrogeochemical behavior of a phreatic aquifer under severe natural and anthropogenic constraints: example of the Djebeniana aquifer (Tunisia) (Origine de la minéralisation et comportement hydrogéochimique d'une nappe phréatique soumise à des contraintes naturelles et anthropiques sévères: exemple de la nappe de Djebeniana (Tunisie)). C R Acad Sci 332:665–671
58. Beitler BB, Benison KC (2009) Geochemical characteristics of naturally acid and alkaline saline lakes in southern Western Australia. Appl Geochem 24:268–284
59. Ben Hamouda MF, Tarhouni J, Leduc C, Zouari K (2011) Understanding the origin of salinization of the Plio-quaternary eastern coastal aquifer of Cap Bon (Tunisia) using geochemical and isotope investigations. Environ Earth Sci 63:889–901
60. Bouderbala A (2015) Groundwater salinization in semiarid zones: an example from Nador plain (Tipaza. Algeria). Environ Earth Sci 73:5479–5496
61. Pulido-Bosch A, Tahiri A, Vallejos A (1999) Hydrogeochemical characteristics of processes in the Temara Aquifer in Northwestern Morocco. Water Air Soil Pollut 114:323–337
62. Jones BF, Vengosh A, Rosenthal E, Yechieli Y (1999) Geochemical investigations. In: Bear J et al (eds) Seawater intrusion in coastal aquifers. Kluwer Academic Publisher, pp 51–71
63. Gimenez E, Bencini A, Pranzini G (2010) Hydrogeochemical considerations about the origin of groundwater salinization in some coastal plains of Elba Island (Tuscany, Italy). Environ Geochem Health 32:243–257
64. Tellam JH, Lloyd JW (1986) Problems in the recognition of seawater intrusion by chemical means: an example of apparent equivalence. Q J Eng Geol Hydroge 19:389–398
65. Revelle R (1941) Criteria for recognition of the sea water in ground-waters. Eos Trans Am Geophys Union 22(3):593–597
66. Todd DK (1959) Annotated bibliography on artificial recharge of groundwater through 1954. US Government Printing Office

67. Parkhurst DL, Appelo CAJ (1999) User's guide to PHREEQC (version 2) a computer program for speciation, batch-reaction, one-dimensional transport, and inverse geochemical calculations: U.S Geological Survey. Water-Resour Investig Rep 99(4259):312
68. Ali S, Shekhar S, Chandrasekhar T, Yadav AK, Arora NK, Kashyap CA, Bhattacharya P, Rai SP, Pande P, Chandrasekharam D (2021) Influence of the water–sediment interaction on the major ions chemistry and fluoride pollution in groundwater of the older Alluvial Plains of Delhi, India. J Earth Syst Sci 130(98):1–16. https://doi.org/10.1007/s12040-021-01585-3
69. Abourida A, Er-rouane S, Bahir M, da Silva MO, Cheggour A (2004) Contribution of environmental isotopes for understanding the functioning of the Miocene-Pliocene-quaternary aquifer of Haouz in Marrakech (Morocco) (contribution des isotopes de l'environnement pour la comprehension du fonctionnement de l'aquiferemio-plioquaternaire du Haouz de Marrakech, Maroc). Estudios Geológicos 60(3–6):161–167. https://doi.org/10.3989/egeol.04603-692
70. Tijani MN, Loehnert EP, Uma KO (1996) Origin of saline groundwaters in the Ogoja area, Lower Benue Trough, Nigeria. J Afr Earth Sci 23(2):237–252
71. Bahir M, El Mountassir O, Ouazar D, Carreira PM (2021) Use of WQI and isotopes to assess groundwater quality of coastal aquifers (Essaouira, Morocco). In: Abrunhosa M, Chambel A, Peppoloni S, Chaminé HI (eds) Advances in geoethics and groundwater management: theory and practice for a sustainable development. Advances in science technology & innovation. Springer, Berlin, pp 251–255. https://doi.org/10.1007/978-3-030-59320-9_51
72. Kaid Rassou K, Fakir Y, Bahir M, Zouari K, Marah M, Monteiro JP (2005) Contribution of isotopic analysis to the understanding of the functioning of coastal aquifers in the hydrological basin of Oualidia lagoon (Apports des analyses isotopiques à la compréhension du fonctionnement des aquifères côtiers du bassin hydrologique de la lagune d'Oualidia). Comun Geol 92:129–142
73. Mennani A (2001) Contribution of hydrochemistry and isotopic studies to the understanding of the functioning of coastal aquifers in the Essaouira coastal zone (Western Morocco) (Apport de l'hydrochimie et de l'isotopie à la connaissance du fonctionnement des aquifères de la zone côtière d'Essaouira (Maroc Occidental)). Thèse de doctorat, Université Cadi Ayyad, Marrakech
74. Rassou K, Fakir Y, Bahir M, Zouari K, Marah M (2005) Origin and dating of groundwater in the hydrological basin of Oualidia lagoon (Origine et datation des eaux souterraines du bassin hydrologique de la lagune d'Oualidia). Estud Geol 61:191–196. https://doi.org/10.3989/egeol.05613-661
75. Geyh M (2000) An overview of ^{14}C analysis in the study of groundwater. Radiocarbon 42 (1):99–114. https://doi.org/10.1017/S0033822200053078
76. Bahir M, Carreira P, Da Silva MO, Fernandes P (2008) Hydrodynamic, hydrochemical, and isotopic characterization of the Kourimat aquifer system (Essaouira Basin, Morocco) (Caractérisation hydrodynamique.hydrochimique et isotopique du système aquifère de Kourimat (Bassin d'Essaouira.Maroc)). Estud Geol 64:61–73. https://doi.org/10.3989/egeol.08641433
77. Craig H (1961) Isotopic variations in meteoric waters. Science 133(3465):1702–1703
78. Carreira PM, Marques JM, Nunes D (2014) Source of groundwater salinity in coastline aquifers based on environmental isotopes (Portugal): natural vs. human interference. A review and reinterpretation. Appl Geochem 41:163–175. https://doi.org/10.1016/j.apgeochem.2013.12.012
79. Carreira PM, Bahir M, Salah O, Galego Fernandes P, Nunes D (2018) Tracing salinization processes in coastal aquifers using an isotopic and geochemical approach: comparative studies in Western Morocco and Southwest Portugal. Hydrgeol J 26(8):2595–2615. https://doi.org/10.1007/s10040-018-1815-1
80. Carreira PM, Nunes D, Marques JM, do Rosário Carvalho M, da Silva MA, Costa A (2023) Environmental isotopes (δ^2H, $\delta^{13}C$, $\delta^{18}O$, ^3H, and ^{14}C) as a diagnostic tool in the appraisal of mineral water management and protection: two case studies – Portugal. Sustain Water Resour Manag 9(4):126
81. Saka D, Akiti TT, Osae S, Appenteng MK, Gibrilla A (2013) Hydrogeochemistry and isotope studies of groundwater in the Ga west municipal area, Ghana. Appl Water Sci 3:588

82. Dansgaard W (1964) Stable isotopes in precipitation. Tellus 16:436468
83. Deshpande RD, Bhattacharya SK, Jani RA, Gupta SK (2003) Distribution of oxygen and hydrogen isotopes in shallow groundwaters from southern India: influence of a dual monsoon system. J Hydrol 271(1–4):226–239. https://doi.org/10.1016/S0022-1694(02)00354-2
84. Bouchaou L, Michelot JL, Chauve P, Mania J, Mudry J (1995) Contribution of stable isotopes to the study of groundwater recharge processes in the aquifers of Tadla (Morocco) under a semi-arid climate (Apports des isotopes stables à l'étude des modalités d'alimentation des aquifères du Tadla (Maroc) sous climat semi-aride). Comptes rendus de l'Académie des sciences Série 2 Sciences de la terre et des planètes 320(2):95–101

Part II
Major Global Contaminants in Groundwater

Geochemical Controls on Fluoride Enrichment in Groundwater of a Geologically Heterogeneous Part of Ghana: Implications for Human Health Risk Assessment

Emmanuel Daanoba Sunkari, Moses Boakye Okyere, Salaam Jansbaka Adams, Musah Saeed Zango, Prosun Bhattacharya, and Shakir Ali

Contents

1 Introduction 299
2 Materials and Methods 300
  2.1 Study Area 300

E. D. Sunkari (✉) and M. B. Okyere
Department of Geological Engineering, Faculty of Geosciences and Environmental Studies, University of Mines and Technology, Tarkwa, Ghana

Department of Geology, Faculty of Science, University of Johannesburg, Johannesburg, South Africa
e-mail: edsunkari@umat.edu.gh; mokyere@uj.ac.za

S. J. Adams
School of Natural and Environmental Sciences, University of Environment and Sustainable Development, PMB – Somanya, Somanya, Eastern Region, Ghana
e-mail: sajansbaka@uesd.edu.gh

M. S. Zango
Department of Environment, Water and Waste Engineering, School of Engineering, University for Development Studies, Tamale, Ghana
e-mail: mzango@uds.edu.gh

P. Bhattacharya
Department of Sustainable Development, Environmental Science and Engineering, KTH Royal Institute of Technology, Stockholm, Sweden
e-mail: prosun@kth.se

S. Ali
Department of Geology, University of Delhi, Delhi, India

Shakir Ali and Abdelazim Negm (eds.), *Groundwater Quality and Geochemistry in Arid and Semi-Arid Regions*, Hdb Env Chem (2024) 126: 297–326, DOI 10.1007/698_2023_1054,

Published online: 21 December 2023

2.2 Sample Collection and Laboratory Analysis 302
2.3 Data Processing 303
2.4 Geospatial Mapping of Hydrochemical Parameters 303
2.5 Multivariate Statistical Analysis 303
2.6 Probabilistic Human Health Risk Assessment 304
3 Results and Discussion 305
3.1 Hydrogeochemistry 305
3.2 Spatial Distribution of Hydrochemical Parameters 308
3.3 Geochemical Evolution of Groundwater 309
3.4 Factors Controlling Groundwater Chemistry 310
3.5 Sources of Groundwater Fluoride Enrichment 311
3.6 Correlation Analysis 315
3.7 Factor Analysis (FA) and Hierarchical Cluster Analysis (HCA) 316
3.8 Human Health Risk Assessment 318
3.9 Sustainable Groundwater Fluoride Remediation 321
4 Summary and Conclusions 321
5 Recommendations 322
References 322

Abstract Fluoride (F^-) contamination in groundwater is a major challenge in many developing countries like Ghana. It is commonly observed that the heterogeneous nature of the geology plays a pivotal role in governing the fluoride mobilization in groundwater. Therefore, this study employed a multi-approach involving hydrogeochemistry and multivariate statistical analysis to investigate the geochemical controls on high F^- in the geologically heterogeneous Bongo District of Ghana. The study also assessed the probable human health risks associated with consuming F^- contaminated groundwater. The study revealed that the dominant water type identified in the area was Na-HCO_3-Cl with other mixed water types. The F^- levels ranged from 0.43 to 3.61 mg/L (average: 1.89 mg/L). Five principal components with eigen values >1.0 explaining a total variance of 88.8% were obtained from factor analysis indicating that both geogenic and anthropogenic sources control the groundwater chemistry. However, F^- mobilization in groundwater is largely because of weathering, ion exchange reactions and dissolution of F^- bearing minerals found in the aquifers. Further, geochemical modelling indicates that the groundwater is undersaturated with respect to calcite, dolomite, fluorite, gypsum, anhydrite, aragonite, halite and quartz. The human health risk assessment showed that children had higher hazard quotient values (0.61–5.11), implying they are more prone to dental fluorosis than the other age groups. Therefore, it is recommended that artificial recharge along with sustainable defluoridation techniques such as absorption, electrodialysis and precipitation should be adopted to remediate the high fluoride menace. In addition, cost-effective and Ghanaian-friendly natural coagulants/absorbents like *Moringa oleifera* and bone char can be promoted for usage at the household level to reduce the fluoride levels in groundwater prior to domestic usage.

Keywords Fluoride contamination in Ghana, Fluoride pollution, Fluorosis in Africa, Geochemical modelling, Non-carcinogenic health risk assessment

1 Introduction

Groundwater is the most significant source of water used for various purposes all over the world. Domestically, it is the most potable and widely used source of drinking water to approximately 95% of rural communities around the world, and thus its quality has a unique set of health benefits, necessitating a high level of focus [1–10].

However, groundwater quality is jeopardized by geogenic and anthropogenic activities that introduce harmful pollutants. The presence of relatively low or high concentrations of these pollutants is a major concern because it renders groundwater unfit for a variety of uses, including drinking. Fluoride is one pollutant that exists in an aqueous solution in an ionic form (F^-) [11, 12]. The lower amount of F^- in groundwater has many advantages, but it is also linked to some health problems. For example, F^- concentrations of at least 0.5 mg/L and less than 1.5 mg/L (WHO guideline) promote healthy bone and tooth growth, significantly reducing the risk of dental caries [11]. On the other hand, highly fluoridated water, above 1.5 mg/L, can significantly cause negative health effects and is evident as dental fluorosis and skeletal fluorosis, respectively, due to long-term consumption [13]. Numerous countries in Asia, Middle East and Africa are mostly associated with high amounts of F^- in groundwater [9, 14–20]. Long-term consumption of F^- contaminated water in most of these areas has resulted in frequent cases of dental fluorosis. For example, Ali et al. [21] and Kumar et al. [22] reported that parts of the Indo-Gangetic Alluvial plains in India have excessive levels of F^- leading to severe cases of bone deformities and dental fluorosis in children. Similarly, in the North East Region of Ghana, Zango et al. [20] recently reported high groundwater F^- concentrations up to 13.29 mg/L, which has significantly increased dental fluorosis in the region, particularly among children.

Bongo District, located in the Upper East Region of Ghana, is primarily dominated by distinctive granitic rocks termed the "Bongo Granite", which have been earlier reported to have high F^- levels [20, 23, 24]. There have been comprehensive researches previously conducted on the levels of F^- in the groundwater of the area. For instance, Atipoka [25], Firempong et al. [26], Alfredo et al. [23] and Sunkari and Abu [27] found high F^- levels between 0.30 and 5.84 mg/L in the groundwater of the Bongo District. Even though these researchers pointed out some significant levels of F^- in the study area, limited interpretations were provided regarding the effects of the underlying rocks on the groundwater, which could be a possible source of the high F^- concentration. There are also limited studies on the assessment of human health risks associated with drinking highly fluoridated water in the Bongo District. In view of this, the study documented the sources of high groundwater F^-

levels, identified geogenic and anthropogenic activities that control groundwater chemistry and assessed the probabilistic human health risks associated with drinking F^- contaminated water in the study area.

2 Materials and Methods

2.1 *Study Area*

2.1.1 Location, Topography and Climate

The Upper East Region of Ghana is divided into 15 districts, including the Bongo District. Bongo, the capital of the district, is located in the central part of the Upper East Region and lies between longitudes of 1°W and 0.62° W and latitudes of 10.83° N and 11.05°N. It has a total population of around 77,885 people and covers an area of about 460 sq.km [28]. The Bongo District is bordered on the west by Kassena-Nankana District, the south by Bolgatanga Municipality and the southeast by Talensi-Nabdam District (Fig. 1).

The Bongo District is typified by a single rainy season, which begins in May/June and ends in September. The average yearly rainfall is between 900 and 1,200 mm [29]. There is a significant dry season marked by cold, dry and dusty harmattan

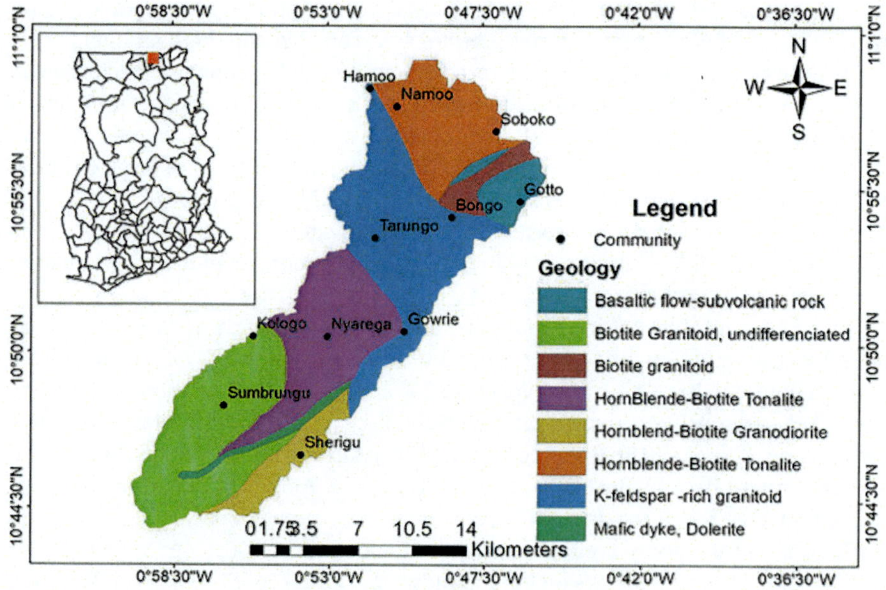

Fig. 1 Location and geological map of the study area

winds from November to mid-February. Temperatures range from as low as 14°C at night to more than 35°C during the day throughout this season. Humidity, on the other hand, is extremely low, making the high temperatures of the day more bearable.

2.1.2 Geology and Hydrogeology

The district is almost flat and low-lying with Paleoproterozoic Basin-type granitoids and Birimian volcanic rocks (Fig. 1). In some parts, the area is covered by phyllites, quartz, sericite schist, greywacke and associated late granitic intrusions. The granites cover the majority of the area and rise from 90 to 300 m above the surrounding terrain on a low, gently rolling relief [28, 30, 31].

The Paleoproterozoic Birimian Supergroup comprising of metasedimentary and metavolcanic rocks, which are intruded by the distinctive late intrusive Bongo granite suite are the dominant rock types in the study area (Fig. 1). The Birimian metavolcanic rocks in the area are predominantly made up of thick layers of basaltic and andesitic lavas, agglomerate beds, tuff and tuffaceous sediments. Low-grade metamorphism has altered the basic volcanics and pyroclastics, as evidenced by chlorite and epidote-bearing rocks that have been loosely clustered together as greenstones [32]. In some communities, the greenstones have been metamorphosed into amphibolites and hornblende schists referred to as Birimian metamorphic rocks due to dynamo-thermal metamorphism. The Bongo granite appears as fine-grained plutons with less quartz as it approaches the Bongo District, where it has locally evolved into biotite-hornblende syenite [33]. The abundance of alkalies in the Bongo granite suggests that it is an alkaline granite suite [34]. The southern and southeastern edges of the study area are covered with some sediments belonging to the Tarkwaian System. The Bongo granite is characterized by high alkali content suggesting that it is an alkaline granite suite [34]. These rocks are mostly made up of eroded Neoproterozoic clastic sediments and carbonates, exposing substantial volcaniclastics, Birimian metavolcanic rocks and belt-type granitoid intrusives [27]. Halite is a prevalent sublimation product in the Birimian metavolcanic rocks. The crystalline basement aquifers of West Africa include the Upper East Region. Basement aquifers are found within the weathered and fragmented bedrock in this location.

The occurrence of groundwater in the Bongo District is determined by the thickness of the overburden, the intensity of fractured bedrocks and the degree of litho-relic decomposition. Another kind of aquifer found in the area is a volcanic aquifer, which is primarily made up of lavas, tuffs and some undifferentiated metamorphic rocks from the Birimian volcanic suite [20]. Precipitation, streams and rivers are the main sources of groundwater recharge in the area. The average recharge rate is estimated to be nearly 4% of annual precipitation, corresponding to almost 40 mm per year. Apambire et al. [31] reported that the water level rise during the rainy season was in the range of 0.1–5.4 m, with an average of 1.8 m, over a three-year period. In most parts of the area, the wells have low yields (0.13–0.32 L/s), but they are suitable for installing handpumps [31]. Although lithologic data were unavailable for the current research, weathering profiles of similar geological terranes in the vicinity show the presence of extensively weathered clay-rich sediments near the surface [35]. There appears to be high amount of sand coupled with the low clay content as the depth increases to the basement granitic rocks. Overburden aquifers have thicknesses up to 100 m, but most wells are installed within the overburden at depth less than 30 m [35].

2.2 *Sample Collection and Laboratory Analysis*

A total of twenty (20) groundwater samples from boreholes in the study area were collected in sterilized 0.5 L polyethylene bottles. The samples were collected by means of a Klyen downhole stainless-steel borehole sampler. They were then placed in the sterilized polyethylene sample bottles, completely filled and tightly capped to avoid contamination or leakage [36]. Accurate labelling was provided for each sample bottle with the collection point identification number, time, date and sample designation and later stored in an ice chest at a temperature of 4°C to prevent chemical reactions in the water. The collected water samples were immediately transported to the Ghana Atomic Energy Commission Laboratory for hydrochemical analysis.

Physical parameters such as pH, temperature, total dissolved solids (TDS) and electrical conductivity (EC) of the samples were determined in situ with pH metres and a portable EC and HACH field titration kit. The samples were filtered using 0.45 µm membranes and major cation concentrations were determined from the filtered and acidified samples (1% v/v HNO3), whereas anion analysis was done from filtered and unacidified samples. A flame photometer was used to analyse $Na^+$and K^+ and EDTA titration was used to determine Ca^{2+} and Mg^{2+}. Dionex ICS 90 ion chromatography system was employed in determining the anions (Cl^-, SO_4^{2-}, HCO_3^-, F^- and NO_3^-) at the Ghana Atomic Energy Commission in Accra. All of the sampling bottles and glassware were immersed in a 10% nitric acid solution for a day and then cleaned with deionized water before use to ensure precision and a high level of confidence in the integrity of the data. Standard chemical solutions made from commercially available chemicals were used to calibrate the instruments, which were then confirmed using Standard Reference Materials (SRM) and Certified Reference Materials (CRM) (CRM). The SRM were examined at predetermined intervals to ensure that the procedure remained in a statistically controlled state. Duplicate samples were also tested and compared, with the results determined to be repeatable within a 5% error margin. The anion–cation balance method was used to assess the validity of the laboratory analysis once again, and only those values within 5% were used for subsequent interpretation.

2.3 Data Processing

Trilinear Piper diagram, Gibbs and bivariate plots using AquaChem software version 4.0 were employed in the processing and interpretation of the hydrochemical data. A study area map, a geological map and spatial distribution maps of the hydrogeochemical parameters were also generated using ArcGIS version 10.7.1 for a better interpretation of the dataset. Multivariate statistical analyses like principal component analysis (PCA) and hierarchical cluster analysis (HCA) were used to understand the interactions and linkages as well as determine the sources of the ions via SPSS Statistics version 25.

2.4 Geospatial Mapping of Hydrochemical Parameters

The inverse distance weighted (IDW) interpolation method was performed via ArcGIS software 10.7.1 in producing fluoride, sodium, total dissolved solids (TDS) and pH spatial distribution maps for the Bongo District. The area was divided into four primary classes using supervised classification with the maximum likelihood approach as the classifier. Inverse distance weighted (IDW) interpolation specifically assumes that variables that are close together are more similar than variables that are farther away [37]. IDW uses the measured values surrounding the prediction location to forecast a value for any unmeasured location. The measured values that are closer to the prediction location have a greater impact on the anticipated value than those that are farthest. IDW presupposes that the measured site has a local influence that decreases as distance increases. It gives more weight to sites that are closest to the predicted location, and the weights decrease as the distance increases [38].

2.5 Multivariate Statistical Analysis

The SPSS statistics software version 25 was used to perform all of the multivariate statistical methods used in this study. The raw datasets were transformed using the centred log-ratio (CLR) transformation to ensure that they are normally distributed, consistent and reliable [39]. The centred log-ratio transformation was performed using the equation below:

$$\text{CLR} = (\log(x_1/g(x))), \ldots, (\log(x_N/g(x))), \tag{1}$$

where x is the hydrochemical parameter, $g(x)$ is the geometric mean of the hydrochemical parameter x and $x_1 \ldots x_N$ are the concentrations of the individual hydrochemical parameters.

Factor analysis is commonly used to identify hidden dimensions that are difficult to interpret using direct analysis. PCA is a technique for reducing the size of a huge dataset while preserving as much information as possible [36, 40]. Understanding the interactions and relationships, as well as identifying the origin of the ions, required the application of Pearson's correlation and R-mode factor analysis. Principal component analysis was employed as the extraction method, while Varimax rotation was used as the rotation method in R-mode factor analysis. In R-mode factor analysis, the Kaiser criterion was used to minimize the number of factors to extract, permitting iteration on only five factors with eigenvalues greater than one [41].

Hierarchical cluster analysis (HCA) was used to group statistically distinct hydrochemical variables. It is an unsupervised pattern detection technique that creates clusters of variables based on their similarities. The spatial relationship between the sampling locations based on their chemical characteristics was determined using Q-mode cluster analysis [42]. In determining the similarity or dissimilarity of the water quality variables and linking the clusters created, the squared Euclidean distance and ward's linkage were utilized. The clusters obtained from the HCA was used to better understand the hydrogeochemical processes taking place in the district.

The PHREEQC software [43] was used to estimate the degree of saturation of the identified minerals phases (calcite, fluorite, aragonite, gypsum, halite, dolomite and quartz) in all the water sampled. The saturation indices (SI), which indicate the thermodynamic tendency of minerals to dissolve or precipitate were computed using the equation below:

$$\mathrm{SI} = \mathrm{Log}\left(\frac{\mathrm{IAP}}{K_{\mathrm{sp}}}\right) \tag{2}$$

where SI means saturation index, IAP is the ion activity product of the dissociated chemical species in solution, and K_{sp} is the equilibrium solubility product for the chemical involved at sample temperature. A positive SI value for a mineral implies saturation, which will result in precipitation, whereas a negative value suggests undersaturation, which will result in mineral dissolution.

2.6 *Probabilistic Human Health Risk Assessment*

The United States Environmental Protection Agency (USEPA) has recommended a quantitative system for determining the possible health risks posed by contaminants in water [44]. According to a previous study [45], people of different ages and genders are more sensitive to contaminants in water than others. Humans are mostly exposed to contaminants in water via oral consumption, dermal contact and inhalation. In the Bongo District the major pathway of contaminants in drinking water is through oral ingestion. Thus, non-carcinogenic risk via oral ingestion for various age groups (infants, children, teenagers and adults) was assessed.

In this study, the assessment of adverse health risks involved in the intake of F^- in drinking water by consumers was done by considering four age groups: infants (<2 years), children (2 to <8 years), teenagers (8 to <18 years) and adults (≥18 years) who were chosen from the population based on their physiological and behavioural changes similar to the studies of Zango et al. [20]. The daily exposure of the various age groups to F^-in drinking water was calculated using the following formula [44]:

$$\text{EDI} = \frac{C_{(f)} \times C_d}{B_w} \tag{3}$$

where EDI expressed in mg/kg/day represents the estimated daily intake; C_f is the concentration of F^- in drinking water; C_d is the daily average consumption of drinking water and B_w represents body weight. The B_w was acquired from community health centres whereas C_d was determined using data from a questionnaire sent out to the target groups. The various age groups were provided C_d and B_w values as follows: infants (0.1 L/day and 5 kg), children (0.85 L/day and 10 kg), teenagers (2 L/day and 50 kg) and adults (2.5 L/day and 75 kg), respectively. The USEPA [44] hazard quotient model was used to calculate the HQ of non-carcinogenic risk of F^- exposure to humans via drinking water using the equation below:

$$\text{HQ} = \frac{\text{EDI}}{\text{R}f\text{D}} \tag{4}$$

R*f*D in the equation above indicates the reference dose of F^- by a given exposure pathway in mg/kg/day. This study used R*f*D of F^- (0.06 mg/kg/day) which was taken from the records of the Integrated Risk Information System (IRIS) provided by the United States Environmental Protection Agency (USEPA). In effect, a population will be exposed to the adverse health effects of F^- when HQ value is (>1) but a population will not be prone to non-carcinogenic health risk of F^- when HQ value is (<1).

3 Results and Discussion

3.1 *Hydrogeochemistry*

The summary statistics of the hydrogeochemical parameters is provided in Table 1, whereas the concentrations of all the major ions are shown in Fig. 2. The pH values range from 7.08 to 7.73 and an average value of 7.33, which is within the WHO recommended range of 6.5 to 8.5 [46]. Such a pH range is an indication of almost neutral to closely alkaline water. The moderately alkaline pH can be attributed to the

Table 1 Summary statistics of the hydrochemical parameters in groundwater of the study area (SAGLV = Samples Above Guideline Values)

Parameter	Unit	No.	Min.	Max.	Average	St. Dev.	WHO [46]	SAGLV
pH		20	7.08	7.73	7.33	7.79	6.5–8.5	None
Temp.	°C	20	24.2	30.6	29.6	1.81	NA	
EC	μS/cm	20	272	1,352	481	232	2,500	None
TDS	mg/L	20	150	744	265	128	NA	None
Na^+	mg/L	20	20.6	210.8	48.5	41.2	200	1
K^+	mg/L	20	0.6	3.1	1.89	0.63	200	None
Mg^{2+}	mg/L	20	1.25	23.3	8.57	5.16	NA	
Ca^{2+}	mg/L	20	1.6	26.8	9.32	6.5	200	None
Cl^-	mg/L	20	14.7	178	42	33.8	250	None
SO_4^{2-}	mg/L	20	2.89	18.3	10.4	3.88	200	None
HCO_3^-	mg/L	20	60	260	103	50.8	NA	
NO_3^-	mg/L	20	3.29	40.1	15.5	11.6	50	None
F^-	mg/L	20	0.43	3.61	1.89	0.88	1.5	15
SI anhydrite		20	−4.43	−2.95	−3.62	0.38		
SI aragonite		20	−2.02	−0.6	−1.26	0.36		
SI calcite		20	−1.88	−0.45	−1.12	0.36		
SI dolomite		20	−3.03	−0.58	−1.87	0.61		
SI fluorite		20	−2.38	−0.77	−1.52	0.41		
SI gypsum		20	−4.21	−2.76	−3.42	0.37		
SI halite		20	−8.02	−6.02	−7.4	0.41		
SI quartz		20	−1.14	−0.83	−0.96	0.08		

dissolution of carbonates and clay formed from metasedimentary rocks, which act as a suitable medium for metals and harmful ions like fluoride adsorption and desorption in groundwater [20]. The temperature varies from 24.2 to 30.6°C and an average of 29.6°C. The values of EC also range from 272 to 1,352 μS/cm with an average of 481 μS/cm falling within the 2,500 μS/cm guideline value [46]. TDS concentrations (150 to 744 mg/L, mean 265 mg/L) were likewise generally low, falling below the WHO guideline range of 1,000 mg/L. The generally low TDS concentrations found in the groundwater of the area suggest that the water can be described as fresh. Again, the low TDS levels imply a rapid time of groundwater occurrence, which limits significant water–rock interaction [32].

Na^+ is the dominating cation in terms of major ion concentrations, with varying concentrations from 20.6 to 210.8 mg/L and an average concentration of 48.5 mg/L. (Table 1). Ca^{2+}, varying from 1.6 to 26.8 mg/L with an average of 9.32 mg/L is the next cation dominating the groundwater, just within the 200 mg/L permissible limit (Table 1). The Mg^{2+} also has concentrations ranging from 1.25 to 23.3 mg/L and a mean value of 8.57 mg/L (Table 1). Among all the cations, K^+ is the least dominated in the groundwater of the area, having concentrations in the range of 0.6 to 3.1 mg/L and a mean value of 1.89 mg/L, which is extremely low relative to the standard guideline value of 200 mg/L (Table 1).

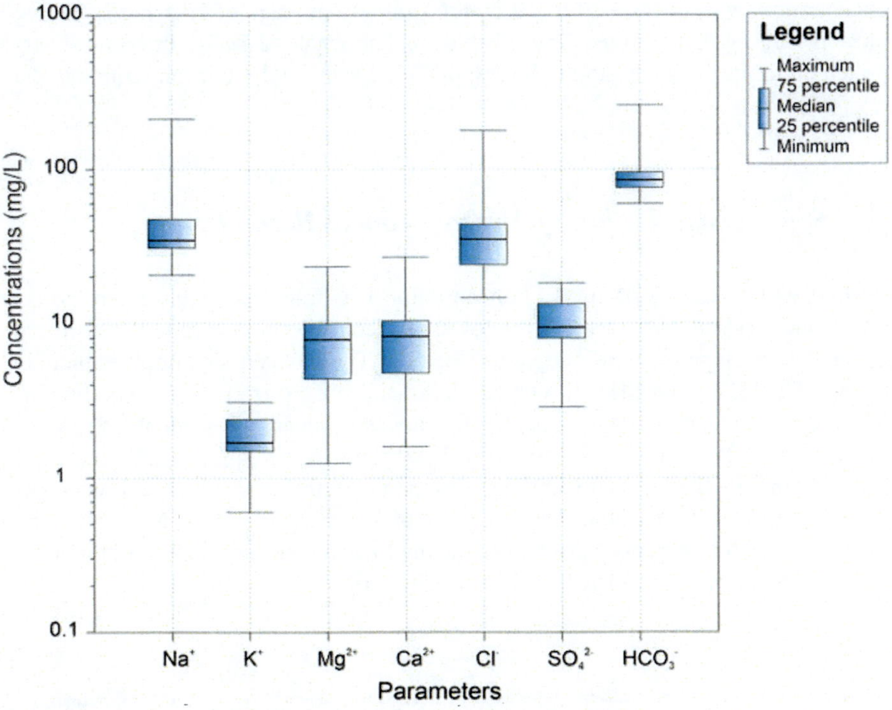

Fig. 2 Box and whisker plot of the major ions in the groundwater

HCO_3^- is the most prevalent major anion found in the water of the area having concentrations ranging from 60 to 260 mg/L with a mean value of 103 mg/L (Table 1). This suggests that carbonate minerals like calcite may have a limited dissolving rate in the groundwater aquifer of the study area (Fig. 1). The second dominant anion is Cl^- which has concentrations in the range of 14.7 to 178 mg/L and an average value of 42 mg/L, being within the 250 mg/L guideline value recommended by WHO (Table 1). Similarly, the SO_4^{2-} concentrations range from 2.89 to 18.3 mg/L with an average value of 10.4 mg/L, all of which fall below the WHO maximum permissible limit of 200 mg/L (Table 1). This could be due to reactions of ion exchange or limited gypsum dissolution in the groundwater.

Anions such as F^- and NO_3^- are of importance in groundwater studies due to adverse effects that could be induced on the health of consumers when taken in higher amounts. The groundwater F^- concentrations are in the range of 0.43 to 3.61 mg/L with an average of 1.89 mg/L (Table 1). This shows that the drinking water is contaminated with F^- because its concentration is above the 1.5 mg/L permissible limit [46]. The average concentration (1.89 mg/L) even outstrips the 1.5 mg/L guideline value and might put people at risk of dental fluorosis. The samples with higher F^- concentrations are generally found in the hornblende-biotite granitoid and K-feldspar-rich granitoid-dominated areas (Fig. 1). The NO_3^-

concentrations vary from 3.29 to 40.1 mg/L with an average of 15.5 mg/L, falling below the 50 mg/L permitted limit (Table 1). The leaching from nitrate fertilizers applied on farms are suggested to be responsible for the NO_3^- concentrations in the area [40].

3.2 *Spatial Distribution of Hydrochemical Parameters*

The highest F^- (2.82–3.61 mg/L) concentration was found in the northern, north-eastern and western parts whereas the lowest F^- concentration is found in northern and southwestern parts of the study area (Fig. 3a). The affected communities include Balungo CHPS, Bongo SHS 1, Vea Kupielga and their surroundings. The highest Na^+ is found in the northern whereas the lowest concentration identified in the western and eastern parts of the area (Fig.3b). Similarly, the highest pH concentration is found in the northern and eastern parts of the area but the lowest pH concentration is found in the western parts of the area (Fig. 3c). Moreover, the highest TDS is found in northern part while the lowest concentration is found in the western and eastern parts of the study area (Fig. 3d).

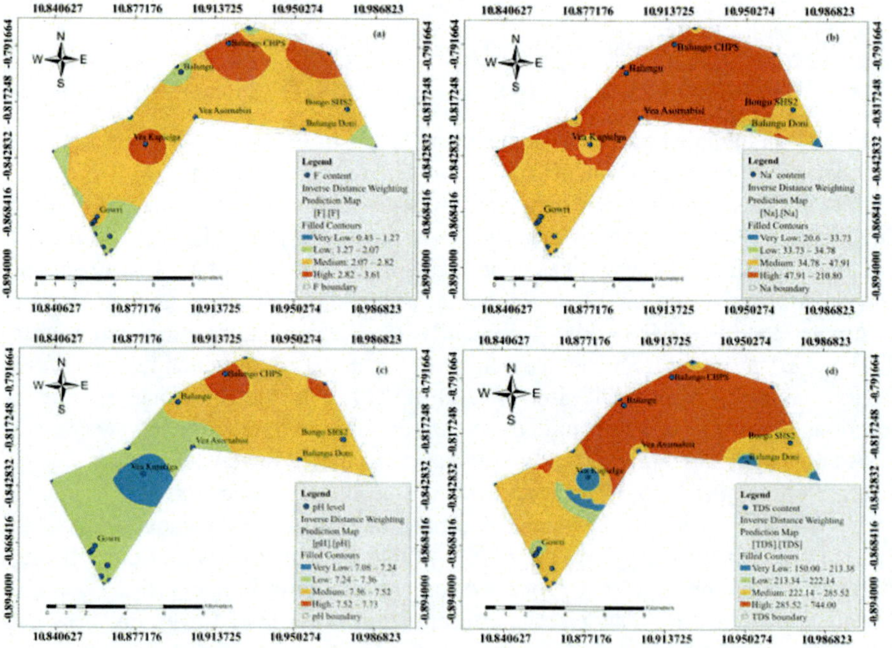

Fig. 3 Trilinear piper diagram showing the water types in the aquifers of the area

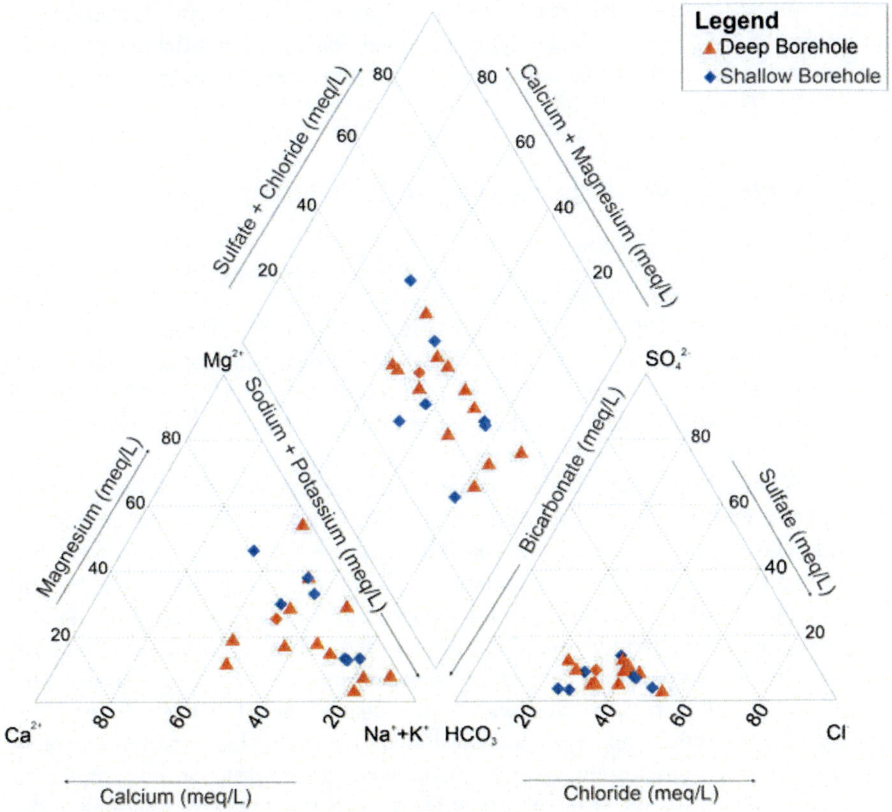

Fig. 4 Spatial distribution maps of (**a**) Na^+, (**b**) F^-, (**c**) pH and (**d**) TDS in the study area

3.3 *Geochemical Evolution of Groundwater*

In groundwater studies, the trilinear Piper diagram [47] is commonly used to discern between different hydrochemical facies and the geochemical evolution of groundwater in aquifer systems. As a result, the type of water in the Bongo District was depicted using a Piper diagram in this study. The principal hydrochemical facies evolve from (Na–Ca–HCO_3 (35%), Na-Mg-HCO_3-Cl (25%), Na-Mg-Ca-HCO_3-Cl (10%), Na-Ca-HCO_3-Cl (10%), Na-Ca-Mg-HCO_3 (5%), Na-Cl-HCO_3 (5%) to Mg-Na-Cl-HCO_3 (5%) (Fig. 4). The Na-Ca-HCO_3 is the most abundant water type in the aquifer which could be attributed to ion exchange reactions and weathering of silicates (Fig. 4). Other water types found in the area may have a similar explanation. The hydrochemistry of some parts of the Bongo District was studied by Zango et al. [20], who found Na-Ca-HCO_3 (85%) as the major type of water in the area. Similarly, the Na-Ca-HCO_3 (85%) was related to the exchange reaction of ions and weathering of silicate minerals, which corroborates the results of this study. The fractured bedrocks of the Birimian metavolcanics and Bongo granitoids serve as a setting for the Na-Ca-HCO_3 water type. This may indicate that silicate weathering and ion exchange are the principal controls of groundwater chemistry in the area.

3.4 Factors Controlling Groundwater Chemistry

Water–rock interaction between groundwater and the underlying rocks may facilitate the release of Na^+ via the dissolution of silicate minerals such as plagioclase feldspars [9, 31, 48, 49]. On Gibbs plots (Fig. 5), the studied samples plot in the rock weathering dominance zone, indicating that water-rock interaction controls the chemistry of the groundwater. Calvi et al. [48] reported that Ca^{2+} and Mg^{2+} always get replaced by Na^+ via cation exchange reaction. The elevated concentrations of Na^+ (an average of 48.5 mg/L) associated with smaller mean values of Ca^{2+} (9.32 mg/L) and Mg^{2+} (8.57 mg/L) (Table 1) may be supported by a likely strong cation exchange reaction that occurred between Ca^{2+} and Na^+ due to the absence of Na-plagioclase-bearing rocks in the area. The bivariate plot of Na^+ against HCO_3^- implies that there is no significant amount of silicate weathering in the area, but the ion exchange reaction is responsible for the enrichment of Na^+in the groundwater chemistry since all the samples are plotting below the 1:1 line (Fig. 6a). Similarly, on the bivariate plot of the Na^+ versus Cl^-, a good percentage of the water samples plot on the 1:1 line (Fig. 6b) which points to the dissolution of halide whereas few samples plotting above and below the equiline indicating both forward and reverse ion exchange reactions for $Na^+$and Cl^-. Moreover, the majority of the samples on the bivariate plot of (Ca^{2+} +Mg^{2+}) against (SO_4^{2-} +HCO_3^-) (Fig. 6c) fall below the equiline suggesting the influence of ion exchange reactions for the enrichment of the

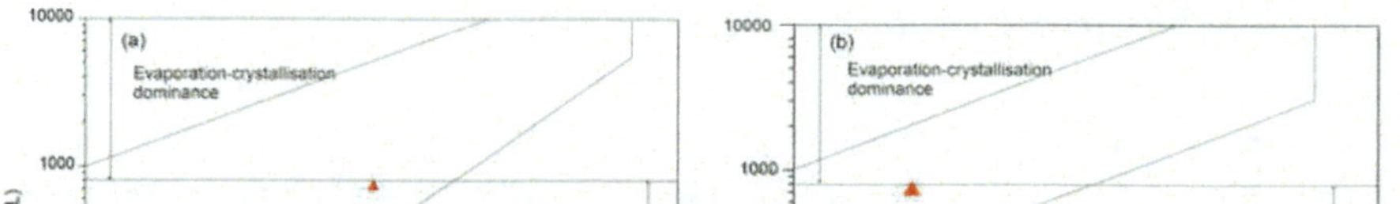

Fig. 5 Gibbs plots showing the mechanisms governing groundwater chemistry

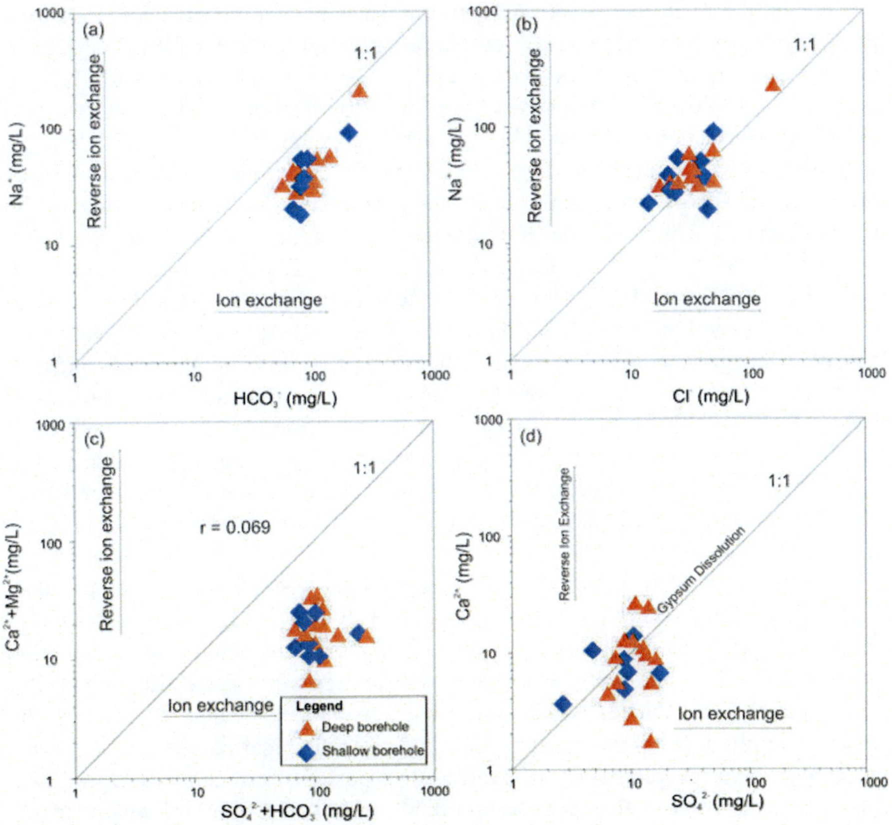

Fig. 6 Bivariate plots (**a**) Ca^{2+}+ Mg^{2+}vs SO_4^{2-} + HCO_3^-, (**b**) Ca^{2+}vs SO_4^{2-}, (**c**) Na^+ vs Cl^- and (**d**) Na^+ vs HCO_3^- showing the major ion relationships

ions in the aquifer. On the bivariate plot of Ca^{2+} versus SO_4^{2-}, some samples plot very close to the 1:1 line indicating the influence of gypsum dissolution whereas most samples plot away from the 1:1 line reflecting both forward and reverse ion reactions (Fig. 6d) and other contributing factors enriching sulphate concentration such as decomposition of leachable sulphates in fertilizers [50].

3.5 *Sources of Groundwater Fluoride Enrichment*

A significant percentage (75%) of the Bongo District boreholes contain F^- levels above the 1.5 mg/L maximum threshold due to their intersection with biotite-rich granitoids (Fig. 6). This compromises the health of inhabitants who rely on groundwater for their daily needs. Alkaline condition (pH = 7.08 to 7.73) is suggested to have played a role in the elevation of groundwater F^-. In recent reports, alkaline settings are known to influence F^- mobilization in groundwater [20]. The Bongo District has many clay minerals; therefore, the alkaline pH associated with the clay might be an influencing mechanism for the mobilization of groundwater F^-. According to Keshavarzi et al. [51], in alkaline water, F^- has a tendency for adsorption on clay mineral surface. This suggests that when alkaline conditions are met, the F^- adsorption from the clay mineral surfaces could be accounting for the increased amounts of F^- in the groundwater in the area. The positive correlation ($r = 0.11$) between F^- and pH confirms this (Table 2). Eventually, the elevated levels of Na^+, pH and HCO_3^- play a role in the enrichment off via similar processes highlighted in the abovementioned Gibbs diagrams, where water–rock interaction dominates the geochemical processes influencing the water in the study area. A similar process is found in semi-arid conditions, where ion exchange interactions between Ca^{2+} and Na^+ result in Ca^{2+}being reduced while Na^+ andF^- are enriched [52]. The abundance of Na-bearing plagioclase indicates that a probable cation exchange process occurred between Ca^{2+} and Na^+ (Fig. 1) and the high content of Na^+ with an average of 48.5 mg/L comparative to the average value of Ca^{2+} (9.32 mg/L) (Table 1).

Moreover, the positive correlations identified between F^- and perhaps most hydrochemical parameters, such as pH, EC, TDS, K^+, Na^+, Mg^{2+}, Cl^- and HCO_3^- (Table 2) suggest that the mechanisms enriching these ions could also enhance the enrichment of F^- in the water. Nevertheless, F^- has a negative correlation with Ca^{2+}, NO_3^- andSO_4^{2-} (Table 2) which infers a reverse ion exchange reaction in the aquifer. The relationship of F^- with other hydrogeochemical parameters was demonstrated on bivariate plots (Fig. 7). pH has a low positive correlation (0.11) with F^- (Table 2). This implies that pH has minimal control over the enrichment of F^- concentration in the aquifer. Similarly, Na^+ low positive correlation with F^- (0.43), indicating that Na plays a minimal role in the enrichment of F^- in the aquifer system of the study area (Table 2). Additionally, the correlation between HCO_3^- and F^- (0.38) (Table 2) suggests HCO_3^- plays a minor role in the enrichment of F^- in the water. SO_4^{2-} correlates negatively with F^- (−0.07) (Table 2), implying that the ions are not enriched by a common source.

Geochemical modelling was employed to understand the geochemical processes that enrich ions in water. A thermodynamic technique was used to calculate the saturation indices (SI) of mineral phases observed in the water. Some of the minerals found in the water are calcite (SI = −1.88 to 0.45, average = −1.12), dolomite (SI = −3.03 to −0.58, average = −1.87), fluorite (SI = −2.38 to −0.77, average = −1.52), gypsum (SI = −4.21 to −2.76, average = −3.42) and halite (SI = −8.02 to −6.02, average = −7.40) aragonite (SI = −2.02 to −0.60), anhydrite (−4.43 to −2.95, average = −3.62), quartz (SI = −1.14 to −0.83, average = −0.96), (Table 1) and this is an indication of undersaturation of the samples with respect to these minerals. On the SI calcite versus F^- (mg/L) plot, all the samples are undersaturated with regard to calcite (Fig. 8a). This means that calcite takes a longer time to dissolve into the water and thus, its minimal

Table 2 Pearson's correlation matrix of the hydrochemical parameters (values in bold indicate correlation coefficients ≥0.7)

	pH	Temp.	EC	TDS	Na^+	K^+	Mg^{2+}	Ca^{2+}	Cl^-	SO_4^{2-}	HCO_3^-	NO_3^-	F^-
pH	1.00												
Temp	−0.08	1.00											
EC	0.39	−0.02	1.00										
TDS	0.39	−0.02	**1.00**	1.00									
Na^+	0.40	−0.05	**0.90**	**0.90**	1.00								
K^+	−0.11	−0.06	0.12	0.12	0.13	1.00							
Mg^{2+}	0.07	−0.08	0.18	0.18	−0.14	0.31	1.00						
Ca^{2+}	−0.05	0.16	−0.15	−0.15	−0.19	−0.43	−0.32	1.00					
Cl^-	0.25	0.19	**0.86**	**0.86**	**0.79**	0.33	0.23	−0.29	1.00				
SO_4^{2-}	0.37	0.53	0.37	0.37	0.38	−0.05	0.05	0.17	0.35	1.00			
HCO_3^-	0.47	−0.26	**0.95**	**0.95**	**0.85**	0.12	0.17	−0.21	**0.72**	0.19	1.00		
NO_3^-	−0.19	0.47	−0.10	−0.10	−0.22	−0.19	−0.12	0.60	−0.15	0.18	−0.23	1.00	
F^-	0.11	−0.40	0.26	0.26	0.43	0.59	0.03	**−0.72**	0.27	−0.07	0.38	**−0.73**	1.00

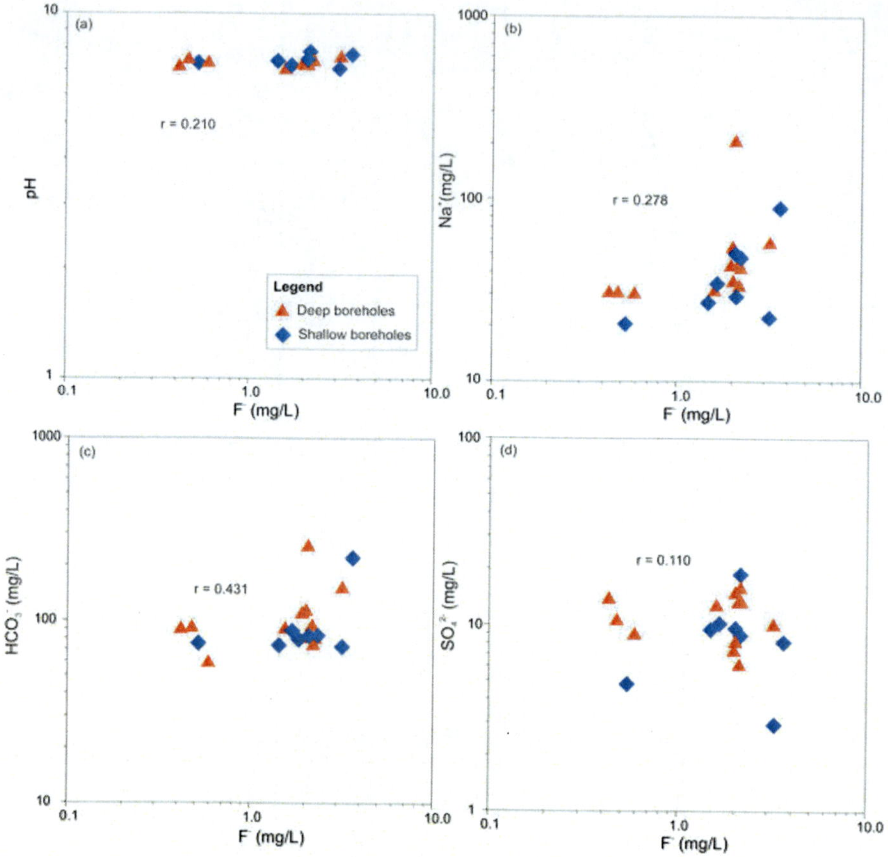

Fig. 7 Bivariate plots **(a)** pH vs F^-, **(b)** Na^+vs F^-, **(c)** HCO_3^- vs F^- and **(d)** SO_4^{2-} vs F^- showing the relationship of F^- with major ions

contribution to the groundwater chemistry. This also further suggests that the undersaturation of the samples could be due to groundwater evolving under alkaline settings. Thereby, enhancing preferential release of F^- and gradual removal of Ca^{2+} in solution. Similarly, the samples are all found in the dolomite undersaturation zone in the SI dolomite versus F^- (mg/L) plot (Fig. 8b). This suggests that dolomite has a minimal contribution to groundwater chemistry. Moreover, on the bivariate plot of fluorite versus F^-, all the samples are undersaturated with respect to fluorite (Fig. 8c). Even though the samples are undersaturated, there is a significant correlation coefficient, which means that there may be remobilization that facilitated the dissolution of fluorite in the water. Therefore, fluorite partly plays a role in the enrichment of F^- in groundwater. However, on the bivariate plot of fluorite versus calcite, all the samples fall with both fluorite and calcite undersaturation zone

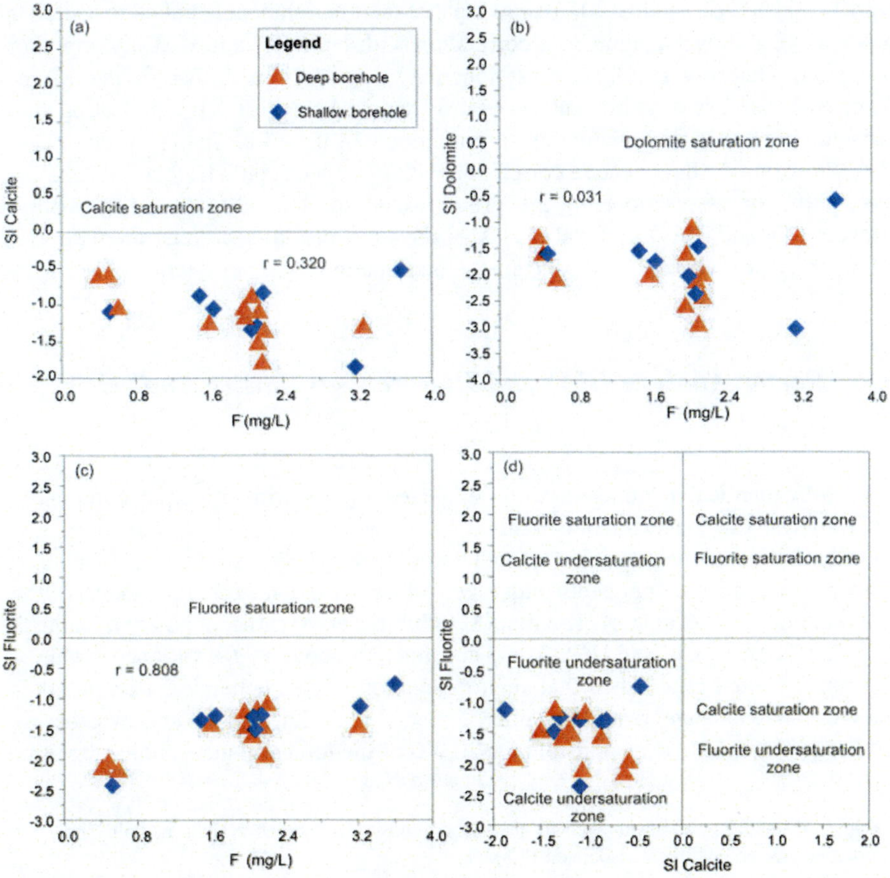

Fig. 8 Bivariate plots showing the saturation indices of mineral pses (**a**) SI calcite vs F^-, (**b**) SI dolomite vs F^-, (**c**) SI fluorite vs F^- and (**d**) SI fluorite vs SI calcite

(Fig. 8d). This suggests that both elements have minimal contributions to groundwater chemistry.

3.6 *Correlation Analysis*

The association between the physicochemical parameters and the major ions employed in this study was shown using Pearson's correlation (Table 2). Correlations with $0.30 \leq p \geq 0.50$ were considered low, $0.50 \leq p \geq 0.70$ were considered as strong whereas those with $p>0.70$ were considered significant correlations. The Pearson's correlation table (Table 2) revealed that EC has a significant correlation with TDS, Na^+, Cl^- and HCO_3^-, showing that their enrichment is linked to a similar process. TDS shows a significant correlation with Na^+, Cl^- and HCO_3^-, implying that the mechanisms enriching these ions may come from a similar source. Moreover, Na^+ displays a significant correlation with Cl^- and HCO_3^- indicating that common process is responsible for the enrichment of these ions in the aquifer. Also, Ca^{2+} shows strong but negative correlation with F^-, which could be that Ca^{2+} and F^- have different associations in the groundwater aquifer. The strong correlation between $Cl^-$and HCO_3^-. Lastly, NO_3^- shows a negative correlation with F^-, which suggests dominance of geogenic contamination.

3.7 *Factor Analysis (FA) and Hierarchical Cluster Analysis (HCA)*

The hydrochemical parameters were subjected to FA with principal component (PC) analysis as the extraction method in order to show their relationships and identify the factors responsible for each association. The Kaiser criterion was used to extract five main components with eigenvalues >1.0 that explained 88.8% of the total variance [53] (Table 3). The first PC exhibits very significant positive loadings on EC, TDS, Na^+, Cl^- and HCO_3^- and accounts for about 35.7% variance (Table 3; Fig. 9). It is well established that such interaction of hydrochemical parameters is due to ion exchange reactions, intense water–rock interaction and weathering [54]. The second PC is responsible for 20.9% of the total variance and has positive

Table 3 Principal component analysis for the groundwater samples (values in bold show the various parameters and their associated PCs)

	PC 1	PC 2	PC 3	PC 4	PC 5
pH	0.306	0.073	0.091	**0.843**	0.069
Temp	−0.054	−0.233	**0.893**	−0.153	−0.023
EC	**0.982**	0.044	0.058	0.118	0.092
TDS	**0.983**	0.046	0.057	0.117	0.092
Na^+	**0.913**	0.209	0.086	0.145	−0.259
K^+	0.142	**0.624**	0.146	−0.461	0.292
Mg^{2+}	0.080	0.120	−0.022	0.033	**0.975**
Ca^{2+}	−0.087	**−0.835**	0.051	0.028	−0.230
Cl^-	**0.856**	0.208	0.270	−0.085	0.155
SO_4^{2-}	0.286	−0.074	**0.782**	0.372	0.002
HCO_3^-	**0.930**	0.126	−0.182	0.208	0.077
NO_3^-	−0.043	**−0.775**	0.326	−0.312	0.038
F^-	0.242	**0.917**	−0.159	−0.004	−0.122
Initial eigen values	5.263	2.770	1.454	1.059	1.003
Percentage of variance	35.654	20.887	13.025	9.869	9.416
Cumulative % of variance	35.654	56.530	69.555	79.424	88.840

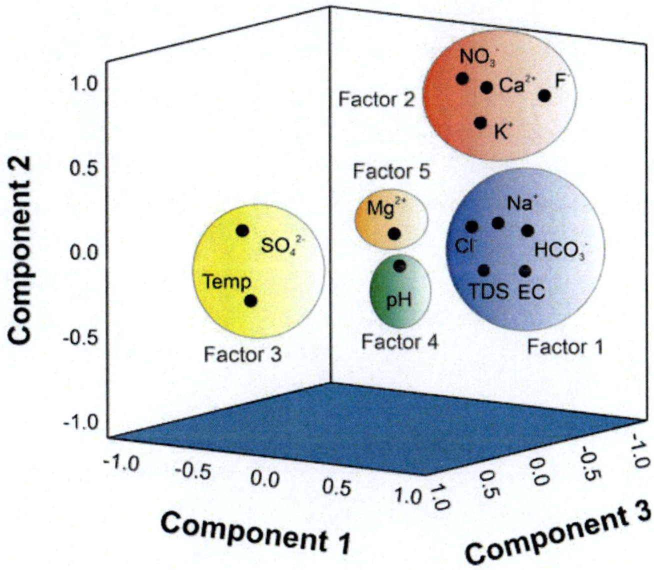

Fig. 9 Principal components extracted from R-mode factor analysis rotated in space using varimax rotation

correlation with Ca^{2+}, K^{+}, NO_3^- and F^- (Table 3; Fig. 9). In the aquifer system, this type of relationship also defines the transition from geogenic to anthropogenic inputs. Dissolution of fluoride-bearing minerals such as fluorite and biotite and ion exchange reactions in the area account for the Ca^{2+}, K^{+} and F^- loading in this PC [50]. However, as the people are mostly farmers, the NO_3^- associated with this PC might be owing to input from agricultural activities contributing to groundwater chemistry by application of nitrate fertilizers on farmlands. The third PC, which accounts for 13.0% of the total variance loads positively on SO_4^{2-} and temperature (Table 3; Fig. 9). This PC further highlights inputs from anthropogenic sources such as the application of sulphate fertilizers on farmlands in the area. The fourth and fifth principal components account for 9.87% and 9.42% of the total variance and correlate positively with pH and Mg^{2+}, respectively (Table 3; Fig. 9). These PCs are subcomponents of PC three, which indicates the role pH plays in mobilization of ions in the groundwater through anthropogenic activities. However, the Mg^{2+} loading may be pointing to inputs from domestic sewage and poor sanitation around the aquifers [55].

The hierarchical cluster analysis (HCA) highlighted three major clusters, with parameters in the same cluster originating from similar processes (Fig. 10). Cluster 1 contains pH, EC, TDS, Na^{+}, Mg^{2+}, Cl^-, SO_4^{2-} and HCO_3^- which corroborates the ion association in factor 1 in the FA with the exception of SO_4^{2-}, pH and Mg^{2+}which confirms the elements in factor 3, factor 4 and factor 5, respectively. This cluster may be as a result of geogenic processes such as water–rock interaction, ion exchange

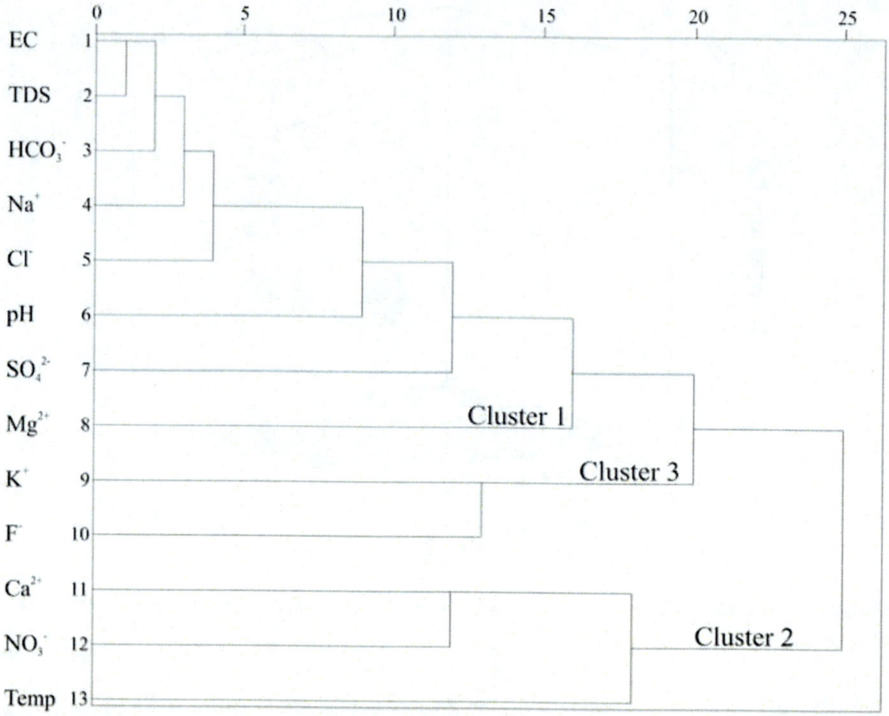

Fig. 10 Dendrogram using average linkage (within groups) criterion to illustrate the clustering behaviour of the major ions

reaction or dissolution of minerals but the SO_4^{2-} may be an input from anthropogenic sources such as the application of sulphate fertilizers in the area [27]. Cluster 2 includes Ca^{2+}, NO_3^- and temperature (Fig. 10), with Ca^{2+} and NO_3^- confirming the element association in factor 2 in the FA except temperature which is found in factor 3. This cluster may be because of both geogenic and anthropogenic processes. Ca^{2+}highlights the geogenic source whereas NO_3^- highlights the inputs from anthropogenic sources such as the use of nitrate fertilizers [56]. The temperature also plays a role in the enrichment of both elements in the water. Cluster 3 includes K^+ and F^- which corroborates the element association in the factor 2 in the FA. This cluster is mainly because of geogenic processes. Certainly, the results of the FA are consistent with that of the HCA.

3.8 *Human Health Risk Assessment*

In this study, non-carcinogenic risk of high F^- in drinking water was assessed. The daily F^- intake values for inhabitants of different ages are estimated (EDI), including

Table 4 Estimated daily intake (EDI) and hazard quotient (HQ) of F^- for four age groups of water consumers in the study area (bold values indicate safety limits for drinking)

Sample	Location		Estimated daily intake (EDI)				Hazard quotient (HQ)			
		F^-	Infants	Children	Teenagers	Adults	Infants	Children	Teenagers	Adults
OBM 1	Bongo	1.49	0.03	0.13	0.06	0.05	0.50	2.11	0.99	0.83
OBM 2	Namoo Amokobisi1	2.02	0.04	0.17	0.08	0.07	0.67	2.86	1.35	1.12
OBM 3	NamooYakabisi	2.00	0.04	0.17	0.08	0.07	0.67	2.83	1.33	1.11
OBM 4	BalunguDoni	2.07	0.04	0.18	0.08	0.07	0.69	2.93	1.38	1.15
OBM 5	Balungu	2.06	0.04	0.18	0.08	0.07	0.69	2.92	1.37	1.14
OBM 6	Bongo Gurogo	0.48	0.01	0.04	0.02	0.02	0.16	0.68	0.32	0.27
OBM 7	SamboliguAyeopia	0.43	0.01	0.04	0.02	0.01	0.14	0.61	0.29	0.24
OBM 8	Bongo Nabisi	1.65	0.03	0.14	0.07	0.06	0.55	2.34	1.10	0.92
OBM 9	ZokoTarongo	1.60	0.03	0.14	0.06	0.05	0.53	2.27	1.07	0.89
OBM 10	NamooAmokobisi	2.13	0.04	0.18	0.09	0.07	0.71	3.02	1.42	1.18
OBM 11	Gowri	2.16	0.04	0.18	0.09	0.07	0.72	3.06	1.44	1.20
OBM 12	VeaAkugrebisi	2.13	0.04	0.18	0.09	0.07	0.71	3.02	1.42	1.18
OBM 13	VeaGunga	2.11	0.04	0.18	0.08	0.07	0.70	2.99	1.41	1.17
OBM 14	VeaAsornabisi	2.10	0.04	0.18	0.08	0.07	0.70	2.98	1.40	1.17
OBM 15	VeaKuplpeliga	0.52	0.01	0.04	0.02	0.02	0.17	0.74	0.35	0.29
OBM 16	VeaNzangongo	0.59	0.01	0.05	0.02	0.02	0.20	0.84	0.39	0.33
OBM 17	Balungo CHPS	3.61	0.07	0.31	0.14	0.12	1.20	5.11	2.41	2.01
OBM 18	Bongo SHS 1	3.20	0.06	0.27	0.13	0.11	1.07	4.53	2.13	1.78
OBM 19	Bongo SHS 2	2.18	0.04	0.19	0.09	0.07	0.73	3.09	1.45	1.21
OBM 20	VeaKupielga	3.18	0.06	0.27	0.13	0.11	1.06	4.51	2.12	1.77
Safety limits		**1.50**	**0.03**	**0.13**	**0.06**	**0.05**	**1.00**	**1.00**	**1.00**	**1.00**

infants (<2 years), children (2 to <8 years), teenagers (8 to <18 years) and adults (≥18 years) in various communities. The calculation of hazard quotient (HQ) was done to identify the level of risk posed to the various age groups for drinking F^- contaminated water (Table 4). As compared to other age groups, children are regarded as the most hypersensitive group in the study area since they have the highest exposure of F^- intake. However, infants are exposed to the intake of higher amounts of F^- (0.04–0.07 mg/kg/day) in 13 communities, exceeding the required safe limit (0.03 mg/kg/day). Fluoride is observed at significant concentrations in the drinking water for infants in the Vea Kupielga, Bongo SHS 1 and Balungo CHPS communities (EDI = 0.06, 0.06 and 0.07 mg/kg/day, respectively). Furthermore, 15 communities, with the exception of five that show minor exposure (0.00–0.13 mg/kg/day) are exposed to high amounts of fluoridated water for children as compared to the 0.13 safe limit. Children from the Bongo SHS 1 (0.27 mg/kg/day), Vea Kupielga community (0.27 mg/kg/day) and Balungo CHPS community (0.31 mg/kg/day) have the highest concentrations of F^- exposure in drinking water. Teenagers show low exposure levels of F^- (0.00–0.06 mg/kg/day) in 6 communities however, the remaining 14 communities have elevated exposure levels reaching an extreme of 0.14 mg/kg/day. The highest exposure is seen in Balungo CHPS community. Similarly, the adult population is exposed to minor F^- levels in 6 communities (0.00–0.05 mg/kg/day), while the remaining 14 communities have greater F^- exposure levels (0.06–0.12 mg/kg/day) when compared to the 0.05 mg/kg/day safe limit. Again, higher F^- exposure levels for adults are observed in Vea Kupielga (0.11 mg/kg/day), Bongo SHS 1 and Balungo CHPS (0.12 mg/kg/day) communities. Consequently, amounts above 0.3 mg/ kg/day may cause acute F^- poisoning [57]. So, in this study, children from only one community are exposed to acute F^- poisoning. It is worth mentioning that children in the study area who consume water containing higher amounts of F^- are also more susceptible to dental and skeletal fluorosis. Higher F^- exposure is shown in most of the population above the 0.122 mg/kg/day maximum threshold suggested by the Health Canada. Again, children show higher levels of F^- exposure, which exceeds the (0.06 mg/kg/day) guideline value of the USEPA even though other age groups have higher exposure too.

The findings of this study are comparable to those of Tekle-Haimanot et al. [58], Huang et al. [59], and Zango et al. [20] who found evidence of potential F^- exposure risk in Ethiopia, China and Ghana, respectively, for children. Most of the communities in the study area have HQ > 1, inferring that the residents are exposed to the health risks accompanying high F^- consumption (Table 4). The children age group show higher HQ values (0.61–5.11) as compared to the other age groups (Table 4). Higher HQ values ≥3 are observed in Namoo Amokobisi, Gowri, Vea Akugrebisi, Bongo SHS 2, Vea Kupielga and Balungo CHPS communities (Table 4). The cause of the affinity of children for non-carcinogenic risk may be because of their moderate body weight (B_w) [59]. The non-carcinogenic risk of F^- in all the communities studied is as follows: children > teenagers > adults > infants (Table 4), showing that children are mainly susceptible to fluorosis.

3.9 Sustainable Groundwater Fluoride Remediation

Curtailing the issue of dental fluorosis in the Bongo District is becoming a quagmire to government and stakeholders. Therefore, it is better to sustainably deal with the root cause than the result. Several studies have proposed sustainable groundwater fluoride remediation methods, which are grouped into in-situ and ex-situ techniques [9, 60, 61]. The in-situ techniques remediate high groundwater F^- in aquifers via artificial recharge [61], whereas the ex-situ methods are employed in reducing groundwater F^- levels at the household or community level [61]. As an in-situ remediation method, it is recommended that the people of the Bongo District and its environs should adopt rainfall harvesting or rainfall recharge as an alternative to deal with the F^- menace. Also, surface waters with low F^- concentrations could be treated and used as alternative sources of potable water.

The best ex-situ F^- remediation methods known in literature that could be adopted by the stakeholders in the Bongo District include absorption technique, electrodialysis technique and precipitation technique. Moreover, more cost-effective and Ghanaian-friendly natural coagulant/absorbent like *Moringa oleifera* and bone char should be used to defluoridate groundwater prior to usage as done elsewhere [62–65]. In the Bongo District, only Kumi et al. [66] recently used integrated bone and biochar to defluoridate the groundwater, which showed almost 100% efficacy in defluoridating the groundwater. This technique should be scaled up and a simple system should be developed using the integrated approach by Kumi et al. [66] to ameliorate the situation.

4 Summary and Conclusions

The study was aimed at identifying the geochemical controls of high groundwater F^- and assessment of the implications for human health risk within the Bongo District of the Upper East Region of Ghana. Areas dominated by hornblende-biotite granitoid and K-feldspar-rich granitoid are proven to be responsible for enriching groundwater with high amounts of F^- and thus most of the samples collected from those sources have high F^- concentrations. The hydrochemical facies are Na-HCO_3-Cl (35%), Na-Mg-HCO_3-Cl (25%), Na-Mg-Ca-HCO_3-Cl (10%), Na-Ca-HCO_3-Cl (10%), Na-Ca-Mg-HCO_3 (5%), Na-Cl-HCO_3 (5%) and Mg-Na-Cl-HCO_3 (5%). The concentrations of F^- range from 0.43 to 3.61 mg/L with an average of 1.89 exceeding the guideline value of 1.5 mg/L as recommended by WHO. A greater percentage (75%) of the samples collected from the boreholes in the study area are affected by F^- contamination. The communities affected are found in the northern, northeastern and southwestern parts of the Bongo District. Groundwater chemistry is principally controlled by water–rock interaction, ion exchange reactions and anthropogenic activities such as the application of nitrate and sulphate fertilizers on farmlands. Bivariate plots and the multivariate statistical techniques used in the study indicate that both geogenic and anthropogenic activities contribute to the increased concentrations of groundwater F^-. People in the Bongo District are exposed to excessive levels of F^- in groundwater, putting their health and livelihood at risk. The study revealed that children in the area are more vulnerable to dental fluorosis than any of the other age groups. The population is likely to be affected by the health risks of F^- in the order: children > teenagers > adults > infants.

5 Recommendations

Based on the key findings of this study, it is recommended that:

1. Groundwater should be treated prior to usage to eliminate any possible contaminant.
2. The people of the Bongo District and its environs should adopt rainfall harvesting or rainfall recharge as an alternative to deal with the fluoride menace.
3. The government should partner with Non-governmental Organizations and researchers to promote the use of the cost-effective household and sustainable defluoridation techniques such as *Moringa oleifera* and bone char.
4. Future research works should focus on assessing the spatiotemporal variation of the groundwater fluoride concentrations in the area.

Acknowledgements The authors acknowledge the financial support from the Government of Ghana through the Book and Research Allowance to faculty members. The first author also thanks the University of Johannesburg, South Africa, for the continuous support as a Postdoctoral Research Fellow in Medical Geology during the time of this research. **Authors' contributions: Emmanuel Daanoba Sunkari:** Conceptualization, Methodology, Investigation, Supervision, Writing – Original draft preparation, Validation, Writing – Reviewing and Editing. **Moses Boakye Okyere:** Methodology, Investigation, Software, Data curation, Writing – Original draft preparation, Visualization. **Salaam Jansbaka Adams:** Methodology, Data curation, Writing – Reviewing and Editing. **Musah Saeed Zango**: Data curation, Validation, Writing – Reviewing and Editing. **Prosun Bhattacharya**: Validation, Visualization, Writing – Reviewing and Editing. **Shakir Ali**: Visualization, Writing – Reviewing and Editing.

References

1. Chen K, Liu Q, Yang T, Ju Q, Yu H (2023) Geochemical characteristics, influencing factors and health risk assessment of groundwater fluoride in a drinking water source area in North Anhui Plain, Eastern China. Stoch Env Res Risk A:1–13. https://doi.org/10.1007/s00477-023-02485-2
2. Choubisa SL, Choubisa D, Choubisa A (2023) Fluoride contamination of groundwater and its threat to health of villagers and their domestic animals and agriculture crops in rural Rajasthan, India. Environ Geochem Health 45(3):607–628. https://doi.org/10.1007/s10653-022-01267-z
3. Çiner F, Sunkari ED, Şenbaş BA (2021) Geochemical and multivariate statistical evaluation of trace elements in groundwater of Niğde Municipality, South-Central Turkey: implications for arsenic contamination and human health risks assessment. Arch Environ Contam Toxicol 80: 164–182. https://doi.org/10.1007/s00244-020-00759-2

4. Furi W, Razack M, Abiye TA, Ayenew T, Legesse D (2011) Fluoride enrichment mechanism and geospatial distribution in the volcanic aquifers of the middle Awash Basin, Northern Main Ethiopian Rift. J African Earth Sci 60(5):315–327. https://doi.org/10.1016/j.jafrearsci.2011.03.004
5. Ijumulana J, Ligate F, Irunde R, Bhattacharya P, Ahmad A, Tomašek I, Maity JP, Mtalo F (2022) Spatial variability of the sources and distribution of fluoride in groundwater of the Sanya alluvial plain aquifers in northern Tanzania. Sci Total Environ 810:152153. https://doi.org/10.1016/j.scitotenv.2021.152153
6. Kanyerere T, Levy J, Xu Y, Saka J (2012) Assessment of microbial contamination of groundwater in Upper Limphasa River Catch- Ment, located in a rural area of northern Malawi. Water SA 38(4):581–596. https://doi.org/10.4314/wsa.v38i4.14
7. Rashed M, Niyazi B (2017) Environmental impact assessment of the former Al-Musk Lake wastewater dumpsite using electromagnetic induction technique. Earth Syst Environ 1(1):10. https://doi.org/10.1007/s41748-017-0010-1
8. Sunkari ED, Abangba T, Ewusi A, Tetteh SEK, Ofosu E (2023) Hydrogeochemical evolution and assessment of groundwater quality for drinking and irrigation purposes in the Gushegu municipality and some parts of east Mamprusi District, Ghana. Environ Monit Assess 195(1): 165. https://doi.org/10.1007/s10661-022-10731-3
9. Sunkari ED, Adams SJ, Okyere MB, Bhattacharya P (2022) Groundwater fluoride contamination in Ghana and the associated human health risks: any sustainable mitigation measures to curtail the long term hazrads? Groundw Sustain Dev 16:100715. https://doi.org/10.1016/j.gsd.2021.100715
10. Sunkari ED, Danladi IB (2016) Assessment of trace elements in selected bottled drinking water in Ghana: a case study of Accra Metropolis. Int J Water Resour Environ Eng 8:137–142. https://doi.org/10.5897/IJWREE2016.0685
11. Ali S, Thakur SK, Sarkar A, Shekhar S (2016) Worldwide contamination of water by fluoride. Environ Chem Lett 14:291–315. https://doi.org/10.1007/s10311-016-0563-5
12. Jabal MSA, Abustan I, Rozaimy MR, Al-Najar H (2014) Fluoride enrichment in groundwater of semi-arid urban area: Khan Younis City, Southern Gaza Strip (Palestine). J African Earth Sci 100:259–266. https://doi.org/10.1016/j.jafrearsci.2014.07.002
13. Ahada CP, Suthar S (2019) Assessment of human health risk associated with high groundwater fluoride intake in southern districts of Punjab, India. Expos Health 11(4):267–275. https://doi.org/10.1007/s12403-017-0268-4
14. Adimalla N, Li P (2019) Occurrence, health risks, and geochemical mechanisms of fluoride and nitrate in groundwater of the Rock-Dominant Semi-Arid Region, Telangana State, India. Hum Ecol Risk Assess Int J 25(1–2):81–103. https://doi.org/10.1080/10807039.2018.1480353
15. Ali S, Shekhar S, Chandrasekhar T, Yadav AK, Arora NK, Kashyap CA, Chandrasekharam D (2021) Influence of the water–sediment interaction on the major ions chemistry and fluoride pollution in groundwater of the Older Alluvial Plains of Delhi, India. J Earth Syst Sci 130(2): 1–16. https://doi.org/10.1007/s12040-021-01585-3
16. Ijumulana J, Ligate F, Bhattacharya P, Mtalo F, Zhang C (2020) Spatial analysis and GIS mapping of regional hotspots and potential health risk of fluoride concentrations in groundwater of northern Tanzania. Sci Total Environ 735:139584. https://doi.org/10.1016/j.scitotenv.2020.139584
17. Kashyap CA, Ghosh A, Singh S, Ali S, Singh HK, Chandrasekhar T, Chandrasekharam D (2020) Distribution, genesis and geochemical modeling of fluoride in the water of tribal area of Bijapur district, Chhattisgarh, Central India. Groundw Sustain Dev 11:100403. https://doi.org/10.1016/j.gsd.2020.100403
18. Nakaya S, Takada R, Yasumoto J, Masuda H, Yoshitani J, Shinjo R, Lugodisha I, Komakech H (2023) Effect of groundwater residence time on geogenic fluoride release into groundwater in the Mt. Meru slope area, Tanzania, the great Rift Valley, East Africa. J Contam Hydrol 253: 104125. https://doi.org/10.1016/j.jconhyd.2022.104125

19. Qiu H, Gui H, Xu H, Cui L, Yu H (2023) Occurrence, controlling factors and noncarcinogenic risk assessment based on Monte Carlo simulation of fluoride in mid-layer groundwater of Huaibei mining area, North China. Sci Total Environ 856:159112. https://doi.org/10.1016/j.scitotenv.2022.159112
20. Zango MS, Pelig-Ba KB, Anim-Gyampo M, Gibrilla A, Sunkari ED (2021) Hydrogeochemical and isotopic controls on the source of fluoride in groundwater within the Vea catchment, northeastern Ghana. Groundw Sustain Dev 12:100–526. https://doi.org/10.1016/j.gsd.2020.100526
21. Ali S, Fakhri Y, Golbini M, Thakur SK, Alinejad A, Parseh I, Shekhar S, Bhattacharya P (2019) Concentration of fluoride in groundwater of India: a systematic review, meta-analysis and risk assessment. Groundw Sustain Dev 9:100224. https://doi.org/10.1016/j.gsd.2019.100224
22. Kumar S, Singh R, Venkatesh AS, Udayabhanu G, Sahoo PR (2019) Medical geological assessment of fluoride contaminated groundwater in parts of indo-Gangetic Alluvial plains. Sci Rep 9(1):1–16. https://doi.org/10.1038/s41598-019-52812-3
23. Alfredo KA, Lawler DF, Katz LE (2014) Fluoride contamination in the Bongo District of Ghana, West Africa: geogenic contamination and cultural complexities. Water Int 39(4): 486–503. https://doi.org/10.1080/02508060.2014.926234
24. Rossiter HMA, Owusu PA, Awuah E, Macdonald AM, Schäfer AI (2010) Chemical drinking water quality in Ghana: water costs and scope for advanced treatment. Sci Total Environ 408: 2378–2386. https://doi.org/10.1016/j.scitotenv.2010.01.053
25. Atipoka FA (2009) Water supply challenges in rural Ghana. Desalination 248(1–3):212–217
26. Firempong CK, Nsiah K, Awunyo-Vitor D, Dongsogo J (2013) Soluble fluoride levels in drinking water-a major risk factor of dental fluorosis among children in Bongo community of Ghana. Ghana Med J 47(1):16–23
27. Sunkari ED, Abu M (2019) Hydrochemistry with special reference to fluoride contamination in groundwater of the Bongo District, Upper East Region, Ghana. Sustain Water Resour Manag 5: 1803–1814. https://doi.org/10.1007/s40899-019-00335-0
28. Sunkari ED, Zango MS, Korboe HM (2018) Comparative analysis of fluoride concentrations in Groundwaters in northern and southern Ghana: implications for the contaminant sources. Earth Syst Environ 2:103–117. https://doi.org/10.1007/s41748-018-0044-z
29. Osumanu IK, Aniah P, Yelfaanibe A (2017) Determinants of adaptive capacity to climate change among smallholder rural households in the Bongo District, Ghana. Ghana J Dev Stud 14(2):142–162. https://doi.org/10.4314/gjds.v14i2.8
30. Abitty EK, Dampare SB, Nude PM, Asiedu DK (2016) Geochemistry and petrogenesis of the K-rich 'Bongo-Type'granitoids in the paleoproterozoic Bole-Nangodi greenstone belt of Ghana. J African Earth Sci 122:47–62. https://doi.org/10.1016/j.jafrearsci.2015.08.011
31. Apambire WB, Boyle DR, Michel FA (1997) Geochemistry, genesis, and health implications of fluoriferous ground waters in the upper regions of Ghana. Environ Geol 33(1):13–24. https://doi.org/10.1007/s002540050221
32. Sunkari ED, Abu M, Bayowobie PS, Dokuz UE (2019) Hydrogeochemical appraisal of groundwater quality in the Ga west municipality, Ghana: implication for domestic and irrigation purposes. Groundw Sustain Dev 8:501–511. https://doi.org/10.1016/j.gsd.2019.02.002
33. Adiamah SS (2016) Petrography and geochemistry of schists and amphibolites from the paleoproterozoic Birimian Suhum Basin, Southeastern Ghana. Doctoral Dissertation, University of Ghana, 112 p
34. Koffi KV, Obuobie E, Banning A, Wohnlich S (2017) Hydrochemical characteristics of groundwater and surface water for domestic and irrigation purposes in Vea catchment, Northern Ghana. Environ Earth Sci 76(4):185. https://doi.org/10.1007/s12665-017-6490-3
35. Craig L, Thomas JM, Lutz A, Decker DL (2018) Determining the optimum locations for pumping low-fluoride groundwater to distribute to communities in a fluoridic area in the upper east region, Ghana. Chem Geol 476:481–492. https://doi.org/10.1016/j.chemgeo.2017.12.001

36. Sunkari ED, Abu M, Zango MS (2021) Geochemical evolution and tracing of groundwater salinization using different ionic ratios, multivariate statistical and geochemical modeling approaches in a typical semi-arid basin. J Contam Hydrol 236:103742. https://doi.org/10.1016/j.jconhyd.2020.103742
37. Setianto A, Triandini T (2013) Comparison of kriging and inverse distance weighted (IDW) interpolation methods in lineament extraction and analysis. J Appl Geol 5:21–29. https://doi.org/10.22146/jag.7204
38. Ouabo RE, Sangodoyin AY, Ogundiran MB (2020) Assessment of ordinary kriging and inverse distance weighting methods for modeling chromium and cadmium soil pollution in E-waste sites in Douala, Cameroon. J Health Pollut 10:20. https://doi.org/10.5696/2156-9614-10.26.200605
39. Reimann C, Filzmoser P, Hron K, Kynčlová P, Garrett RG (2017) A new method for correlation analysis of compositional (environmental) data – a worked example. Sci Total Environ 607:965–971. https://doi.org/10.1016/j.scitotenv.2017.06.063
40. Sunkari ED, Abu M, Zango MS, Wani AML (2020) Hydrogeochemical characterization and assessment of groundwater quality in the Kwahu-Bombouaka group of the VoltaianSupergroup, Ghana. J African Earth Sci 169:103899. https://doi.org/10.1016/j.jafrearsci.2020.103899
41. Stanton MC, Diggle PJ (2013) Geostatistical analysis of binomial data: generalized linear or transformed Gaussian modelling. Environmetrics 24:158–171. https://doi.org/10.1002/env.2205
42. Amanah TRN, Putranto TT, Helmi M (2019) Application of cluster analysis and principal component analysis for assessment of groundwater quality – a study in Semarang, Central Java, Indonesia. IOP Conf Ser Earth Environ Sci 248(1):12–63. https://doi.org/10.1088/1755-1315/248/1/012063
43. Parkhurst DL, Appelo CAJ (1999) User's guide to PHREEQC (version 2): a computer program for speciation, batch-reaction, one-dimensional transport, and inverse geochemical calculations. Water-Res Investig Rep 99. 312 p. https://acamedia.info/sciences/J_G/references/PHREEQC_Manual.pdf
44. USEPA (1989) Risk assessment guidance for superfund volume I: human health evaluation manual (part A), vol 1. U.S. Environmental Protection Agency, Office of Emergency and Remedial Response, Washington, pp 20–50
45. Yin S, Xiao Y, Han P, Hao Q, Gu X, Men B, Huang L (2020) Investigation of groundwater contamination and health implications in a typical semi-Arid Basin of North China. Water 12(4):1137. https://doi.org/10.3390/w12041137
46. World Health Organization, WHO (2017) WHO drinking water guideline (fourth). WHO Library Cataloguing-in-Publication Data. https://apps.who.int/iris/bitstream/handle/10665/254636/9789241550017-eng.pdf
47. Piper AM (1953) A graphic representation in the geochemical interpretation of groundwater analyses. Am Geophys Union Trans U S A 25:914–923
48. Calvi C, Daniel M, Cristina D, Florencia G (2016) Abundance and distribution of fluoride concentrations in groundwater: La Ballenera catchment, southeast of Buenos Aires Province, Argentina. Environ Earth Sci 75:534. https://doi.org/10.1007/s12665-015-4972-8
49. Ewusi A, Sunkari ED, Seidu J, Coffie-Anum E (2022) Hydrogeochemical characteristics, sources and human health risk assessment of heavy metal dispersion in the mine pit water–surface water–groundwater system in the largest manganese mine in Ghana. Environ Technol Innov 26:102312. https://doi.org/10.1016/j.eti.2022.102312
50. Laxmankumar D, Satyanarayana E, Dhakate R, Saxena PR (2019) Hydrogeochemical characteristics with respect to fluoride contamination in groundwater of Maheshwarm Mandal, RR District, Telangana State, India. Groundw Sustain Dev 8:474–483. https://doi.org/10.1016/j.gsd.2019.01.008
51. Keshavarzi B, Moore F, Esmaeili A, Rastmanesh F (2010) The source of fluoride toxicity in Muteh Area, Isfahan, Iran. Environ Earth Sci 61(4):777–786. https://doi.org/10.1007/s12665-009-0390-0

52. Karro E, Uppin M (2013) The occurrence and hydrochemistry of fluoride and boron in carbonate aquifer system, central and Western Estonia. Environ Monit Assess 185(5): 3735–3748. https://doi.org/10.1007/s10661-012-2824-5
53. Kaiser HF (1960) The application of electronic computers to factor analysis. Educ Psychol Meas 20(1):141–151. https://doi.org/10.1177/001316446002000116
54. Yidana SM, Banoeng-Yakubo B, Akabzaa TM (2010) Analysis of groundwater quality using multivariate and spatial analysis in the Keta Basin, Ghana. J African Earth Sci 58:220–234. https://doi.org/10.1016/j.jafrearsci.2010.03.003
55. Berhe BA, Dokuz UE, Celik M (2017) Assessment of hydrogeochemistry and environmental isotopes of surface and Groundwaters in the Kutahya Plain, Turkey. J African Earth Sci 134: 230–240. https://doi.org/10.1016/j.jafrearsci.2017.06.015
56. Miyittah MK, Tulashie SK, Tsyawo FW, Sarfo JK, Darko AA (2020) Assessment of surface water quality status of the aby lagoon system in the Western region of Ghana. Heliyon 6(7): 44–66. https://doi.org/10.1016/j.heliyon.2020.e04466
57. Akiniwa K (1997) Re-examination of acute toxicity of fluoride. Fluoride 30(2):89–104. https://www.fluoridealert.org/wp-content/uploads/akiniwa-1997.pdf
58. Tekle-Haimanot R, Melaku Z, Kloos H, Reimann C, Fantaye W, Zerihun L, Bjorvatn K (2006) The geographic distribution of fluoride in surface and groundwater in Ethiopia with an emphasis on the Rift Valley. Sci Total Environ 367:182–119. https://doi.org/10.1016/j.scitotenv.2005.11.003
59. Huang D, Yang J, Wei X, Qin J, Ou S, Zhang Z, Zou Y (2017) Probabilistic risk assessment of Chinese residents' exposure to fluoride in improved drinking water in endemic fluorosis areas. Environ Pollut 222:118–125. https://doi.org/10.1016/j.envpol.2016.12.074
60. Abugri DA, Pelig-Ba KB (2011) Assessment of fluoride content in tropical surface soils used for crop cultivation. Afr J Environ Sci Technol 5(9):653–660. https://www.ajol.info/index.php/ajest/article/view/72064
61. Brindha K, Elango L (2011) Fluoride in groundwater: causes, implications and mitigation measures. In: Monroy SD (ed) Fluoride properties, applications and environmental management, pp 111–136. https://www.novapublishers.com/catalog/product_info.php?products_id=15895
62. Aziz NA, Jayasuriya N, Fan L, Al-Gheethi A (2021) A low-cost treatment system for underground water using Moringaoleifera seeds and Musa cavendish peels for remote communities. J Chem Technol Biotechnol 96(3):680–696. https://doi.org/10.1002/jctb.6581
63. daConceição VM, Ambrosio Ugri MCB, Silveira C, Nishi L, Vieira MF, de Jesus BF, Vieira AMS, Bergamasco R (2015) Removal of excess fluoride from groundwater using natural coagulant Moringa oleifera Lam and microfiltration. Can J Chem Eng 93(1):37–45. https://doi.org/10.1002/cjce.22101
64. daConceição VM, Yamaguchi NU, de Jesus BF, Bergamasco R (2021) Process performance combining natural coagulant Moringa oleifera Lam and ultrafiltration for groundwater defluoridation. Water Air Soil Pollut 232(6):1–13. https://doi.org/10.1007/s11270-021-05120-4
65. Gayathri G, Beulah M, Pallavi HJ, Sarath Chandra K (2021) Defluoridation of drinking water–fluoride wars. In: Advances in energy and environment. Springer, Singapore, pp 179–187. https://doi.org/10.1007/978-981-33-6695-4_17
66. Kumi M, Anku WW, Govender PP, Obiri-Nyarko F (2023) Bench-scale integrated bone and biochar bed treatment of geogenic fluoride contaminated groundwater from Bongo in Ghana. Groundw Sustain Dev 21:100929. https://doi.org/10.1016/j.gsd.2023.100929

Uncovering Fluoride Contamination in Groundwater of Arid and Semi-Arid Regions: Stigma to Solutions

Shakir Ali

Contents

1 Introduction 328
2 Sources 329
3 Background 329
3.1 Permissible Limit 329
3.2 Why Do We Need to Study Fluoride? 329
3.3 Boon or Bane? 330
4 Prominent Fluoride-Polluted Areas in Arid and Semi-Arid Regions 330
5 Methods of Defluoridation 332
6 Challenges 332
7 Conclusions 333
8 Recommendations 334
References 334

Abstract Fluoride contamination in the groundwater is a threat to water security. This is highly dangerous in arid and semi-arid regions which comprises one-third of the world region where more than 20% of the population resides. Elevated levels of fluoride in the groundwater are largely reported from these areas which is widely used for drinking without any prior treatment. It is of great concern as the consumption of elevated levels of fluoride causes dental fluorosis and skeletal fluorosis to humans. Furthermore, insufficient rainfall and high evapotranspiration vis-à-vis longer water–rock interactions are noticeably responsible for the elevated levels of fluoride in these regions. Unluckily, most of these regions fall in the developing world, and thus, governing policies are many times inefficient in providing sustainable water supply for drinking. Therefore, more feasible and practical solutions in conjunction with strong policy intervention are required to solve this global contaminant issue. This chapter highlights various facets of fluoride contamination and

S. Ali (✉)
Department of Geology, University of Delhi, Delhi, India

Shakir Ali and Abdelazim Negm (eds.), *Groundwater Quality and Geochemistry in Arid and Semi-Arid Regions*, Hdb Env Chem (2024) 126: 327–338, DOI 10.1007/698_2023_1060,

Published online: 29 December 2023

discusses various challenges in providing fluoride-free water in contaminated arid and semi-arid regions.

Keywords Arid and semi-arid regions, Fluoride, Fluoride-contaminated groundwater, Fluoride pollution, Fluorosis

1 Introduction

Arid and semi-arid areas cover roughly one-third of the global regions where nearly 20% (~1.6 billion) of the population lives (www.un.org; [1]). In these areas, groundwater is largely consumed for drinking and widely used for irrigation. Therefore, drinking water should be free from any contaminants to avoid diseases. Contaminants such as fluoride are predominantly found in these regions due to limited rainfall and high temperature resulting in longer water-sediments/rock interaction which promotes fluoride release into the groundwater. Thus, the Asian and African regions were largely investigated and reported to be highly contaminated [2, 3]. However, fluorosis is also frequently reported from few European countries like Estonia [4] and Poland [5] and also from various regions of Latin America [6, 7].

Fluoride is a major significant contaminant commonly found in the groundwater globally [2]. In fact, it is fluorine, one of the lightest halogen elements, that is highly mobile at higher temperature and in aqueous solution forms fluoride ion (F^-; [8]). Fluoride is now considered as a global groundwater contaminant due to its occurrence globally [8, 9]. Fluoride in the groundwater is primarily governed by various factors such as fluoride content in the sediments/rocks, water–rock interaction, and mean residence time of water. In South Asia, drinking wells installed either in both unconsolidated aquifers or crystalline aquifers were largely reported to be contaminated with fluoride [10, 11].

Numerous researchers observed that the pH of the groundwater plays a pivotal role in the mobilization of fluoride from sediments/rocks into the groundwater [10, 12]. In general, it was found that fluoride in the groundwater shows positive correlation with ions like arsenic, sodium, and bicarbonate, whereas fluoride shows negative correlation with calcium ion [9]. However, few authors also have distinct observations [13]. Fluoride in the groundwater is primarily controlled by the presence and content of fluorine-bearing minerals [14], water–rock interaction (i.e., substitution of fluoride ions with hydroxyl ions), mean residence time of the water, and the local prevailing hydro-meteoric conditions such as temperature and rainfall [15, 16]. However, studies conducted on the investigation of geogenic sources of fluoride are limited globally; thus, the mobilization of fluoride from the sediments/rocks is still unclear. However, the substitution of fluoride ions with hydroxyl ions (OH^-) due to their similar ionic radii is mainly found to be possibly responsible for the release of fluoride into the water [17, 18]. Furthermore, the historic investigation on fluoride can be found in a study conducted by Nordstrom and Smedley [9].

In this chapter, fluoride contamination in arid and semi-arid regions was investigated and the sources, mobilization, and consequences of fluoride consumption were documented. Furthermore, this study discusses knowledge gaps in research and suggests possible remedies and various challenges in achieving fluoride-free water in the arid and semi-arid regions.

2 Sources

Fluoride can be derived from both geogenic and anthropogenic sources. Geogenic sources are widespread, whereas anthropogenic sources are mostly localized in nature [9]. Fluoride occurs in various rocks like igneous, sedimentary, and metamorphic with different proportions [19]. For example, igneous rocks enriched in felsic minerals in particular contain plausible fluoride. Numerous natural fluorine-bearing minerals were found to occur on Earth with varying concentration. Nordstrom and Smedley [9] reported that nearly 300 minerals contain fluoride in varying proportions. However, the most commonly occurring fluorine-bearing minerals are fluorite (CaF_2), biotite [$K(Mg, Fe)_3$ ($AlSi_3O_{10}$) (OH, F)], phlogopite [K $Mg_3(AlSi_3O_{10})$ (F, OH)], apatite [CaF_2 $3Ca_3(PO_4)_2$], cryolite (Na_3AlF_6), etc. Interestingly, frequently occurring minerals like micas only contain up to 1% of fluoride, while the accessory minerals like fluorite may have up to 50% of fluoride [9, 10, 18]. A detailed investigation of fluoride content in various rocks is undocumented. Besides groundwater, food beverages [20] and various types of tea [21] also contribute a noticeable amount of fluoride to the human body.

Anthropogenic sources include fluorine emission from volcanic eruption (www.usgs.gov), mining of fluorine-rich minerals, plenty and unscientific use of fertilizers in agriculture, brick production [22], industrial wastes, etc. [10].

3 Background

3.1 *Permissible Limit*

The WHO (2011) and BIS (Bureau of Indian Standards; 2012) have set the maximum permissible limit as 1.5 mg/L for fluoride in drinking water. BIS has also set 1 mg/L as the maximum desirable limit [16].

3.2 *Why Do We Need to Study Fluoride?*

The advancement of fluoride studies is increased due to its potential health impacts on humans, and thus, studies on fluoride contamination have exponentially increased

Table 1 Fluoride consumption and related health effects in humans [2]

Fluoride concentration (mg/L)	Health effect
<0.5	Helpful in promoting dental caries
0.5–1.5	Strongly required for bones and teeth
1.5–4.0	Dental fluorosis especially among children
>4.0	Dental fluorosis and early sign of skeletal fluorosis
>10	Strong crippling skeletal fluorosis

during the last two decades. Fluoride has potentially dangerous effects on human health and is evident in the form of dental fluorosis and skeletal fluorosis depending upon the duration, amount of consumption, and age of the consuming groups [19]. Consumption of fluoride above the maximum permissible limit (more than 1.5 mg/L) is well connected to fluorosis and the children are more vulnerable to dental fluorosis ([8]; Table 1). In the long term, it is also associated with irreversible skeletal fluorosis [9]. It was also found to be linked to liver and kidney failure, hypertension, thyroid problems, and defects in the reproduction system [8, 23, 24]. Various studies conducted on probabilistic assessment of non-carcinogenic risk assessment of fluoride revealed that children are more vulnerable to fluorosis than other age groups [19, 25]. This could be attributed to their lower body weight and high demand for food and water.

3.3 *Boon or Bane?*

It was mentioned earlier that the consumption of fluoride-bearing water above the maximum permissible limit is associated with dental fluorosis and skeletal fluorosis. However, fluorine below the desirable limit (<1 mg/L) is pivotal for tooth enamels and bone mineralization [2]. Therefore, fluoride is also considered as a sword with two edges. There is no immediate effect of fluoride on humans; thus, contaminated water is often consumed without any prior treatment. Few studies also correlate negatively with high fluoride consumption to the development of children's intelligence [26]. Ding et al. [27] observed a good correlation between fluoride consumption and children's intelligence. However, this relationship is not well documented.

4 Prominent Fluoride-Polluted Areas in Arid and Semi-Arid Regions

Numerous countries falling in the arid and semi-arid regions were found to be highly contaminated by fluoride (Fig. 1). The arid and semi-arid region receives limited rainfall and high evapotranspiration rates which are likely responsible for the high fluoride concentration in the groundwater.

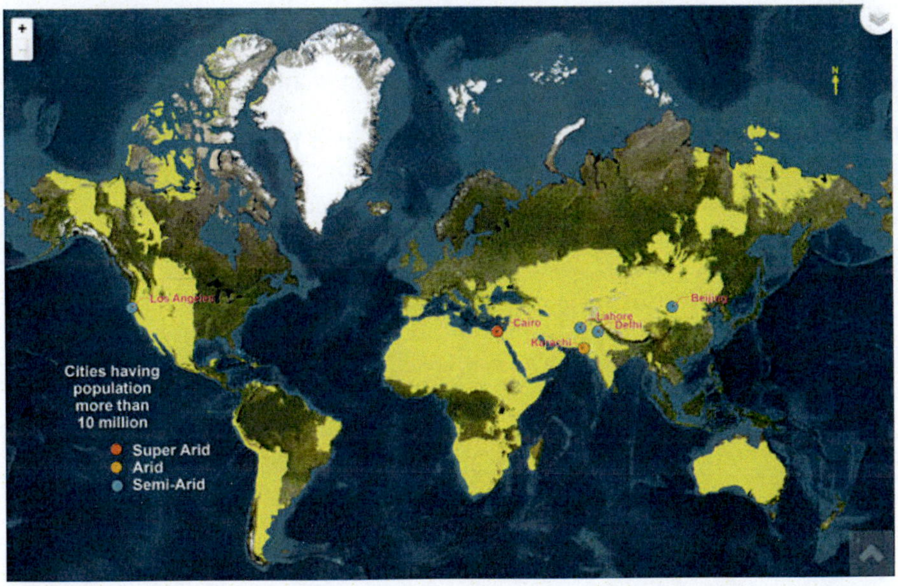

Fig. 1 Regions falling in the arid and semi-arid areas showing cities with more than 10 million populations in the region (highlighted in yellow color. World atlas desertification; https://wad.jrc.ec.europa.eu/; open access)

Therefore, countries such as India [10], China [28], Iran [25, 29], Brazil [7], Argentina [30], Mexico [6], Ghana [31], Pakistan [32], Turkey [33], Sri Lanka [34], Yemen [35], Afghanistan [36], and Tanzania [37] were mainly found to be highly contaminated with fluoride [2].

In India, many states falling in arid and semi-arid regions such as Andhra Pradesh, Rajasthan, Orissa, Punjab, Haryana, Delhi, Maharashtra, and Chhattisgarh were found to have plausible fluoride in the groundwater [2, 10, 16, 38]. For instance, more than 18 mg/L of fluoride was reported from the Haryana, western India [39]. Brindha and Elango [40] reported high fluoride concentrations from the crystalline rocks of southern India. Shaji et al. [41] also reported elevated fluoride levels from Kerala, southern India. Naaz et al. [42] reported high fluoride values in the arid and semi-arid regions of central India. Senthilkumar et al. [43] and Kadam et al. [44] also reported elevated levels of fluoride from western India. The agricultural dominated Indo-Gangetic plains were also reported to be contaminated with fluoride [19, 45, 46]. Furthermore, groundwater in the western Thar desert of India including Pakistan is also reported to be highly contaminated [47, 48].

In Africa, countries in the vicinity of the East African Rift Valley are repetitively reported to be contaminated by fluoride as the geothermal water often contains high fluoride [49–53]. Considerably high fluoride values are also observed by Mwiathi et al. [54] from Kenyan rift valley. Bianchini et al. [55] also obtained high fluoride from waters of Ethiopian Rift Valley. Egbueri et al. [56] and Sunkari and Abu [31] studied fluoride contamination in Nigeria and Ghana and reported elevated levels of fluoride in numerous drinking wells. In China, Liu et al. [57] reported fluoride median value higher than the maximum permissible limit (1.5 mg/L). Li et al. [58] also observed high concentration of fluoride in the vicinity of industrial areas of China. Wu et al. [28] also reported high fluoride values from the arid areas of China. Fluoride-contaminated groundwater was also reported from many drinking wells especially the semi-arid western part of the USA [59].

In Pakistan, Durrani and Farooqi [60] investigated Quetta region and found fluoride levels higher than the safe limit. Khattak et al. [32] also investigated the Punjab Province of Pakistan along with India and reported fluoride values higher than the safe limit in many drinking wells. Chandrajith et al. [34] studied numerous drinking wells of Sri Lanka and also reported elevated fluoride levels. Aqeel et al. [35] reported higher fluoride values in Yemen. Ijumulana et al. [37] investigated various sources and fluoride distribution in the groundwater of Tanzania.

Considerable studies also show that groundwater along the coastal areas is also contaminated by fluoride due to seawater intrusion and reported mainly from China [61, 62] and Bangladesh [63] and from few other localities. However, this association needs in-depth investigation.

5 Methods of Defluoridation

Numerous defluoridation methods such as ion exchange, reverse osmosis, coagulation and precipitation, adsorption, and membrane separation were investigated and found to be efficient in fluoride removal from groundwater at various capacities (Table 2; [12, 67, 68]).

In general, the reverse osmosis method is widely used; however, the method is expensive and ineffective in providing drinking water to large communities. Therefore, newly developed and privileged societies are only benefitted. Traditional Nalgonda defluoridation technique generated huge sludge and requires highly skilled labor. However, an environmentally friendly and cost-effective method of providing water to large community is still challenging. From this perspective, other methods such as bioremediation and phytoremediation are environmentally friendly and cost-effective methods. These methods involve the use of microbes to remove toxic substances; however, their application on a larger scale, i.e., scalability, is often challenging and time-consuming.

6 Challenges

Wrong perception, socio-economically disadvantaged societies, lack of awareness, acceptability, and scalability are some major hindrances to tackle the fluoride problem especially in arid and semi-arid regions particularly in the developing countries [8, 69]. Ineffective and weak policies and implementations of policies

Table 2 Various technologies used for defluoridation in arid and semi-arid regions (compiled from [8, 64–66])

Defluoridation method	Description	Limitations
Adsorption	Require use of adsorbents for filtration. Several adsorbents such as laterite, amberlite resin, hydrous ferric oxide, and biochars are used. Naturally occurring low-cost adsorbents used for adsorption are fruit peel, seeds, tulsi, ragi powder, red mud, Multani matti, orange peel powder, chalk powder rice husk, etc.	Only functional in a specific range of pH (5–6)
Ion exchange	Fluoride is removed using an anion exchange resin	Expensive technology
Coagulation–precipitation/Nalgonda	Addition of chemicals such as lime and alum which leads to precipitation of fluoride	High chemical dose is required. This method also requires skilled labor. Generate huge amount of sludge
Membrane separation (reverse osmosis/nanofiltration/electrodialysis)	Fluoride is removed through semi-permeable membrane. Reverse osmosis (RO) is more commonly used over nanofiltration (NF) due to its high efficiency in removing dissolved solids	Expensive technology

are major challenges in these regions. Strong policy intervention in the hotspot areas is warranted [8, 70, 71]. In few countries such as India, Pakistan, Sri Lanka, and Turkey, there are numerous unlicensed wells owned by individuals, and water from these private wells is largely consumed for drinking without any prior treatment. This is due to unreliability of the drinking water supply authorities. Furthermore, lack of awareness, education, and consciousness toward health especially in deprived societies is responsible for many diseases including fluorosis. Therefore, the cases of fluorosis are common in these regions.

7 Conclusions

This chapter documents the sources, mobilization, and consequences of fluoride and addresses the potential remedies and challenges in arid and semi-arid regions globally. The study suggests that fluoride contamination is predominant in the areas due to limited rainfall and high insolation vis-à-vis largely governed by local meteorological conditions. The study further suggests that fluoride content is influenced by the pH of the water. Furthermore, this study highlighted the defluoridation techniques and possible remediation. Due to scalability issues of available defluoridation techniques, providing safe drinking water to large communities living in these regions is always challenging. Implementation and adaptation of strong policies of providing safe water supply in arid and semi-arid regions is warranted. This chapter will help policymakers to revisit the existing policies in contaminated regions of arid and semi-arid areas.

8 Recommendations

Tackling fluoride contamination in drinking water in regions like arid and semi-arid areas should be critically studied. In this perspective, on a local level, rejuvenation of water bodies and development of recharge structures is recommended to dilute in situ fluoride. Regular monitoring of public drinking wells for water quality should be prioritized. However, providing fluoride-free drinking water access to considerable community is always challenging. The research on fluoride should be advanced so that the problem can be sorted out with a well-defined scientific and more practical approach. Based on the socio-economic situation, community-based solutions will be highly effective particularly in South Asia like India, Bangladesh, Pakistan, and Sri Lanka. Furthermore, awareness toward the problem, willingness of the concerned governing authorities, and a strong decisive policy intervention and learning from the past are recommended.

Acknowledgements It is duly acknowledged that no funding is involved for conducting this study. The author thanks the entire team of Springer for their support.

References

1. Abdelhak M (2022) Soil improvement in arid and semiarid regions for sustainable development. In: Natural resources conservation and advances for sustainability. Elsevier, pp 73–90. https://doi.org/10.1016/B978-0-12-822976-7.00026-0
2. Ali S, Thakur SK, Sarkar A, Shekhar S (2016) Worldwide contamination of water by fluoride. Environ Chem Lett 14:291–315. https://doi.org/10.1007/s10311-016-0563-5
3. Yadav KK, Kumar S, Pham QB (2019) Fluoride contamination, health problems and remediation methods in Asian groundwater: a comprehensive review. Ecotoxicol Environ Saf 183:109362. https://doi.org/10.1016/j.ecoenv.2019.06.045
4. Indermitte E, Saava A, Karro E (2009) Exposure to high fluoride drinking water and risk of dental fluorosis in Estonia. Int J Environ Res Public Health 6(2):710–721. https://doi.org/10.3390/ijerph6020710
5. Razowska-Jaworek L, Cudak J (2022) Occurrence, origin and health hazards of high fluoride waters in the western part of the Neogene multilayered sedimentary basin in Opole Province, Poland. Univ J Geosci 9(1):1–20. https://doi.org/10.13189/ujg.2022.090101
6. Gutiérrez M, Alarcón-Herrera M, Gaytán-Alarcón AP (2023) Arsenic and fluorine in northern Mexico: spatial distribution and enrichment factors. Environ Monit Assess 195(1):1–12. https://doi.org/10.1007/s10661-022-10818-x

7. Martins VTS, Pino DS, Bertolo R, Hirata R, Babinski M, Pacheco DF, Rios AP (2018) Who to blame for groundwater fluoride anomaly in São Paulo, Brazil? Hydrogeochemistry and isotopic evidence. Appl Geochem 90:25–38
8. Kumar R, Ali S, Sandanayake S, Islam MA, Ijumulana J, Maity JP, Vithanage M, Armienta MA, Sharma P, Hamisi R, Kimambo V (2024) Fluoride as a global groundwater contaminant. In: Inorganic contaminants and radionuclides. Elsevier, pp 319–350
9. Nordstrom D, Smedley P (2022) Fluoride in groundwater – the groundwater project. https://gw-project.org/books/fluoride-in-groundwater/
10. Ali S, Shekhar S, Kumar R, Brindha K, Li P (2023) Genesis and mobilization of fluoride in groundwater of India: statistical evaluation, health impacts, and potential remedies. J Hazard Mater Adv:100352. https://doi.org/10.1016/j.hazadv.2023.100352
11. Saha D, Shekhar S, Ali S, Elango L, Vittala S (2020) Recent scientific perspectives on the Indian hydrogeology. Proc Indian Natl Sci Acad 86(1):459–478. https://doi.org/10.16943/ptinsa/2020/49790
12. Edmunds, W.M., Smedley, P.L. (2013). Fluoride in natural waters. In: Olle, S. (eds) Essentials of medical geology. pp. 311–336. Springer, Dordrecht. ISBN: 978-94-007-4374-8. https://doi.org/10.1007/978-94-007-4375-5_13
13. Dehbandi R, Moore F, Keshavarzi B (2018) Geochemical sources, hydrogeochemical behavior, and health risk assessment of fluoride in an endemic fluorosis area, central Iran. Chemosphere 193:763–776. https://doi.org/10.1016/j.chemosphere.2017.11.021
14. Mukherjee I, Singh UK (2018) Groundwater fluoride contamination, probable release, and containment mechanisms: a review on Indian context. Environ Geochem Health 40(6):2259–2301. https://doi.org/10.1007/s10653-018-0096-x
15. Ali S (2022) Process based insight into the groundwater contamination and fluoride pollution in parts of Delhi. PhD thesis, Department of Geology, University of Delhi
16. Ali S, Shekhar S, Chandrasekhar T, Yadav AK, Arora NK, Kashyap CA, Bhattacharya P, Rai SP, Pande P, Chandrasekharam D (2021) Influence of the water–sediment interaction on the major ions chemistry and fluoride pollution in groundwater of the Older Alluvial Plains of Delhi, India. J Earth Syst Sci 130(98):1–16. https://doi.org/10.1007/s12040-021-01585-3
17. Ali, S., Shekhar, S., Chandrasekhar, T. (2022). Non-carcinogenic health risk assessment of fluoride in groundwater of the river Yamuna flood plain, Delhi, India. In: Hamid R. P. (eds) Computers in earth and environmental sciences. Elsevier. ISBN: 978-0-323-89861-4. 455–466. https://doi.org/10.1016/B978-0-323-89861-4.00046-4
18. Vithanage M, Bhattacharya P (2015) Fluoride in the environment: sources, distribution and defluoridation. Environ Chem Lett 13:131–147. https://doi.org/10.1007/s10311-015-0496-4
19. Ali S, Fakhri Y, Golbini M, Thakur SK, Alinejad A, Parseh I, Shekhar S, Bhattacharya P (2019) Concentration of fluoride in groundwater of India: a systematic review, meta-analysis and risk assessment. Groundw Sustain Dev 9(100224):1–10. https://doi.org/10.1016/j.gsd.2019.100224
20. Mridha D, Priyadarshni P, Bhaskar K, Gaurav A, De A, Das A, Joardar M, Chowdhury NR, Roychowdhury T (2021) Fluoride exposure and its potential health risk assessment in drinking water and staple food in the population from fluoride endemic regions of Bihar, India. Groundw Sustain Dev 13:100558
21. Essebbahi I, Ouazzani C, Moustaghfir A, Er-ramly A, El Baroudi Y, Dami A, Balouch L (2023) Analysis of the fluoride levels of well water and tea consumed by the Moroccan population in different rural areas. Mater Today Proc 72:3347–3350
22. Fuge R (2019) Fluorine in the environment, a review of its sources and geochemistry. Appl Geochem 100:393–406. https://doi.org/10.1016/j.apgeochem.2018.12.016
23. Ahmad S, Singh R, Arfin T, Neeti K (2022) Fluoride contamination, consequences and removal techniques in water: a review. Environ Sci Adv 1:620–661. https://doi.org/10.1039/D1VA00039J
24. Kheradpisheh Z, Mirzaei M, Mahvi AH, Mokhtari M, Azizi R, Fallahzadeh H, Ehrampoush MH (2018) Impact of drinking water fluoride on human thyroid hormones: a case-control study. Sci Rep 8:1–7. https://doi.org/10.1038/s41598-018-20696-4

25. Keramati H, Miri A, Baghaei M, Rahimizadeh A, Ghorbani R, Fakhri Y, Bay A, Moradi M, Bahmani Z, Ghaderpoori M, Mousavi Khaneghah A (2019) Fluoride in Iranian drinking water resources: a systematic review, meta-analysis and non-carcinogenic risk assessment. Biol Trace Elem Res 188:261–273. https://doi.org/10.1007/s12011-018-1418-7
26. Karimzade S, Aghaei M, Mahvi A (2014) Investigation of intelligence quotient in 9–12-year-old children exposed to high- and low-drinking water fluoride in West Azerbaijan Province, Iran. Fluoride 47:9–14
27. Ding Y, Gao Y, Sun H, Han H, Wang W, Ji X, Liu X, Sun D (2011) The relationships between low levels of urine fluoride on children's intelligence, dental fluorosis in endemic fluorosis areas in Hulunbuir, Inner Mongolia, China. J Hazard Mater 186:1942–1946. https://doi.org/10.1016/j.jhazmat.2010.12.097
28. Wu J, Li P, Qian H (2015) Hydrochemical characterization of drinking groundwater with special reference to fluoride in an arid area of China and the control of aquifer leakage on its concentrations. Environ Earth Sci 73:8575–8588. https://doi.org/10.1007/s12665-015-4018-2
29. Amiri V, Ali S, Sohrabi N et al (2023) Hydrogeochemical evaluation with emphasis on nitrate and fluoride in urban and rural drinking water resources in western Isfahan province, central Iran. Environ Sci Pollut Res. https://doi.org/10.1007/s11356-023-30001-0
30. Alcaine AA, Schulz C, Bundschuh J, Jacks G, Thunvik R, Gustafsson JP, Mörth CM, Sracek O, Ahmad A, Bhattacharya P (2020) Hydrogeochemical controls on the mobility of arsenic, fluoride and other geogenic co-contaminants in the shallow aquifers of northeastern La Pampa Province in Argentina. Sci Total Environ 715:136671
31. Sunkari ED, Abu M (2019) Hydrochemistry with special reference to fluoride contamination in groundwater of the Bongo District, Upper East Region, Ghana. Sustain Water Resour Manag 5:1803–1814. https://doi.org/10.1007/s40899-019-00335-0
32. Khattak JA, Farooqi A, Hussain I, Kumar A, Singh CK, Mailloux BJ et al (2022) Groundwater fluoride across the Punjab plains of Pakistan and India: distribution and underlying mechanisms. Sci Total Environ 806:151353
33. Baba A, Tirol N (2023) Understanding the challenges: sustainable usage of groundwater resources in Türkiye. In: Ali S, Armanuos AM (eds) Groundwater in arid and semi-arid areas. Earth and environmental sciences library. Springer, Cham. https://doi.org/10.1007/978-3-031-43348-1_5
34. Chandrajith R, Diyabalanage S, Dissanayake CB (2020) Geogenic fluoride and arsenic in groundwater of Sri Lanka and its implications to community health. Groundw Sustain Dev 10:100359. https://doi.org/10.1016/j.gsd.2020.100359
35. Aqeel A, Al-Amry A, Alharbi O (2017) Assessment and geospatial distribution mapping of fluoride concentrations in the groundwater of Al-Howban Basin, Taiz-Yemen. Arab J Geosci 10:312. https://doi.org/10.1007/s12517-017-3069-y
36. Hayat E, Baba A (2017) Quality of groundwater resources in Afghanistan. Environ Monit Assess 189:318. https://doi.org/10.1007/s10661-017-6032-1
37. Ijumulana J, Ligate F, Irunde R, Bhattacharya P, Ahmad A, Tomašek I, Maity JP, Mtalo F (2022) Spatial variability of the sources and distribution of fluoride in groundwater of the Sanya alluvial plain aquifers in northern Tanzania. Sci Total Environ 810:152153
38. Kashyap CA, Ghosh A, Singh S, Ali S, Singh HK, Chandrasekhar T, Chandrasekharam D (2020) Distribution, genesis and geochemical modeling of fluoride in the water of tribal area of Bijapur district, Chhattisgarh, central India. Groundw Sustain Dev 11:1–11. https://doi.org/10.1016/j.gsd.2020.100403
39. Ali S, Shekhar S, Bhattacharya P, Verma G, Chandrasekhar T, Chandrashekhar AK (2018) Elevated fluoride in groundwater of Siwani Block, Western Haryana, India: a potential concern for sustainable water supplies for drinking and irrigation. Groundw Sustain Dev 7:410–420. https://doi.org/10.1016/j.gsd.2018.05.008
40. Brindha K, Elango L (2013) Geochemistry of fluoride rich groundwater in a weathered granitic rock region, southern India. Water Qual Expo Health 5:127–138. https://doi.org/10.1007/s12403-013-0096-0

41. Shaji E, Gómez-Alday J, Hussein S et al (2018) Salinization and deterioration of groundwater quality by nitrate and fluoride in the Chittur Block, Palakkad, Kerala. J Geol Soc India 92:337–345. https://doi.org/10.1007/s12594-018-1017-4
42. Naaz A, Kumar B, Narayan C et al (2015) Assessment of fluoride pollution in groundwaters of arid and semi-arid regions of Tonalite–Trondjhemite series in central India. Water Qual Expo Health 7:545–556. https://doi.org/10.1007/s12403-015-0171-9
43. Senthilkumar M, Mohapatra B, Gnanasundar D et al (2021) Identifying fluoride endemic areas and exposure pathways for assessment of non-carcinogenic human health risk associated with groundwater fluoride for Gujarat state, India. Environ Sci Pollut Res 28:50188–50203. https://doi.org/10.1007/s11356-021-14156-2
44. Kadam A, Wagh V, Umrikar B et al (2020) An implication of boron and fluoride contamination and its exposure risk in groundwater resources in semi-arid region, Western India. Environ Dev Sustain 22:7033–7056. https://doi.org/10.1007/s10668-019-00527-w
45. Nizam S, Virk HS, Sen IS (2022) High levels of fluoride in groundwater from northern parts of Indo-Gangetic plains reveals detrimental fluorosis health risks. Environ Adv 8:100200. https://doi.org/10.1016/j.envadv.2022.100200
46. Sulaiman MA, Divya MMZ, Anjum S, Kumari A (2023) Groundwater contamination by fluoride and mitigation measures for sustainable management of groundwater in the Indo-Gangetic Plains of India. In: Ali S, Armanuos AM (eds) Groundwater in arid and semi-arid areas. Earth and environmental sciences library. Springer, Cham. https://doi.org/10.1007/978-3-031-43348-1_12
47. Rafique T, Naseem S, Ozsvath D, Hussain R, Bhanger MI, Usmani TH (2015) Geochemical controls of high fluoride groundwater in Umarkot sub-district, Thar Desert, Pakistan. Sci Total Environ 530:271–278. https://doi.org/10.1016/j.scitotenv.2015.05.038
48. Singh CK, Kumari R, Singh N, Mallick J, Mukherjee S (2013) Fluoride enrichment in aquifers of the Thar Desert: controlling factors and its geochemical modelling. Hydrol Process 27 (17):2462–2474. https://doi.org/10.1002/hyp.9247
49. Addison MJ, Rivett MO, Robinson H, Fraser A, Miller AM, Phiri P, Mleta P, Kalin RM (2020) Fluoride occurrence in the lower east African rift system, southern Malawi. Sci Total Environ 712:136260. https://doi.org/10.1016/j.scitotenv.2019.136260
50. Ijumulana J, Ligate F, Irunde R, Bhattacharya P, Maity JP, Ahmad A, Mtalo F (2021) Spatial uncertainties in fluoride levels and health risks in endemic fluorotic regions of northern Tanzania. Groundw Sustain Dev 14:100618. https://doi.org/10.1016/j.gsd.2021.100618
51. Kimambo V, Bhattacharya P, Mtalo F, Mtamba J, Ahmad A (2019) Fluoride occurrence in groundwater systems at global scale and status of defluoridation–state of the art. Groundw Sustain Dev 9:100223. https://doi.org/10.1016/j.gsd.2019.100223
52. Kut KMK, Sarswat A, Srivastava A, Pittman Jr CU, Mohan D (2016) A review of fluoride in African groundwater and local remediation methods. Groundw Sustain Dev 2-3:190–212. https://doi.org/10.1016/j.gsd.2016.09.001
53. Olaka LA, Wilke FD, Olago DO, Odada EO, Mulch A, Musolff A (2016) Groundwater fluoride enrichment in an active rift setting: central Kenya Rift case study. Sci Total Environ 545:641–653. https://doi.org/10.1016/j.scitotenv.2015.11.161
54. Mwiathi NF, Gao X, Li C, Rashid A (2022) The occurrence of geogenic fluoride in shallow aquifers of Kenya Rift Valley and its implications in groundwater management. Ecotoxicol Environ Saf 229:113046. https://doi.org/10.1016/j.ecoenv.2021.113046
55. Bianchini G, Brombin V, Marchina C, Natali C, Godebo TR, Rasini A, Salani GM (2020) Origin of fluoride and arsenic in the main Ethiopian rift waters. Fortschr Mineral 10(5):453. https://doi.org/10.3390/min10050453
56. Egbueri JC, Ezugwu CK, Unigwe CO, Onwuka OS, Onyemesili OC, Mgbenu CN (2021) Multidimensional analysis of the contamination status, corrosivity and hydrogeochemistry of groundwater from parts of the Anambra Basin, Nigeria. Anal Lett 54(13):2126–2156. https://doi.org/10.1080/00032719.2020.1843049

57. Liu S, Gao B, Qin H et al (2022) Concentration mechanism of fluorine, arsenic, and uranium in groundwater of the Hailar Basin, China. Environ Earth Sci 81:444. https://doi.org/10.1007/s12665-022-10560-0
58. Li Y, Wang F, Feng J, Lv JP, Liu Q, Nan FR, Zhang W, Qu WY, Xie SL (2019) Long term spatial-temporal dynamics of fluoride in sources of drinking water and associated health risks in a semiarid region of northern China. Ecotoxicol Environ Saf 171:274–280. https://doi.org/10.1016/j.ecoenv.2018.12.090
59. McMahon PB, Brown CJ, Johnson TD, Belitz K, Lindsey BD (2020) Fluoride occurrence in United States groundwater. Sci Total Environ 732(139217):1–15. https://doi.org/10.1016/j.scitotenv.2020.139217
60. Durrani TS, Farooqi A (2021) Groundwater fluoride concentrations in the watershed sedimentary basin of Quetta Valley, Pakistan. Environ Monit Assess 193:1–18. https://doi.org/10.1007/s10661-021-09365-8
61. Chen Q, Hao D, Gao Z, Shi M, Wang M, Feng J, Deng Q, Xia L, Zhang C, Yu Y (2020) The enrichment process of groundwater fluorine in sea water intrusion area of Gaomi City, China. Groundwater 58(6):882–891. https://doi.org/10.1111/gwat.12990
62. Cao W, Zhang Z, Guo H, Fu Y, Gao Z, Nan T, Ren Y, Li Z (2023) Spatial distribution and controlling mechanisms of high fluoride groundwater in the coastal plain of Bohai Rim, North China. J Hydrol 617:128952. https://doi.org/10.1016/j.jhydrol.2022.128952
63. Rahman MM, Bodrud-Doza M, Siddique T, Zahid A, Islam ARMT (2020) Spatiotemporal distribution of fluoride in drinking water and associated probabilistic human health risk appraisal in the coastal region, Bangladesh. Sci Total Environ 724:138316. https://doi.org/10.1016/j.scitotenv.2020.138316
64. Jagtap S, Yenkie MK, Labhsetwar N, Rayalu S (2012) Fluoride in drinking water and defluoridation of water. Chem Rev 112(4):2454–2466. https://doi.org/10.1021/cr2002855
65. Kashyap SJ, Sankannavar R, Madhu GM (2021) Fluoride sources, toxicity and fluorosis management techniques – a brief review. J Hazard Mater Lett 2:1–8. https://doi.org/10.1016/j.hazl.2021.100033
66. Mohapatra M, Anand S, Mishra BK, Giles DE, Singh P (2009) Review of fluoride removal from drinking water. J Environ Manage 91(1):67–77. https://doi.org/10.1016/j.jenvman.2009.08.015
67. Ingle NA, Dubey HV, Kaur N, Sharma I (2014) Defluoridation techniques: which one to choose. J Health Res Rev 1:1–4
68. Kumar R, Sharma P, Rose PK, Sahoo PK, Bhattacharya P, Pandey A, Kumar M (2023) Co-transport and deposition of fluoride using rice husk-derived biochar in saturated porous media: effect of solution chemistry and surface properties. Environ Technol Innov 30:103056
69. Gutierrez L, Nocella G, Ghiglieri G, Idini A (2021) Willingness to pay for fluoride-free water in Tanzania: disentangling the importance of behavioural factors. Int J Water Resour Dev:1–20. https://doi.org/10.1080/07900627.2021.1996341
70. Benzian H, Guarnizo-Herreño CC, Kearns C, Muriithi MW, Watt RG (2021) The WHO global strategy for oral health: an opportunity for bold action. Lancet 398:192–194. https://doi.org/10.1016/S0140-6736(21)01404-5
71. Lowery G, Flinders M, Gibson BJ (2021) When evidence alone is not enough: the problem, policy and politics of water fluoridation in England. Evid Policy 17:507–523. https://doi.org/10.1332/174426420X16079614941921

Nitrate Contamination in Groundwater of Arid and Semi-Arid Regions, Ecotoxicological Impacts, and Management Strategies

Subhash Chander, Sangita Yadav, and Asha Gupta

Contents

1 Introduction ... 340
2 Detection and Analysis of Nitrate ... 341
2.1 Ion Chromatography ... 342
2.2 Colorimetry ... 342
2.3 Ion-selective Electrode ... 342
2.4 Nitrate Test Strip ... 342
3 Sources of Nitrate Contamination in Groundwater of Arid and Semi-arid Regions ... 344
3.1 Natural Sources ... 344
3.2 Anthropogenic Sources ... 344
4 Drinking Water Standards ... 345
5 Nitrate as a Global Groundwater Pollutant in Arid and Semi-Arid Regions ... 345
6 Identification of Various Nitrate Sources in Groundwater ... 351
7 Nitrogen Transformation Processes ... 352
8 Effects of Nitrate on Human Health and Environment ... 354
8.1 Effects on Human Health ... 354
8.2 Environmental Health Effects ... 355
9 Technologies for Nitrate Remediation from Groundwater ... 355
10 Management Strategies for Safe Water Supply in Arid and Semi-arid Regions ... 356
10.1 Effective Framework for the Management of Groundwater ... 356
10.2 Nitrate Contamination Management Strategies ... 358
10.3 Options for Safe Drinking Water Supply ... 361
11 Summary and Future Perspective ... 362
12 Conclusion ... 362
13 Recommendations ... 363
References ... 364

S. Chander (✉), S. Yadav, and A. Gupta
Department of Environmental Science and Engineering, Guru Jambheshwar University of Science & Technology, Hisar, India
e-mail: ashagupta@gjust.org

Shakir Ali and Abdelazim Negm (eds.), *Groundwater Quality and Geochemistry in Arid and Semi-Arid Regions*, Hdb Env Chem (2024) 126: 339–370, DOI 10.1007/698_2023_1047,

Published online: 19 December 2023

Abstract Groundwater is the primary source of drinking and irrigation in arid and semi-arid regions. In the last few decades, groundwater contamination by nitrate has reached its maximum levels. Several geogenic and anthropogenic sources were found to be responsible for the nitrate contamination. Studies around the globe show that the extensive use of nitrogen-based fertilizers is the principal cause of nitrate contamination in arid and semi-arid aquifers. Nitrate in the drinking water can harm human health by resulting in methemoglobinemia, infectious diseases, thyroid problems, and increased risk of colorectal cancer. Therefore, the growing demand for groundwater, especially in arid and semi-arid regions, necessitates the development of effective nitrate removal strategies. Several existing technologies, such as reverse osmosis, ultrafiltration, chemical and biological denitrification, ion exchange, adsorption, and electrodialysis, can remove nitrate from groundwater. However, their applicability is contingent on several variables, including necessary infrastructure, the cost-effectiveness of the technology, scalability, and its widespread acceptance. Management of nitrate-contaminated groundwater entails source reduction, removal or transformation technologies, groundwater conservation, education, legislation, and guiding principles. Thus, this chapter focuses on nitrate contamination in groundwater, health and environmental impacts, management strategies, and options for safe water supply in arid and semi-arid regions worldwide.

Keywords Arid and semi-arid regions, Groundwater nitrate, Groundwater quality management, Human health effects, Methemoglobinemia, Remediation technologies

1 Introduction

Sustainable development goals created by the UN general assembly require the provision of high-quality drinking water. According to the WHO, one-third of the global population lacks access to clean and safe drinking water. Groundwater is a principal source of fresh water, which provides almost 50% of the world's drinking water and around 43% of irrigation water. This resource is under threat from several factors, including climate change, land use, and rapid population growth [1]. The quality and quantity of many aquifers in arid and semi-arid regions worldwide are degrading, especially where groundwater is the only source of drinking and irrigation. A decline in the water table and deterioration of groundwater, especially with nitrate contamination, is the major problem in arid and semi-arid regions [2]. Several natural and anthropogenic nitrate sources can contaminate groundwater. Some of the primary reasons for an elevated level of nitrate in aquifers of arid and semi-arid regions include mineralization of organic plants, agricultural activities (mainly inorganic fertilizers), industrial activities, human waste disposal (septic and sewage disposal), and nitrification of soil organic nitrogen [2–5]. Agricultural irrigation return flows in arid and semi-arid regions often contain elevated levels of salts, nitrate, and pesticides [6]. Numerous studies have shown that groundwater nitrate is driven majorly by the extensive use of fertilizers or manure in agro-based activities in these regions [7, 8].

The WHO [9] has established 50 mg/L as the safe drinking water level for nitrate, while the Bureau of Indian Standards (BIS) has set this limit to 45 mg/L (IS: 10500-2012). Nitrate levels in drinking water that exceed this limit can impair ecosystems and human health. Blue baby syndrome or methemoglobinemia is one of drinking water's most visible side effects with nitrate concentrations above the WHO-recommended limit [7]. Furthermore, the elevated levels of nitrates can cause infectious diseases, thyroid issues, increased risk of colorectal cancer, methemoglobinemia, congenital disabilities, possibly stomach cancer, and low birth weight [1, 10, 11]. The overgrowth of aquatic plants and algae due to excess nitrates in surface water causes eutrophication [12]. It can cause permanent damage to aquatic ecosystems, even to the point of causing mass fish mortality. Likewise, irrigation with nitrate-polluted groundwater may harm crop production. The Food and Agriculture Organization (FAO) has established a threshold value of 22 mg/L for irrigation water; a level above this may damage sensitive crops like sugar beet or grapes [1].

In arid and semi-arid regions, alternative water supply sources are becoming scarcer while groundwater demand is rising. There is an urgent need to develop technologically and economically sustainable, accessible, and practical solutions for mitigating nitrate pollution [7]. Several existing technologies, such as reverse osmosis, ultrafiltration, chemical and biological denitrification, ion exchange, adsorption, and electrodialysis are capable of removing nitrate from groundwater [7, 13]. However, their applicability depends on several variables, including necessary infrastructure, the cost-effectiveness of the technology, and its widespread acceptance and scalability [11]. It is also imperative to develop and implement nitrate management measures for groundwater. Nitrogen source inventories, basin management plans, and identifying and quantifying primary sources and their loads to groundwater are some strategies for reducing nitrate pollution. The management of nitrate-contaminated groundwater in arid and semi-arid regions should include source reduction measures, removal or transformation technologies, groundwater conservation, educational actions, legislative efforts, and practical guidelines [10, 14–16]. Therefore, this chapter aims to focus on nitrate contamination in groundwater, their health and environmental impacts, management strategies, and options for safe water supply in arid and semi-arid regions globally.

2 Detection and Analysis of Nitrate

Numerous techniques can be utilized to detect and analyze nitrate in groundwater. Before analysis, it is necessary to consider some common factors, such as proper sampling, storage conditions, interference ions, etc. The sample must be filtered through 0.45 μm membranes to remove turbidity and bacteria. Those samples that cannot be analyzed immediately should be refrigerated at 4°C and must not acidify because rapid oxidation of nitrite to nitrate happens at lower pH. Several widely known analytical methods for nitrate determination and their fundamental features are discussed here.

2.1 Ion Chromatography

Ion chromatography is the most extensively used analytical technique for analyzing nitrate in groundwater. This technique is based on ion exchange and conductivity-based detection. It also permits the analysis of additional anions in water samples, such as nitrite, chloride, fluoride, sulfate, and nitrate. Ion chromatography utilizes ion exchange resins to separate atomic or molecule ions based on their interaction with the specific resin. The advantages include being free from ionic interference, high accuracy and precision, a variety of detection modes, high separation efficiency, selectivity, and speed and detection thresholds ranging from 0.01 to 1 mg/L [12, 17–19]. However, a disadvantage of the technique is that organic acids may affect analytical procedures.

2.2 Colorimetry

Many colorimetric methods are available for nitrate analysis in the water samples; they use copper-treated cadmium metal to reduce nitrate to nitrite. Nitrite is then combined with additional regents to produce a highly colored diazonium dye that can be detected at 520 nm. However, cadmium and hydrazine used in these techniques generate toxic by-products; hence waste disposal must be regulated [2]. For nitrate analysis, similar enzymatic approaches may utilize hydrazine or nitrate reductase. The enzymatic approach has the benefit of avoiding the harmful effects of cadmium and hydrazine.

2.3 Ion-selective Electrode

Ion-selective electrodes can detect nitrate in groundwater samples with high precision. Potentiometric measurements of nitrate using ion-selective electrodes allow relatively rapid measurement of NO_3^- -N concentration ranging from 0.14 to 1,400 mg/L. However, this method is susceptible to significant interferences and requires linear calibration and controlled conditions for reliable results [2].

2.4 Nitrate Test Strip

A sample can be screened for nitrate interferences before analysis using test strips. Test strips are easy and quick but inaccurate in the evaluation process. For example,

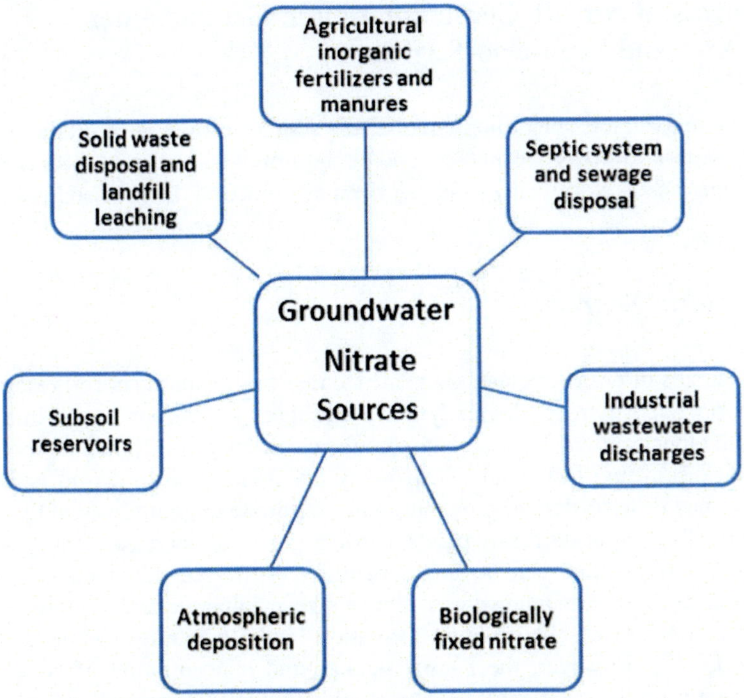

Fig. 1 Possible groundwater nitrate sources in arid and semi-arid regions

Table 1 Sources of nitrate in arid and semi-arid regions

Sources	Descriptions	Examples
Point	Single identifiable source and high concentration at a particular location	Concentrated animal confinement areas, leaky septic tanks, manure storage areas, accidental spills of nitrogen-rich chemicals, and dairy lagoons
Diffuse	Multiple sources dispersed around a region emit pollutants and have long-term impacts on human health and the ecosystem	Nitrogenous fertilizer, manure, and chemicals in agriculture, sewage pipe leaks, inappropriate household waste disposal, mining activities, dissolved nitrogen in precipitation, and return flow after irrigation

(Source: Adopted from [20]; [21])

Hach™ test strips are widely used based on the color change in response to the nitrate concentration and allow rapid evaluation of nitrate [2].

3 Sources of Nitrate Contamination in Groundwater of Arid and Semi-arid Regions

According to available scientific literature, the sources of nitrate in groundwater in arid and semi-arid areas are natural and/or anthropogenic (Fig. 1). As stated in Table 1, these sources can also be categorized as point and diffused sources.

3.1 Natural Sources

Natural sources of nitrate include geogenic (nitrate from natural subsoil reservoirs), atmospheric deposition, biologically fixed nitrogen, and groundwater-immanent input from other aquifers that may be hydraulically connected [2]. Nitrate reservoirs have been discovered in the subsoil of many dry regions of the world, and these reservoirs may be a substantial geogenic source of nitrate in groundwater [12]. Additionally, the fixation of nitrate by plants in arid regions can increase nitrate levels in groundwater [6]. Nitrate can be found naturally in nitrate salt deposits such as sodium nitrate. The continuing interaction between minerals and bacteria located in fissures and crevices in geologic formation leads to nitrate contamination of groundwater [7]. However, the natural background concentration of NO_3^--N in groundwater is far below 10 mg/L due to precipitation infiltration and mineralization of organic plants and animals; if these concentrations rise, it could be due to agricultural, industrial, or human waste disposal [22].

3.2 Anthropogenic Sources

Human actions, directly and indirectly, affect the quality of groundwater. Many anthropogenic factors affect the augmentation of nitrate in groundwater, like excessive use of fertilizers, septic systems, and human-induced wastes [23, 24]. Over-application and unscientific use of nitrogen-based fertilizers is the primary culprit of nitrate pollution in arid and semi-arid aquifers [2, 25]. Ammonium in inorganic fertilizers converts to the more mobile nitrate form in an oxidizing soil environment. Enzyme urease converts urea into nitrate, which is then utilized by plants or leaches into shallow aquifers [22]. Further, irrigated agriculture on heavily fertilized sandy soils is more susceptible to nitrate leaching. A variety of sources, including agriculture (primarily inorganic fertilizers, livestock manure, etc.), industry (untreated and poorly treated industrial wastewater), human waste disposal (septic and sewage disposal), landfill leaching, manure ponds, and polluted river and aquifer interactions, all contribute to nitrate contamination in groundwater [2–6, 8, 26]. Regarding nitrogen-related water quality indicators (nitrate, nitrite, and ammonia), agriculture sector pollution exceeds that of urban and industrial sources [6]. The primary causes

Table 2 Limit of nitrate concentration in drinking water permitted by various agencies

Organizations/ agencies	Conc. as NO_3^- (mg/L)	Con. as NO_3^--N (mg/L)	References
WHO	50	10	Zendehbad et al. [28], WHO [9]
BIS	45	–	Singh et al. [13]; IS:10500-2012 [29]
US-EPA	45	10	Xin et al. [15], EPA [30]
EDWD	50	–	Xin et al. [15], Agarwal et al. [31]
MEP, China	–	10	Agarwal et al. [31]

MEP Ministry of Environmental Protection, *EDWD* European Drinking Water Directive

of nitrate pollution in developing nations are low living standards, inadequate sanitation, leaking septic tanks, and improper sewage disposal [1]. Similarly, nitrate concentrations are higher in many urban areas due to increasing human and animal waste [23]. Furthermore, stable isotope studies indicate that most nitrate in groundwater of arid and semi-arid regions is due to fertilizers and human waste [12]. A small contribution of nitrate may be from the industrial sectors that use nitric acid, urea, and anhydrous ammonia. In addition, as the forest has a high capacity for nitrogen transfer, deforestation also results in nitrate leaching into groundwater [27].

4 Drinking Water Standards

Primary drinking water regulations are intended to safeguard public health from specific contaminants such as nitrate. High nitrate levels in drinking water can pose several health risks; consequently, various agencies worldwide have established safe nitrate levels in drinking water. Environmental protection agencies set a limit of 10 mg/L for NO_3^--N in drinking water, below which no adverse effects on human health due to methemoglobinemia were observed [22]. A comparison of the nitrate concentration standards established by various agencies is shown in Table 2.

5 Nitrate as a Global Groundwater Pollutant in Arid and Semi-Arid Regions

Nitrate is a tasteless, odorless form of nitrogen and is naturally produced in the soil and other mediums, such as groundwater. It is an essential component of the nitrogen cycle and is used by most plants as a macronutrient. Nitrate can leach easily into the aquifers from the unsaturated soil zone because of high solubility and mobility in water [21]. Due to its significant solubility, it is known as the most prevalent pollutant in groundwater. Nitrate may be represented in drinking water as nitrate and nitrate-nitrogen [15]. The aridity index classifies arid lands into a desert (i.e., hyper-arid and arid) and semi-desert (i.e., semi-arid). These regions are characterized by fluctuating precipitation, high evaporation rates, and an annual wet and dry season [32]. About one-third of the world's population resides in drylands, which account for about 41% of the planet's surface area [33]. Most people in these regions rely on the groundwater supply for daily requirements. Additionally, a considerable proportion of the population relies on agricultural activities for survival. Over the past several decades, unsustainable agrarian practices have increased the potential of groundwater pollution with nitrates [14]. Agricultural irrigation return flows contain high salts and nitrate concentrations, eventually leaching and contaminating groundwater [6]. In addition, urbanization, industrialization, and waste disposal can contribute significantly to groundwater nitrate contamination worldwide [2]. These anthropogenic activities demonstrate that nitrate is the most prevalent pollutant in the groundwater of arid and semi-arid regions.

Studies have shown that nitrate is the most prevalent pollutant in the aquifers of arid and semi-arid regions worldwide. Alsabti et al. [34] found that 68% of groundwater samples of Kuwait Bay had nitrate concentrations above WHO standards, ranging from 22.7 to 803.9 mg/L due to anthropogenic factors such as fertilizer use and urbanization. From 1991 to 2003, a total of 5,101 groundwater wells were sampled in 51 research studies across the United States; more than 4% of the sampled wells had nitrate levels above the EPA [30] limit of NO_3^--N [35]. Shukla and Saxena [27] pointed out that San Joaquin Valley (United States) is the nitrate's epicenter and affects over 275,000 people. Rahmati et al. [36] reported that 12.9% of samples from the Ghorveh-Dehgelan aquifer in Kurdistan (Iran) surpassed the maximum permissible level set by WHO [9]. Antiguedad et al. [37] observed the presence of nitrate concentrations in many alluvial floodplains in Europe. According to Beutel et al. [38], nitrate concentrations exceeding 10 mg/L as NO_3^--N are most common in the eastern alluvial fans subregion Central Valley of California. Nawale et al. [39] point out that the Wardha sub-basin (India) has a high health risk of non-carcinogenic disease due to drinking nitrate-contaminated groundwater. Adimalla [40] demonstrates that the aquifers of Telangana (India) have a concentration of nitrate (NO_3^-) ranging from 17 to 120 mg/L, and around 57% of samples were above the BIS permissible limits for drinking water. Zendehbad et al. [28] found that the urban aquifer of Mashhad (Iran) has excessive nitrate in 110 wells out of 261 wells due to sewage contamination. Jandu et al. [41] found that 86% of samples had nitrate content higher than the WHO maximum safe limit and found to be in the range of 10.2 to 519.6 mg/L in Jhunjhunu, Rajasthan (India). Ahadal and Suthar [42] studied the Malwa region of Punjab (India) and found that over 92% of sites have higher nitrate than the WHO recommendation. Waste dump sites, animal waste, nitrogen-based fertilizers, and industrial effluents are the foremost reasons for contamination. Further, Table 3 demonstrates the groundwater nitrate, possible sources, and sample percentages exceeding various drinking water standards worldwide in arid and semi-arid regions. In addition, Fig. 2 depicts sampling locations/ regions of reported nitrate in arid and semi-arid regions of the world and Fig. 3 gives

Table 3 Regions/locations demonstrating groundwater nitrate concentration along with possible sources and sample percentages exceeding various drinking water standards in arid and semi-arid regions worldwide

Location/ region no.	Region/location	Climate	Reported value as NO_3^- (mg/L)	Percentage of sample > DWS (mg/L as NO_3^-)	Possible nitrate sources	References
1	Churu Rajasthan, India	Arid and Semi-arid	Min = 0.8 Max = 498.7 Mean = 44.7	NS = 515 28.54% > BIS	Anthropogenic activities	Tanwer et al. [43]
2	South Kuwait's Bay, Kuwait	Arid	Min = 22.7 Max = 803.9 Mean = 143.8	NS = 19 68.42% > WHO	Agricultural fertilizers, sewage disposal and landfills	Alsabti et al. [34]
3	Cuddapah, South A.P, India	Semi-arid	Min = 23.2 Max = 110.8	NS = 30 86% > BIS	Anthropogenic activities	Sunitha et al. [44]
4	Maadher central parts, Hodna, Algeria	Semi-arid	Min = 12 Max = 407 Mean = 173	NS = 33 67.64% > WHO	Agricultural activities and dumping sites	Selmane et al. [45]
5	Djelfa region, Algeria	Semi-arid	Min = 3 Max = 336	NS = 19 58% > WHO	Agricultural fertilizers and livestock waste	Ali Rahmani and Chibane [46]
6	Coastal parts, Southern Saudi Arabia	Arid	Min = 7 Max = 124 Mean = 48	NS = 80 45% > WHO	–	Masoud et al. [47]
7	North-eastern, Iran	Semi-arid	–	NS = 82 48% > HWC	Agricultural fertilizers and sewage effluents	Atabati et al. [48]
8	Nile Valley, Qena City, Egypt	Arid	Min = 0.1 Max = 257 Mean = 53	NS = 41 30% > WHO	Chemical fertilizers and manure	Mohammed et al. [49]
9	Shekhawati region, Northern India	Semi-arid	Min = 2 Max = 1803	NS = 163 65% > BIS	–	Singhal et al. [50]
10	Tiruppur region, Southern India	Semi-arid	Min = 10 Max = 290 Mean = 83.45	NS = 40 58% > WHO	Agricultural fertilizers, manure and septic tanks	Karunanidhi et al. [51]

(continued)

Table 3 (continued)

Location/ region no.	Region/location	Climate	Reported value as NO_3^- (mg/L)	Percentage of sample > DWS (mg/L as NO_3^-)	Possible nitrate sources	References
11	Telangana, India	Semi-arid	Min = 12 Max = 212 Mean = 69	NS = 105 71% > BIS	–	Adimalla et al. [52]
12	Kano plains, Kisumu city, Kenya	Arid and Semi-arid	Min < 0.04 Max = 90.6	NS = 62 63% > WHO	Sewage and manure application	Nyilitya et al. [53]
13	Telangana, India	Semi-arid	Min = 17 Max = 120 Mean = 58.74	NS = 35 57% > WHO	Agricultural fertilizers, human and livestock waste	Adimalla and Li [23]
14	Telangana, India	Semi-arid	Min = 4 Max = 440 Mean = 73	NS = 194 52% > WHO	Agricultural activities	Adimalla and Wu [10]
15	Yamuna sub-basin Panipat, Haryana, India	Semi-arid	Min = 0.5 Max = 69 Mean = 12.8	NS = 74 6.6% > WHO	–	Kaur et al. [54]
16	Guanzhong basin, China	Semi-arid	Min = 0 Max = 90 Mean = 18.26	NS = 191 24.61% > WHO	Agricultural fertilizers, manures, septic tanks and organic effluent	Zhang et al. [55]
17	Bardaskan, Southeast Iran	Arid	Min = 0 Max = 77.37 Mean = 17.57	NS = 30 6.6% > WHO	–	Radfarda et al. [56]
18	Sylhet, Bangladesh	Semi-arid	Min = 0 Max = 25.86 Mean = 5.70	NS = 23 -	–	Ahmed et al. [57]
19	Varamin aquifer, Tehran, Iran	Semi-arid	Min = 0.22 Max = 74.14	NS = 70 11.42% > WHO	Agricultural fertilizers, domestic waste and sewage	Nejatijahromi et al. [58]

20	Nirmal Province, Telangana, India	Semi-arid	Min = 0.8 Max = 80 Mean = 36.91	NS = 34 26% > BIS	Agricultural fertilizers, sewage disposal and septic system	Adimalla et al. [59, 60]
21	Malwa region, Punjab, India	Semi-arid	Min = 38.45 Max = 198.05 Mean = 33.45	NS = 76 92% > BIS	Agricultural fertilizers	Ahada and Suthar [42]
22	Scopia basin, Central Greece	Semi-arid	Min = 6.2 Max = 134.6	NS = 41 29% > WHO	Geogenic, fertilizers and wastewater	Charizopoulos et al. [61]
23	Halayieb, Egypt	Arid	Min = 11.7 Max = 131.27	NS = 11 54.54% > WHO	–	Zaki et al. [62]
24	Bou-Areg, Nador region, Morocco	Semi-arid	Min = 16 Max = 300	NS = 94 50% > WHO	Agricultural activities	Re and Sacchi [63]
25	Zhongning, Ningxia, Northwest China	Arid	Min = 59.6 Max = 103 Mean = 17.9	NS = 50 60% > WHO (NO_3^- -N; 10 mg/L)	Agricultural fertilizers and wastewater	Chen et al. [64]
26	Lar area, Southern Iran	Arid	Min = 1.5 Max = 70.7	NS = 34 5.9% > WHO	Agricultural activities	Rezaei et al. [65]
27	Kharkiv City, East Ukraine	Semi-arid	Mean = 31.8 –	NS = 25 30% > WHO	Sewage disposal	Vystavna et al. [66]
28	Bahira plain aquifer, Central Morocco	Arid	Min = 0.5 Max = 584	NS = 62 18% > WHO	Agricultural activities	Karroum et al. [67]
29	Ghorveh Dehgelan aquifers, Western Iran	Semi-arid	Min = 12.4 Max = 111.6 Mean = 32.15	NS = 93 12.9% > WHO	Agricultural fertilizers	Rahmati et al. [36]
30	Basin of Seversky Donets River, Russia/Ukraine	Semi-arid	Min = 37.2 Max = 189.9	NS = 39 7.69% > WHO	Natural and anthropogenic	Vystavna et al. [68]
31	Southeast, Granada, Peninsula	Semi-arid	Min = 4 Max = 561 Mean = 74.5	NS = 175 -	–	Rodriguez-Galiano et al. [69]

(continued)

Table 3 (continued)

Location/ region no.	Region/location	Climate	Reported value as NO_3^- (mg/L)	Percentage of sample > DWS (mg/L as NO_3^-)	Possible nitrate sources	References
32	Southern Hodna, Algeria	Arid	Min = 6 Max = 158 Mean = 76.28	NS = 18 61% > WHO	Agricultural fertilizers	Abdesselam et al. [25]
33	Basin and ranges of USA	Semi-arid	Min = 0.1 Max = 130	NS = 3,539 8% > WHO	Agricultural fertilizers or urban waste	Anning et al. [70]
34	Toyserkan, western Iran	Semi-arid	Min = 1 Max = 162 Mean = 30	NS = 95 9.5% > WHO	Manure and inorganic fertilizers	Jalali [71]
35	Castilla-La Mancha, Southeast Spain	Semi-arid	Min = 0.4 Max = 125	NS = 684 –	Agricultural activities	Moratalla et al. [72]
36	Tumkur Taluk, Karna-taka, India	Semi-arid	Min = 0.44 Max = 261 Mean = 54.3	NS = 269 48.5% > BIS	–	Ramakrishnaiah et al. [73]
37	Badain Jaran desert, Northwest China	Arid	Min = 0.1 Max = 113	NS = 52 32%, > WHO	Agricultural activities nitrification, subsistence farming and grazing	Gates et al. [74]
38	Serowe, Botswana	Semi-arid	Min = 0 Max = 219 Mean = 22.8	NS = 51 –	Agricultural fertilizers, manures, septic tanks and precipitation	Stadler et al. [75]
39	Livermore, USA	Semi-arid	Min = 5.2 Max = 60.2	NS = 35 17% > WHO	Agricultural fertilizers and livestock waste	Moore et al. [76]

No. of Samples (NS); Drinking Water Standards (DWS): (BIS, 45 mg/L; WHO, 50 mg/L; HWC, NO_3^--N; max., 10 mg/L)

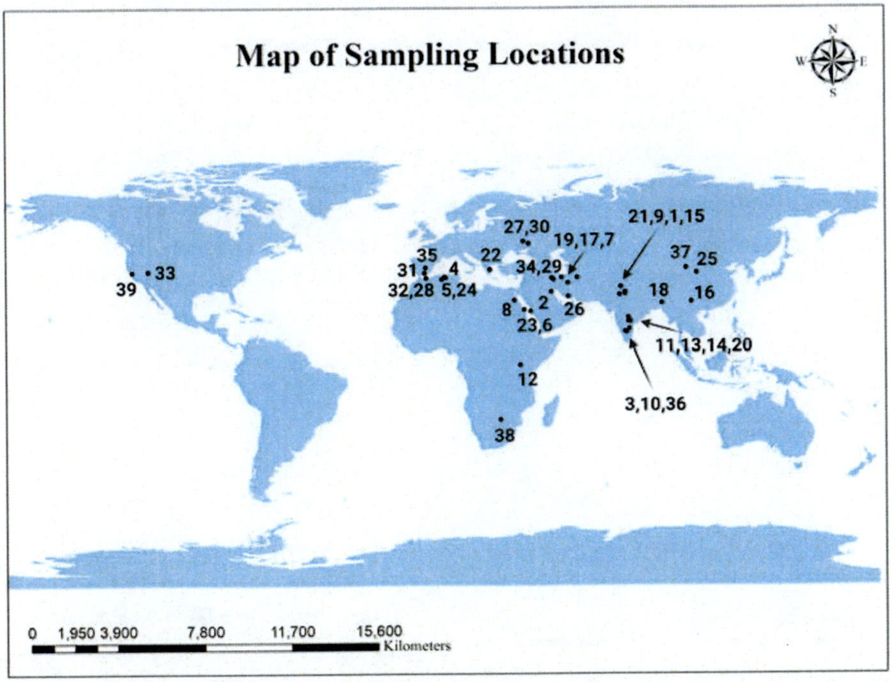

Fig. 2 Sampling locations/regions of reported nitrate in groundwater of arid and semi-arid regions

a visual representation of sampling locations together with the percentage of samples exceeding various nitrate drinking water guidelines.

6 Identification of Various Nitrate Sources in Groundwater

Although there are several approaches for identifying nitrate sources in groundwater, the stable dual isotopes (nitrogen and oxygen) approach is extensively used and widely accepted to identify agricultural fertilizers, manure, human waste, and other sources. Many scientific studies globally successfully used $\delta^{15}N$ and $\delta^{18}O$ isotope composition of NO_3^- to identify different sources, fate, and their related contributions to nitrate in aquifers [24, 28, 77]. The numerous sources (e.g., atmospheric, agriculture fertilizer and sewage, or manure) have distinct compositions of nitrogen ($^{15}N/^{14}N$) and oxygen ($^{18}O/^{16}O$) isotopes, which are widely used for source identification of Nitrate [77, 78]. However, a homogeneous signal of dual isotopes in aquifers reveals naturally occurring nitrate [75]. Nitrate derived from fertilizers and sewage has a distinct range of ^{15}N- NO_3^-, whereas soil microbial and atmospheric source has a different range of ^{18}O- NO_3^- [78]. When numerous nitrate sources are

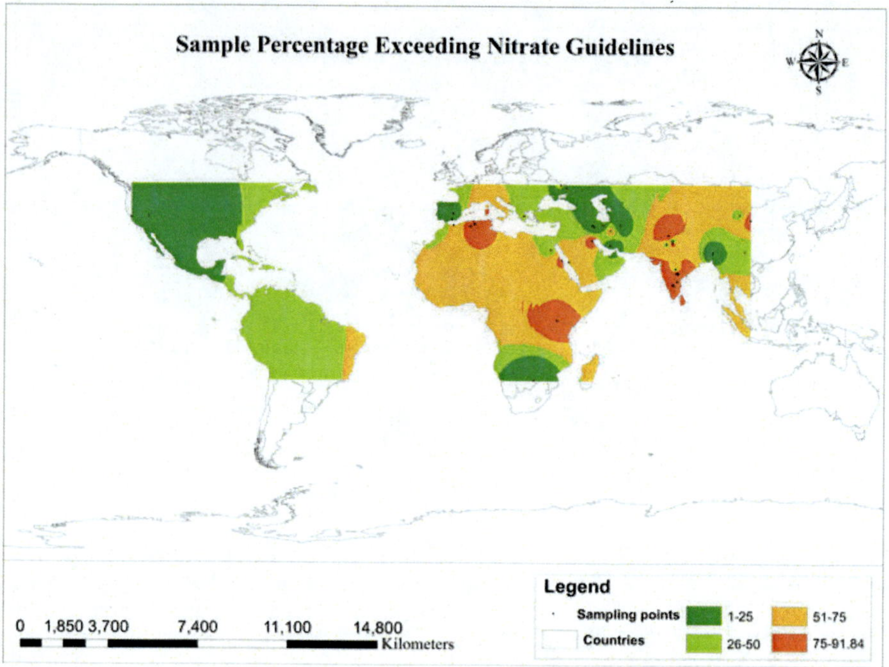

Fig. 3 Locations along with the percentage of samples exceeding various nitrate drinking water guidelines

present, isotopic quantification is also accompanied by evaluation uncertainty and lacking [77].

7 Nitrogen Transformation Processes

Nitrogen is accessible to plants through ammonium and nitrate via nitrification or nitrogen fixation activities within the root zone. Some bacterial species, including those that interact with the roots of higher plants and those that are free-living, can assimilate atmospheric nitrogen. Some fungi and blue-green algae species can also assimilate atmospheric nitrogen. Under aerobic conditions, Nitrosomonas species convert organic nitrogen, ammonium ion (NH_4^+), or ammonia to NO_2^- (nitrite), which is then converted into nitrate by nitrite-oxidizing bacteria such as Nitrobacter species [2]. Nitrate is created when soil-dwelling aerobic and anaerobic bacteria decompose dead plants and other organic remains into ammonium ions, which are then changed into nitrate. The soil biota quickly converts ammonium to nitrate in soils under aerated or oxidizing conditions. Globally, around 193 million tons of biological nitrogen fixation (land and seas) and 94 million non-biological (atmospheric lightning and industrial) fixations occur [27, 79]. There are many ways to reduce nitrate levels, such as plant absorption, mineralization-immobilization

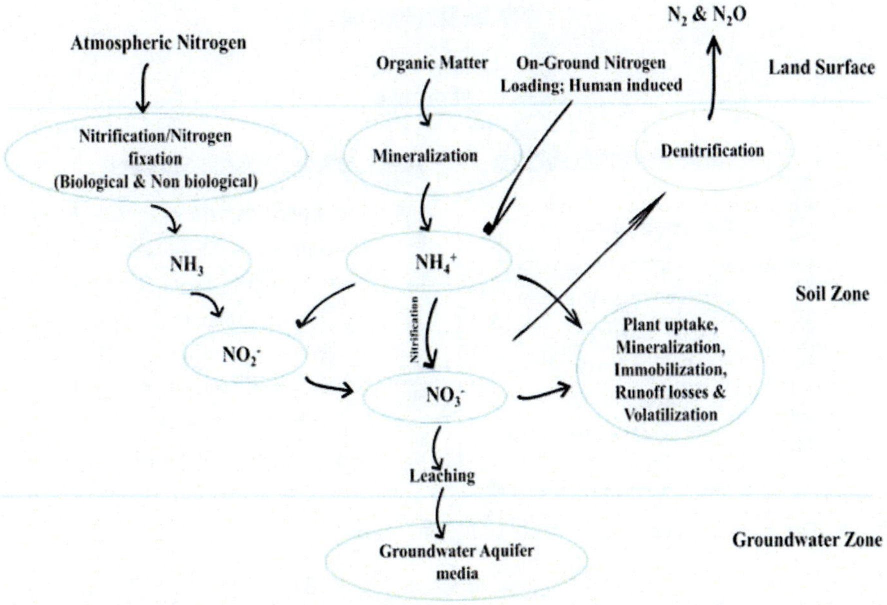

Fig. 4 Subsurface nitrogen transformation processes and nitrate leaching

processes, volatilization, runoff losses, and denitrification. These processes limit the nitrate flux into groundwater either individually or in combination.

However, nitrate ions are weakly bound to soil particles (negatively charged) and may percolate into the aquifer. When oxygen is scarce in soil for microbial respiration, microbial denitrification is frequently observed with greater than 60% pore saturation. Nitrate or nitrite is employed as the terminal electron acceptor in the respiratory process of microbial reduction of nitrate ions when oxygen is scarce. As a result, high energy molecule adenosine tri-phosphate is produced. The electron transfer during this phase provides energy to the denitrifying bacteria to stimulate new cell biomass [7]. Autotrophic and heterotrophic denitrification is essential for converting nitrate into nitrogen gas to reduce nitrate leaching in groundwater. Several factors influence nitrate leaching, including land use patterns, on-ground nitrogen loading, groundwater recharge, soil nitrogen dynamics, soil properties, and groundwater level [21]. Different ecosystems have varying capacities for nitrogen accumulation and transmission, which can be used to estimate the probability of nitrate contamination. An ecosystem's ability to accumulate nitrate is referred to as the accumulation potential, whereas the ability to transfer nitrate to another ecosystem is referred to as the transfer potential. The atmosphere and agricultural systems have substantial transmission potential, increasing groundwater pollution likelihood [27]. Figure 4 demonstrates the subsurface nitrogen transformation processes and nitrate leaching into groundwater.

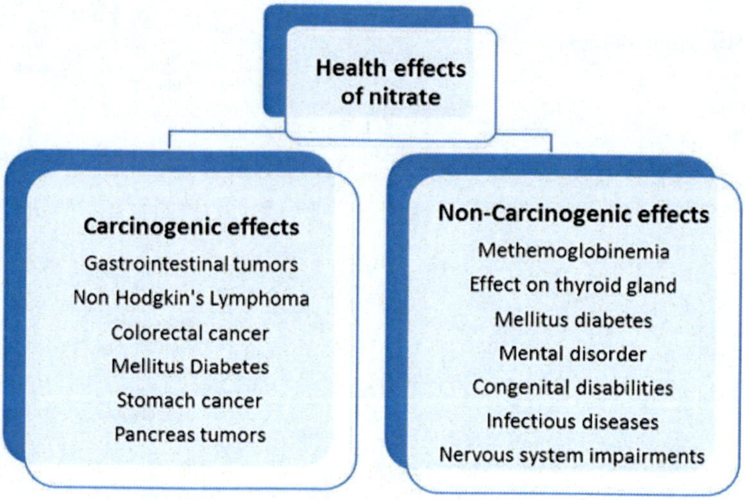

Fig. 5 The major health effects of nitrate in drinking water

8 Effects of Nitrate on Human Health and Environment

Nitrate, a prevalent groundwater pollutant in arid and semi-arid regions, can harm ecosystems and human health. The major effects of nitrate in drinking water are depicted in Fig. 5.

8.1 Effects on Human Health

Nitrate in drinking water has adverse health effects if consumed excessively for an extended time period. Humans usually consume nitrate through the consumption of drinking water and food beverages. Still, when the maximum contamination level in drinking water is exceeded, it can account for up to 50% of human nitrate consumption [27]. It can enter the bloodstream from the stomach and upper intestines via drinking water [20]. Most of the nitrite absorption into the bloodstream appears in the intestines. Blue baby syndrome, or methemoglobinemia, is one of the prominent health effects of drinking water with nitrate concentrations greater than the upper safe limit of WHO [9] for an extended period in infants under 6 months of age [7]. Bacteria in the infant gastrointestinal tract convert nitrate to nitrite. Nitrite oxidizes the iron of hemoglobin to generate methemoglobinemia, decreasing the blood's oxygen-carrying capacity. The babies have an unusual blue-grey skin tone associated with 10% or higher methemoglobin levels. If the illness is not diagnosed and treated promptly, it might result in shortness of breath, a heart attack, and mortality [17]. According to the USEPA, the Hazard Quotient (HQ) value for non-carcinogenic human health risks associated with nitrate in drinking groundwater is unity, with a value greater than unity reflecting an individual's susceptibility to non-carcinogenic health risk [54]. Studies also reported that consuming elevated levels of nitrates can cause weakness, vomiting, mental disorder, abdominal disorder, hypertension, dizziness, infectious diseases, nervous system impairments, thyroid issues, gastrointestinal tumors, non-Hodgkin's lymphoma, mellitus diabetes, stomach cancer, pancreas tumors, increased risk of colorectal cancer, congenital disabilities, possible stomach cancer (adults), and low birth weight in humans [1, 5, 10, 11, 24, 80, 81]. Nitrate has been identified as a potential human carcinogen that can produce N-nitroso compounds through endogenous nitrosation [82]. Nitrate has also been associated with chronic digestive diseases and an increased risk of digestive cancer [6].

8.2 *Environmental Health Effects*

Many streams and rivers rely on groundwater for base flow, and increased nitrate concentrations in groundwater can pollute these resources. When there is an abundance of nitrate in surface water, aquatic plants and algae grow more quickly, causing eutrophication [12]. Eutrophication is commonly associated with anthropogenic nitrate sources. When numerous algae die and decompose, the decomposers consume a substantial amount of oxygen, altering the aquatic ecosystem. The adverse effects of eutrophication include reduced light penetration, decreased plant productivity in deeper waters, and decreased oxygen content in the water body [20]. It can considerably contribute to the eutrophication of coastal and marine environments [2, 83]. Nitrates can cause permanent damage to aquatic ecosystems, even to the point of causing mass fish mortality. Nitrate contamination harms humans by lowering environmental quality, increasing health risks, and increasing environmental management costs. Irrigating with nitrate-contaminated groundwater may damage sensitive crops like sugar-beet or grapes. As a result, the FAO established a 22 mg/L threshold value for irrigation water for sensitive crops [1]. The nitrate-nitrogen concentration in water between 100 and 200 mg/L reduces livestock appetite [84].

9 Technologies for Nitrate Remediation from Groundwater

Technological and economically viable, accessible, and practical solutions are required to mitigate nitrate pollution. The increasing demand for groundwater necessarily involves the development of efficient nitrate removal strategies. Various technologies efficiently removed nitrate from groundwater worldwide depending on infrastructure, affordability, and acceptability. Furthermore, energy and cost-efficient nitrate removal technologies are required to achieve global sustainable development goals and quality standards. Researchers for removing nitrate from groundwater have proposed a wide range of in-situ and ex-situ technologies. The in-situ treatment method involves nitrate treatment at the site, while the ex-situ option primarily involves the pump and treatment method away from the site. However, the ex-situ method is most effective when the contaminant plume is well-defined. The limitations of this method include co-contaminant availability, operation and maintenance, and scale of operation for water treatment. The treatment technologies may be categorized into nitrate reduction and removal methods. Some globally accepted techniques for nitrate removal are ion exchange, reverse osmosis, adsorption, electrodialysis, chemical denitrification using zerovalent iron, and biological denitrification [1, 7, 13, 85–87]. Some of these techniques can be combined for increased effectiveness and offset other technologies' drawbacks. A few conventional nitrate removal techniques are summarized in Table 4.

10 Management Strategies for Safe Water Supply in Arid and Semi-arid Regions

In arid and semi-arid regions, groundwater must be managed sustainably because it is an essential resource for irrigation and drinking water. Developing and implementing management strategies is necessary to reduce the elevated nitrate concentration in aquifers. Also, technological and policy reforms are required to mitigate its effects on humans and the environment. An effective management system should include a well-abandonment strategy and source reduction measures. However, source reduction activities like best agriculture management practices, domestic wastewater treatment, municipal solid waste management, etc., improve groundwater quality over the years to decades, so in-situ remediation may also be considered for hotspot sites with short-term objectives. The management comprises non-structural measures in addition to structural measures like physical activities and construction projects. The non-structural measures include laws, regulations, funding, education, and policies.

10.1 Effective Framework for the Management of Groundwater

Groundwater management involves collecting and analyzing data to identify nitrate-contaminated areas and quantify the scope of the problem. The essential management consideration is the fate and transport of nitrate in unsaturated and saturated zones. The potential sources of contamination are identified to establish available management options that reduce nitrate levels below the established standards. Then, examine the environmental and economic aspects of the available options. Soil and

Table 4 Various conventional nitrate removal techniques, along with their description, benefits, and drawbacks

Techniques	Descriptions	Benefits	Drawbacks	References
Reverse osmosis	In reverse osmosis, groundwater is forced through a cell membrane at a pressure of 300 to 1,500 psi, leaving contaminants on one side and water on another	Continuous operation, used for nitrate-affected saline groundwater, can separate 0.1 to 1 nm pollutants size, post-treatments are not required	High costs of operation, membrane fouling and deterioration, maintenance of membrane and issues of brine effluent disposal	Singh et al. [13], Huno et al. [7]
Ion-exchange	A strong base anion exchange resin is used for NO_3^- exchange with Cl^- and CO_3^{2-} from groundwater.	Regeneration and reuse of exhausted resin, effectiveness, simple to operate, economical method, especially trimethylamine used for nitrate exchange	SO_4^{2-} ions reduce resin's nitrate removal ability, brine disposal, and pretreatment required	Singh et al. [13], Tokazhanov et al. [8]
Electrodialysis	In this technique, ions are transferred from a less concentrated to a more concentrated solution using direct electric voltage and membranes	May simultaneously remove contaminants and desalinate, with greater precision and simple operation	Alkaline conditions reduced the efficiency of nitrate separation, more energy demand and pretreatment required	Abascal et al. [1], Sharma and Bhattacharya [85]
Adsorption	It is a surface phenomenon in which various natural and synthesized sorbents, agri-waste by-products, and industrial wastes are used for pollutant remediation	Convenience, cost-effective, lower energy demands, used for removal of both organic and inorganic pollutants	Removal depends upon initial nitrate concentrations, a dose of adsorbent, reaction time, pH, and operating temperature	Singh et al. [13], Huno et al. [7], Chander et al. [87], Yadav et al. [88]
Biological denitrification	Biological denitrification involves the reduction of nitrate under anaerobic conditions by using bacterial species	Environment-friendly, cost-effective, used for in-situ and ex-situ remediation	Higher levels of nitrate are challenging to eliminate, long time required, need optimum carbon-to-nitrogen ratio, bacterial sludge, high monitoring needs, sensitivity to environmental conditions, risk of	Huno et al. [7]

(continued)

Table 4 (continued)

Techniques	Descriptions	Benefits	Drawbacks	References
			nitrite formation, and post-treatment required	
Chemical denitrification	In this method, chemicals like zerovalent iron, elemental sulfur, zinc, and aluminum are used to reduce nitrate from water	Complete nitrate ion reduction may be achieved using zerovalent iron under controlled acidic conditions	Condition-dependent, ammonia stripping and post-treatment are necessary	Singh et al. [13], Huno et al. [7]
Catalytical reduction	This method removes nitrite and nitrate from water using catalysts such as lead, copper $A1_2O_3$, palladium-alumina, etc.	Complete nitrate removal may be possible	Cost-effectiveness and ammonia formation issue	Tokazhanov et al. [8]
Photo-catalytical method	The method is based on the acceleration of photodegradation of organic pollutants, pathogens, and other pollutants in the presence of a catalyst	High selectivity for a particular pollutant	Formation of nitrite and ammonium, Reusability of the catalyst as it is unchanged during the process	Sharma and Bhattacharya [85], Zhang et al. [89]

groundwater models may be analyzed before decision implementation [21]. Figure 6 depicts the management framework for groundwater resources.

10.2 Nitrate Contamination Management Strategies

Globally, legislative measures are crucial requirements for the management of groundwater resources. Maintaining groundwater quality and preventing future nitrate pollution requires understanding the variables and processes influencing nitrate occurrence, transport, and fate. Nitrogen source inventories and basin management plans are essential for reducing nitrate from aquifers [20]. Preventive measures should be taken to avoid nitrate contamination. Land use planners, decision-makers, and environmental regulators must identify areas with high nitrate loads to implement preventative measures like manure storage in concrete pits to reduce leaching [90]. Furthermore, continuous seasonal groundwater quality monitoring is essential for implementing these measures. Numerous researchers such as

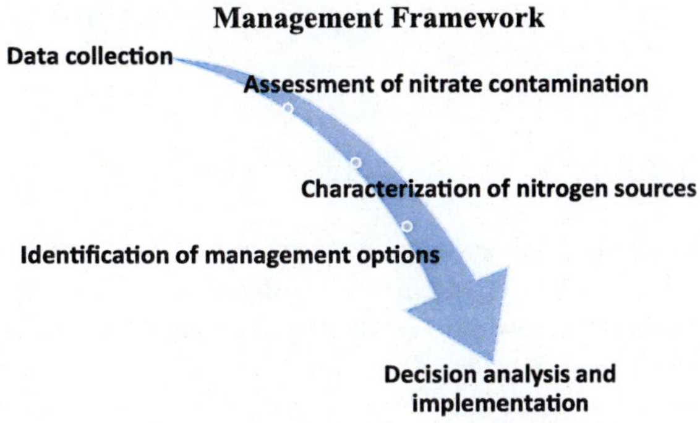

Fig. 6 Management framework for the groundwater resources (Permission from [21])

Singh et al. [13], Rahman et al. [14], Zhang et al. [16], Adimalla and Wu [10], Bastani and Harter [90], Li et al. [91], Han et al. [6], and Almasri [21] proposed nitrate management solutions such as source reduction, removal or transformation technologies, groundwater conservation, educational actions, legislative efforts, and guidelines, among others.

10.2.1 Agricultural Source Management

In agricultural areas, multiple sources may control the dynamics and occurrence of nitrate in groundwater. Here, management should be based on applying fertilizer and manure, cultivation techniques, and irrigation methods. Increasing fertilizer use efficiency, application quantity, and time and implementing integrated nutrient management will help farmers save money on fertilizer application and prevent long-term nitrate contaminations [92, 93]. To reduce reliance on fertilizers and the risk of fertilizer, new strains of nitrogen-fixing microorganisms (like Rhizobium and blue-green algae) with increased nitrogen-fixing capacity should be developed. Furthermore, long-term field research must be conducted to compile an up-to-date list of the best management techniques and application guidelines for fertilizers.

Additionally, various optimization models should be utilized to determine the optimal irrigation and groundwater storage options. Furthermore, each country must enact legislation for agricultural groundwater management, similar to the European Union's nitrates directive for reducing nitrate sources (EC 1991). Online resources for agricultural advice should be made available to decrease nitrate pollution. In 2010, the Chinese Ministry of Agriculture and Rural Affairs issued "Guidance for Scientific Fertilization of Major Crops", which included detailed irrigation and fertilization recommendations [94]. Suitable denitrification models should be developed for groundwater management; these models will reduce nitrate leaching. Implementing and maintaining artificial recharge schemes must involve non-governmental organizations and local governments. Society should be educated on groundwater quality and its proper management through seminars, short films, etc. Several mitigation tactics, such as balanced fertilization, crop rotation, adopting improved irrigation techniques, and implementing environmental legislation, can avert nitrate problems.

10.2.2 Domestic Wastewater Management

Expanding the sewerage network and centralizing the wastewater treatment system will mitigate the detrimental effects of improperly treated domestic wastewater discharge. However, providing complete sewer coverage to all rural and semi-urban areas in arid and semi-arid regions is not feasible due to economic constraints. Domestic wastewater in rural and semi-urban areas is a source of nitrate in groundwater; this issue can be resolved by implementing a decentralized or on-site wastewater treatment system. The wastewater must be collected, treated, and disposed of or reused close to the point of generation in a decentralized treatment system [15]. This technique typically settles solids in a septic tank, followed by treatment in secondary treatment facilities, such as anaerobic lagoons or constructed wetlands.

10.2.3 Solid Waste Management

The top priority of municipal solid waste management should establish a legal framework for regulating landfills and eliminating illegal dumpsites. These regulations typically address location restrictions, liner requirements, leachate collection and removal, and groundwater monitoring requirements from the standpoint of groundwater management. If waste is collected in properly designed, built, and maintained landfills, there is a low chance that contaminants will seep into the groundwater.

10.2.4 Treatment of Drinking Water

Groundwater is the principal source of domestic drinking water in arid and semi-arid regions of the world. It is expensive and time-consuming to treat highly nitrate-contaminated groundwater, so it is recommended to use alternate drinking water sources if they are available. Nitrate treatment technology should be deployed at drinking water treatment plants to improve the quality of nitrate-contaminated groundwater in regions without alternative water sources [9]. The polluted groundwater can be reused using water treatment technologies. Every country, mainly the developing world must set drinking water standards and provide water within these limits. Several conventional nitrate removal techniques and methods are outlined in Table 4, and they can be implemented in treatment plants based on the requirements.

10.2.5 Other Measures

Groundwater management and its use in conjunction with surface water are essential in arid and semi-arid regions. Recharging aquifers during abundant rainfall is one method of promoting this conjunctive use. Indigenous water management techniques may be used due to their local adaptability compared to more sophisticated and advanced techniques. The nitrate concentration of a particular region must be depicted on several regional or local maps and these maps should be digitized to effectively manage nitrate pollution in groundwater aquifers. Further, GIS should be used to assess the effectiveness of various management strategies because it significantly improves data collection and processing, evaluation of the nitrate leaching risk index, identification of diverse vulnerability zones, model development, and scenario planning for management options. The only appropriate nitrate standard has been set for groundwater; managers should handle these within the scope of the profile from the surface to groundwater. Furthermore, mathematical models of nitrogen transport must be developed to quantify the outcomes of management options before their actual implementations at various spatial and temporal scales. Water experts should increase their research on water quantity and quality to aid government decision-making and achieve sustainable development of the world's water resources. Water specialists and scholars should conduct more research on water quantity and quality to help governments make decisions and accomplish the long-term development of the world's water resources.

10.3 Options for Safe Drinking Water Supply

The drinking water in arid and semi-arid regions is already in poor condition; based on global scientific research data, the following solutions are suggested for safe water supply:

(a) For safe drinking water in arid and semi-arid regions, collecting rainwater and taking precautions against contaminants in rainwater storage tanks is necessary. Local governments should implement rainwater harvesting practices to ensure a safe water supply in the short and long term.
(b) To provide potable water to residential areas of these regions, protected water supply schemes and treatment plants to remove contaminants should be implemented. Furthermore, nitrate pollution must be addressed by installing distillation plants or implementing appropriate removal techniques.
(c) The local government should take immediate action to reduce groundwater nitrate pollution and ensure the availability of potable water from alternate sources (i.e., rivers and canals) in arid and semi-arid regions.
(d) Promote cost-effective, sustainable seawater desalination and ensure a source-to-tap approach to water supply management.
(e) Promoting organic manure over nitrogen-based fertilizers in arid and semi-arid regions.

(f) The use of groundwater in conjunctive with surface water is another option for a safe drinking water supply in arid and semi-arid regions. Mixing contaminated water with clean water decreases nitrate concentration; however, this method is unsafe for infants but safe for animals and adults.

11 Summary and Future Perspective

Nitrate is one of the principal pollutants found in the groundwater globally; excessive levels have adversely damaged ecosystems and human health. Therefore, technological and economically viable, accessible, and practical solutions will be required to mitigate nitrate pollution. Also, policy reforms are needed to minimize its effects on humans and the environment. Nitrogen source inventories, basin management plans, and identifying and quantifying primary sources and their loads to groundwater are some strategies for reducing nitrate pollution. Furthermore, various technologies like reverse osmosis, ultrafiltration, chemical and biological denitrification, ion exchange, adsorption, and electrodialysis have been widely used to eliminate nitrate from groundwater. However, the by-products of these technologies have significant limits; therefore, hybrid methods will be required in the future to combat the nitrate threat. Improved and ongoing communication between scientists, water managers, and water consumers is essential for achieving the sustainability of groundwater resources. The management of nitrate-contaminated groundwater in arid and semi-arid regions should include source reduction measures, removal or transformation technologies, groundwater conservation, educational actions, legislative efforts, and guidelines. Likewise, we can choose appropriate management alternatives with the help of the multicriteria decision analysis approach. In addition to structural measures like physical activities and construction projects, the management includes non-structural measures such as policies, guidance, and funding. Regional actions will be strengthened in the short term to decrease nitrate contamination. However, future research must develop enhanced ways to eliminate nitrate from the environment efficiently.

12 Conclusion

This chapter compiles information on the quality of groundwater aquifers, ecotoxicological impacts, and management options for arid and semi-arid regions worldwide. It has been determined that agricultural fertilizers and septic systems are the principal contributors to nitrate in most arid and semi-arid locations. The existence of nitrate concentrations that exceed WHO standards necessitates an immediate management strategy in order to prevent ecotoxicological effects. Therefore, the region's groundwater requires "Treatment" before consumption and must be safeguarded against additional contamination. The present removal and transformation